T0231056

BLIND IMAGE DECONVOLUTION

Theory and Applications

Blind Image Deconvolution

Theory and Applications

Edited by
Patrizio Campisi
Karen Egiazarian

CRC Press
Taylor & Francis Group
Boca Raton   London   New York

CRC Press is an imprint of the
Taylor & Francis Group, an **informa** business

CRC Press
Taylor & Francis Group
6000 Broken Sound Parkway NW, Suite 300
Boca Raton, FL 33487-2742

International Standard Book Number-10: 0-8493-7367-0 (Hardcover)
International Standard Book Number-13: 978-0-8493-7367-1 (Hardcover)

Library of Congress Cataloging-in-Publication Data

Campisi, Patrizio, 1968-
    Blind image deconvolution : theory and applications / Patrizio Campisi and Karen Egiazarian.
        p. cm.
    Includes bibliographical references and index.
    ISBN 978-0-8493-7367-1 (alk. paper)
    1. Image processing--Digital techniques--Mathematics. 2. Spectrum analysis--Deconvolution--Mathematics. I. Egiazarian, K. (Karen), 1959- II. Title.

TA1632.C35 2007
621.36'7--dc22   2006101025

Visit the Taylor & Francis Web site at
http://www.taylorandfrancis.com

and the CRC Press Web site at
http://www.crcpress.com

Contents

2 Blind Image Deconvolution Using Bussgang Techniques:
Applications to Image Deblurring and Texture Synthesis   43
Patrizio Campisi, Alessandro Neri, Stefania Colonnese, Gianpiero
Panci, Gaetano Scarano

3 Blind Multiframe Image Deconvolution Using Anisotropic
Spatially Adaptive Filtering for Denoising and Regularization   95
Vladimir Katkovnik, Karen Egiazarian, Jaakko Astola

André Jalobeanu, Josiane Zerubia, Laure Blanc-Féraud

Preface

Image deconvolution is receiving an always increasing attention from the academic as well the industrial world due to both its theoretical and practical implications. In many imaging applications an observed image can be thought of as a degraded replica of the original one. This degradation is usually nonlinear, thus taking into account phenomena such as quantization, and spatially and temporally varying as in nonuniform motion or imperfect optics. However, for the sake of mathematical tractability the observation model is linearized, thus assuming that the observed image is obtained as the output of a linear, shift invariant system, namely the *Point Spread Function* (PSF), fed by the original image and contaminated by additive noise.

The term "deconvolution" refers to the process of reconstructing the original image from either one or multiple degradated observations, using the information about both the imaging system and the original image. Unfortunately, there are applications where the PSF is either unknown or partially known, and little information is available about the original image. In these cases the original image estimation process is called *blind image deconvolution* (BID).

The field of image deconvolution has several applications in different areas such as image restoration, microscopy, medical imaging, biological imaging, remote sensing, astronomy, nondestructive testing, geophysical prospecting, and many others. There are several factors which motivate the use of blind deconvolution in image processing. In fact, in many practical scenarios it may be expensive, dangerous, or physically impossible either to obtain *a priori* information about the image to be acquired or information about the PSF. For example, in applications such as astronomy and satellite imaging, the atmospheric turbulence, which partially determines the PSF, is difficult to characterize also because of its time-varing nature. In image photography the PSF of the imaging system (e.g. the lens) is only approximately known. Also in medical imaging applications the PSF of the instruments has to be measured and therefore it is subject to errors. In all these situations as well as in many others, blind deconvolution can be used to generate images of improved quality, which makes this field of research a stimulating and a challenging one.

This book is intended to put together the latest research in BID with specific attention to applications, thus filling a gap in the available literature on blind image deconvolution. Specifically, in this book we introduce the problem of blind deconvolution for images, we provide an overview of the existing algorithms which represent the state of the art for BID with reference to specific applications such as image restoration, microscopy, medical imaging, biological imaging, remote sensing, and astronomy. Current trends in this research area are also described.

The book is organized as follows. In Chapter 1, a survey on the field of blind image deconvolution with emphasis on the Bayesian techniques is provided. In Chapter 2, single channel and multichannel blind image deconvolution are presented in the unifying framework of the Bussgang deconvolution algorithm, with application to texture analysis and synthesis and to image restoration. In Chapter 3, an adaptive method for blind deblurring of noisy multiframe images is presented and a recursive projection gradient algorithm is developed for the minimization of an energy criterion. In Chapter 4, the blind image deconvolution problem is addressed in the Bayesian framework by using a methodology based on variational approximations which can be viewed as a generalization of the expectation maximization algorithm. In Chapter 5, the issue of blind deconvolution of medical images is taken into account. An exhaustive overview of the existing techniques for blindly deconvolving medical ultrasound images is provided along with an assessment of future deconvolution tools for processing biological data acquired by some of the new macro and micro systems used in small organs and cells imaging. In Chapter 6, blind deconvolution for remote sensing imaging is investigated and some methods are provided in a Bayesian framework. In Chapter 7, the problems of deconvolution and blind deconvolution in astronomy are investigated. Specifically, some basic methods used in radio astronomy are described as well as wavelet-based deconvolution approaches. Some insights in blind deconvolution of astronomical images are eventually given. Multiframe blind image deconvolution, that is, the capture of the same scene by different sensors and the fusion of the so obtained degraded acquisitions in order to achieve an image of better quality, is investigated in Chapters 8, 9, and 10. Specifically, in Chapter 8, a superresolution-based method, which allows estimation of blurs with a transitory behavior, such as the ones due to atmospheric turbulence, camera motion, and so on, is detailed. In Chapter 9, two methods, relying on the merge of classification and fusion of the available multiple frames in order to produce a more regularized result, are presented. In Chapter 10, methods for the reconstruction of high-resolution images from multiple undersampled images of a scene that is obtained by using a charge-coupled device (CCD) or a CMOS detector array of sensors which are shifted relative to each other by subpixel displacements are described.

Patrizio Campisi
Roma

Karen Egiazarian
Tampere

Editors

Patrizio Campisi is currently an Associate Professor with the Department of Applied Electronics, University of Rome, Roma TRE, Italy. He received the M.Sc. in Electronic Engineering from the University of Rome La Sapienza, Italy and the Ph.D. degree in Electronic Engineering from the University of Rome, Roma TRE. He has held visiting positions at the University of Toronto, Canada, at the Beckman Institute, University of Illinois at Urbana-Champaign, USA, at the Tampere University of Technology, Finland, and at the Ecole Polytechnique de L'Univesité de Nantes, IRRCyN, France. His research interests have been focused on digital image and video processing with applications to multimedia. He has been working on image deconvolution, image restoration, image analysis, texture coding, classification, and synthesis, multimedia security, data hiding, and biometrics. He is co-recipient of a IEEE ICIP 2006 best student paper award for the paper titled "Contour detection by multiresolution surround inhibition."

Karen Egiazarian received M.Sc. in mathematics from Yerevan State University (1981), Ph.D. in physics and mathematics from Moscow M.V. Lomonosov State University (1986) and Doctor of Technology from Tampere University of Technology, Finland (1994). He is currently a Professor of the Institute of Signal Processing of Tampere University of Technology, Tampere, Finland, leading a group of Transforms and Spectral Techniques. His research interests include image and video analysis, nonlinear methods, image compression, digital logic, transform-based technique. He is a coauthor of three books, several patents and over 400 publications.

Contributors

Dan R. Adam
Department of Biomedical Engineering, Technion University, Haifa, Israel.

Bruno Amizic
Department of EECS, Northwestern University, Evanston, IL, USA.

Jaakko Astola
Institute of Signal Processing, Tampere University of Technology, Tampere, Finland.

S. Derin Babacan
Department of EECS, Northwestern University, Evanston, IL, USA.

Tom E. Bishop
School of Engineering and Electronics, The University of Edinburgh, Edinburgh, Scotland, UK.

Laure Blanc-Féraud
Ariana research group (INRIA-I3S), Sophia Antipolis, France.

Patrizio Campisi
Dipartimento di Elettronica Applicata, Università degli Studi Roma Tre, Roma, Italy.

Tony Chan
Department of Mathematics, UCLA, Los Angeles, CA, USA.

Stefania Colonnese
Dipartimento INFOCOM, Università "La Sapienza" di Roma, Italy.

Gabriel Cristóbal
Instituto de Óptica, Consejo Superior de Investigaciones Científicas, Madrid, Spain.

Karen Egiazarian
Institute of Signal Processing, Tampere University of Technology, Tampere, Finland.

Jan Flusser
Institute of Information Theory and Automation, Academy of Sciences of the Czech Republic, Prague, Czech Republic.

Nikolas P. Galatsanos
Department of Computer Science, University of Ioannina, Ioannina, Greece.

Alexia Giannoula
Department of Electrical and Computer Engineering, University of Toronto, Toronto, Canada.

Jianxin Han
Department of Electrical and Computer Engineering, University of Toronto, Toronto, Canada.

Dimitrios Hatzinakos
Department of Electrical and Computer Engineering, University of Toronto, Toronto, Canada.

André Jalobeanu
PASEO research group, MIV team (LSIIT UMR 7005 CNRS-ULP), Illkirch, France.

Vladimir Katkovnik
Institute of Signal Processing, Tampere University of Technology, Tampere, Finland.

Aggelos K. Katsaggelos
Department of EECS, Northwestern University, Evanston, IL, USA.

Aristidis Likas
Department of Computer Science, University of Ioannina, Ioannina, Greece.

Oleg V. Michailovich
Department of Electrical and Computer Engineering, University of Alberta, Edmonton, Canada.

Rafael Molina
Departamento de Ciensas de la Computacion e I. A., Universidad de Granada, Spain.

Fionn Murtagh
Department of Computer Science, Royal Holloway, University of London, Egham, UK

Alessandro Neri
Dipartimento di Elettronica Applicata, Università degli Studi Roma Tre, Roma, Italy.

Michael K. Ng
Department of Mathematics, Hong Kong Baptist University, Hong Kong.

Gianpiero Panci
Dipartimento INFOCOM, Università "La Sapienza" di Roma, Italy.

Eric Pantin
Service d'Astrophysique, SAP/SEDI, CEA-Saclay, Gif-sur-Yvette, France.

Robert J. Plemmons
Department of Mathematics and Computer Science, Wake Forest University, Winston-Salem, NC, USA.

Gaetano Scarano
Dipartimento INFOCOM, Università "La Sapienza" di Roma, Italy.

Filip Šroubek
Institute of Information Theory and Automation, Academy of Sciences of the Czech Republic, Prague, Czech Republic.

Jean-Luc Starck
Service d'Astrophysique, SAP/SEDI, CEA-Saclay, Gif-sur-Yvette, France.

Josiane Zerubia
Ariana Research Group (INRIA-I3S), Sophia Antipolis, France.

1

Blind Image Deconvolution: Problem Formulation and Existing Approaches

Tom E. Bishop

School of Engineering and Electronics, The University of Edinburgh, Edinburgh, Scotland, UK
e-mail: t.e.bishop@ed.ac.uk

S. Derin Babacan, Bruno Amizic, Aggelos K. Katsaggelos

Department of EECS, Northwestern University, Evanston, IL, USA
e-mail: (sdb, amizic)@northwestern.edu, aggk@ece.northwestern.edu

Tony Chan

Department of Mathematics, UCLA, Los Angeles, CA, USA
email: chan@math.ucla.edu

Rafael Molina

Departamento de Ciensas de la Computacion e I. A., Universidad de Granada, Granada, Spain
e-mail: rms@decsai.ugr.es

1.1 Introduction

Images are ubiquitous and indispensable in science and everyday life. Mirroring the abilities of our own human visual system, it is natural to display observations of the world in graphical form. Images are obtained in areas ranging from everyday photography to astronomy, remote sensing, medical imaging, and microscopy. In each case, there is an underlying object or scene we wish to observe; the *original* or *true* image is the ideal representation of the observed scene.

Yet the observation process is never perfect: there is uncertainty in the measurements, occurring as blur, noise, and other degradations in the recorded images. Digital image restoration aims to recover an estimate of the original image from the degraded observations. The key to being able to solve this *ill-posed* inverse problem is proper incorporation of prior knowledge about the original image into the restoration process.

Classical image restoration seeks an estimate of the true image assuming the blur is known. In contrast, *blind* image restoration tackles the much more difficult, but realistic, problem where the degradation is unknown. In general, the degradation is nonlinear (including, for example, saturation and quantization) and spatially varying (non-uniform motion, imperfect optics); however, for most of the work, it is assumed that the observed image is the output of a Linear Spatially Invariant (LSI) system to which noise is added. Therefore it becomes a Blind Deconvolution (BD) problem, with the unknown blur represented as a Point Spread Function (PSF).

Classical restoration has matured since its inception, in the context of space exploration in the 1960s, and numerous techniques can be found in the literature (for recent reviews see [1, 2]). These differ primarily in the prior information about the image they include to perform the restoration task. The earliest algorithms to tackle the BD problem appeared as long ago as the mid-1970s [3, 4], and attempted to identify known patterns in the blur; a small but dedicated effort followed through the late 1980s (see for instance [5–9]), and a resurgence was seen in the 1990s (see the earlier reviews in [10, 11]). Since then, the area has been extensively explored by the signal processing, astronomical, and optics communities. Many of the BD algorithms have their roots in estimation theory, linear algebra, and numerical analysis.

An important question one may ask is why is BD useful? Could we not simply use a better observation procedure in the first place? Perhaps, but there always exist physical limits, such as photonic noise, diffraction, or an observation channel outside of our control, and often images must be captured in suboptimal conditions. Also there are existing images of unique events that cannot be retaken that we would like to be able to recover (for instance with forensics or archive footage); furthermore in these cases it is often infeasible to measure properties of the imaging system directly. Another reason is that of cost. High-quality optics and sensing equipment are expensive. However, processing power is abundant today and opens the door to the application of increasingly sophisticated models. Thus BD represents a valuable tool that can be used for improving image quality without requiring complicated calibrations of the real-time image acquisition and processing system (i.e., in medical imaging, video conferencing, space exploration, x-ray imaging, bio-imaging, and so on).

The BD problem is encountered in many different technical areas, such as astronomical imaging [12, 13], remote sensing [14], microscopy [15], medical imaging [16], optics [17, 18], photography [19, 20], superresolution applications [21], and motion tracking applications [22], among others.

For example, astronomical imaging is one of the primary applications of BD algorithms [12, 13]. Ground-based imaging systems are subject to blurring due to the rapidly changing index of refractions of the atmosphere. Extraterrestrial observations of the Earth and the planets are degraded by motion blur as a result of slow camera shutter speeds relative to the rapid spacecraft motion.

BD is used for improving the quality of the Poisson distributed film grain noise present in blurred x-rays, mammograms, and digital angiographic images. In such applications, most of the time the degradations are unavoidable because the medical imaging systems limit the intensity of the incident radiation in order to protect the patient's health [23].

In optics, BD is used to restore the original image from the degradation introduced by a microscope or any other optical instrument [17, 18]. The Hubble Space Telescope (HST) main mirror imperfections have provided an inordinate amount of images for the digital image processing community [12].

In photography, depth-of-field effects and misfocusing frequently result in blurred images. Furthermore, motion blur and camera shake are also problems during long exposures in low lighting. The parameters describing these processes are generally unknown; however, BD can enable restoration of these images [19, 20].

As a final example, in tracking applications the object being tracked might be blurred due to its own motion, or the motion of the camera. As a result, the track is lost with conventional tracking approaches and the application of BD approaches can improve tracking results [22].

In the rest of this chapter, we survey the field of blind image deconvolution. In conducting this review though, we develop and present most of the techniques within the Bayesian framework. Whilst many methods were originally derived by other means, this adds consistency to the presentation and facilitates comparison among the different methods. Even today, due to the difficulty of simultaneously estimating the unknown image and blur in the presence of noise, the problem still provides a very fertile ground for novel processing methods.

The chapter is organized as follows. In Section 1.2 we mathematically define the BD problem. In Section 1.3 we provide a classification of the existing approaches to BD. In Section 1.4 we formulate the BD problem within the Bayesian framework, and in Section 1.5 we survey the probabilistic models for the observation, the original image, the blur, and their associated unknown parameters. In Section 1.6 we discuss solutions to the BD problem as inference models under the Bayesian framework. Finally, in Section 1.7 we briefly review BD models which appeared in the literature and cannot be easily obtained from the Bayesian formulation. Conclusions are presented in

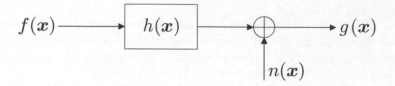

FIGURE 1.1:   Linear space invariant image degradation model.

Section 1.8.

1.2   Mathematical Problem Formulation

In digital image processing, the general, discrete model for a linear degradation caused by blurring and additive noise is given by

$$g\left(\mathbf{x}\right) = \sum_{\mathbf{s} \in S_h} h(\mathbf{x}, \mathbf{s}) f(\mathbf{s}) + n(\mathbf{x}), \qquad \mathbf{x} = (x_1, x_2) \in S_f \qquad (1.1)$$

where $f\left(\mathbf{x}\right)$, $g\left(\mathbf{x}\right)$, $h\left(\mathbf{x}, \mathbf{s}\right)$, and $n\left(\mathbf{x}\right)$ represent the original image, the observed image, the blur or PSF, and the observation noise, respectively, $S_f \subset \mathbb{R}^2$ is the support of the image, and $S_h \subset \mathbb{R}^2$ is the support of the PSF. The additive noise process $n\left(\mathbf{x}\right)$ may originate during image acquisition, processing, or transmission. Common types of noise are electronic, photoelectric, film grain, and quantization noise. It is common to assume $n(\mathbf{x})$ is White Gaussian Noise (WGN), uncorrelated with the image (although certain types of noise may in practice be signal dependent).

The general objective of blind restoration is to estimate $\mathbf{f}$ and $\mathbf{h}$. The difficulty in solving this problem with a spatially varying blur motivates the use of a space-invariant model for the blur. This leads to the following expression for the degradation system:

$$g(\mathbf{x}) = (f * h)(\mathbf{x}) + n(\mathbf{x}) = \sum_{\mathbf{s} \in S_h} h(\mathbf{x} - \mathbf{s}) f(\mathbf{x}) + n(\mathbf{x}), \qquad (1.2)$$

where the operator $(*)$ denotes 2-D convolution. The block diagram of the LSI degradation model presented in Equation (1.2) is shown in Figure 1.1.

The image degradation model described by Equation (1.1) or Equation (1.2) is often represented in terms of a matrix-vector formulation, that is,

$$\mathbf{g} = \mathbf{Hf} + \mathbf{n}, \qquad (1.3)$$

where the vectors $\mathbf{g}$, $\mathbf{f}$, and $\mathbf{n}$ represent the observed image, the original image, and the observation noise ordered lexicographically by stacking either the rows or the columns of each image into a vector. $\mathbf{H}$ is a Block Toeplitz with Toeplitz Blocks (BTTB) matrix when Equation (1.2) is used, which can be approximated by a Block Circulant with Circulant Blocks (BCCB) matrix [24]. Since BCCB matrices are diagonalized using the 2-D Discrete Fourier Transform (DFT), this allows (1.2) to be expressed in the discrete frequency domain. Note that it will also be useful to write $\mathbf{Hf} = \mathbf{Fh}$, where $\mathbf{F}$ is a matrix formed from the image data and $\mathbf{h}$ is a vector parameterizing the blur.

The BD problem refers to finding estimates $\hat{f}(\mathbf{x})$ and $\hat{h}(\mathbf{x})$ for $f(\mathbf{x})$ and $h(\mathbf{x})$ based on $g(\mathbf{x})$ and prior knowledge about $f(\mathbf{x})$, $h(\mathbf{x})$, and $n(\mathbf{x})$. It should be noted that although the degradation model is LSI, the deconvolution algorithm may be nonlinear or spatially varying or both.

Blind deconvolution is an ill-posed problem: the solution of Equation (1.2) may not depend continuously on the data, may not be unique, or it may not even exist [1, 25]. With practical approaches, an approximate solution to the problem is estimated, so that the existence of the solution can be disregarded; however, the nonuniqueness and the sensitivity of the solution to the noise are still serious problems.

1.3 Classification of Blind Image Deconvolution Methodologies

We may classify BD approaches into two categories according to the stage at which we identify the blur: *a priori* or jointly with the image.

A priori **blur identification methods** With this approach, the PSF is identified separately from the original image, and later used in combination with one of the classical image restoration algorithms in order to restore the original image. A parametric blur model may be used, for example, one of the general models to be described in Section 1.5.2; then the objective is to identify the most likely blur parameters $\mathbf{h}$ from the observation. This approach has been used in [3, 4], for example, and in a Bayesian context in [26].

Experimental approaches are also possible: images of one or more point sources are collected and used to obtain an estimate of the PSF (this, for example, was done with the HST). Furthermore if a good understanding of the imaging system is available for a specific application, we may make an *a priori* prediction of the blur; however unless we use this model as a prior in one of the other algorithms, this is not a blind procedure as such. This may be possible in microscopy, medical ultrasound, remote sensing, or optical telescope systems (e.g., Tiny Tim modeling of the HST).

Joint identification methods The majority of existing methods fall into this class, where the image and blur are identified simultaneously. However, in practice many methods in this category use an alternating approach to estimate **f** and **h** rather than truly finding the joint solution. Prior knowledge about the image and blur is typically incorporated in the form of models like those presented in Section 1.5.2 and Section 1.5.3. Parameters describing such models are also required to be estimated from the available data; often this is performed before image and blur identification, although simultaneous identification is possible, e.g., see [27].

In the following section, we use the Bayesian framework to present the different BD approaches proposed in the literature. As we will see, most of them can be seen as special cases of an application of Bayes' theorem: the main differences between them are the choices of the function to be optimized, and the prior distributions used to model the original image and the degradation process. Using the Bayesian framework allows us to describe the general BD problem in a systematic way, and to identify the similarities and differences between the proposed approaches in the above two categories. Each category may be seen as a particular inference model in the Bayesian paradigm.

1.4   Bayesian Framework for Blind Image Deconvolution

A fundamental principle of the Bayesian philosophy is to regard all parameters and observable variables as unknown stochastic quantities, assigning probability distributions based on subjective beliefs. Thus in BD, the original image **f**, the blur **h**, and the noise **n** in Equation (1.3) are all treated as samples of random fields, with corresponding *prior* Probability Density Functions (PDFs) that model our knowledge about the imaging process and the nature of images. These distributions depend on parameters which will be denoted by Ω. The parameters of the prior distributions are termed *hyperparameters*.

Often Ω is assumed known (or is first estimated separately from **f** and **h**). Alternatively we may adopt the *hierarchical* Bayesian framework whereby Ω is also assumed unknown, in which case we also model our prior knowledge of its values. The PDFs for the hyperparameters are termed *hyperprior* distributions. This abstraction allows greater robustness to error when there is uncertainty, and is essential when we are less confident in the observed data (due to a lower Signal-to-Noise Ratio [SNR]). This *hierarchical* modeling allows us to write the joint global distribution

$$p(\Omega, \mathbf{f}, \mathbf{h}, \mathbf{g}) = p(\Omega)p(\mathbf{f}, \mathbf{h}|\Omega)p(\mathbf{g}|\Omega, \mathbf{f}, \mathbf{h}), \qquad (1.4)$$

where $p(\mathbf{g}|\Omega, \mathbf{f}, \mathbf{h})$ is termed the *likelihood* of the observations. Typically, we assume that $\mathbf{f}$ and $\mathbf{h}$ are *a priori* conditionally independent, given Ω, i.e., $p(\mathbf{f}, \mathbf{h} \mid \Omega) = p(\mathbf{f} \mid \Omega) p(\mathbf{h} \mid \Omega)$. Then the task is to perform inference using the posterior

$$p(\mathbf{f}, \mathbf{h}, \Omega \mid \mathbf{g}) = \frac{p(\mathbf{g} \mid \mathbf{f}, \mathbf{h}, \Omega) p(\mathbf{f} \mid \Omega) p(\mathbf{h} \mid \Omega) p(\Omega)}{p(\mathbf{g})}. \quad (1.5)$$

Note that this corresponds to the joint estimation method described in Section 1.3. We can also marginalize either $\mathbf{f}$, to describe the *a priori* blur identification method as

$$p(\mathbf{h}|\mathbf{g}) = \int \cdots \int p(\mathbf{f}, \mathbf{h}, \Omega \mid \mathbf{g}) \, d\mathbf{f} \cdot d\Omega, \quad (1.6)$$

or marginalize $\mathbf{h}$ to obtain

$$p(\mathbf{f}|\mathbf{g}) = \int \cdots \int p(\mathbf{f}, \mathbf{h}, \Omega \mid \mathbf{g}) \, d\mathbf{h} \cdot d\Omega. \quad (1.7)$$

Note that by marginalizing $\mathbf{h}$ this approach seeks to bypass the blur identification stage and estimate the image $\mathbf{f}$ directly, by using appropriate constraints or prior knowledge, although this is less common in practice.

In the following sections we study first the various prior models for the image, blur, and hyperparameters that have appeared in the literature. We then analyze the estimation of these unknown quantities as inference models under the Bayesian framework.

1.5 Bayesian Modeling of Blind Image Deconvolution

1.5.1 Observation Model

The first stage of the Bayesian formulation is specifying the likelihood of the observed image, $\mathbf{g}$. Due to the model in Equation (1.3), the PDF of $\mathbf{g}$ is related to that of the observation noise, $\mathbf{n}$. A typically used model for $\mathbf{n}$ is zero mean independent WGN with distribution $\mathcal{N}(\mathbf{n} \mid 0, \beta^{-1}\mathbf{W}^{-1})$, where β^{-1} denotes the variance, and $\mathbf{W}$ is a diagonal weights matrix included for generality. It allows us to represent spatially varying noise statistics, and it also introduces flexibility in the energy minimization formulation, as will be described later. For stationary noise, $\mathbf{W} = \mathbf{I}$. Thus we have

$$p(\mathbf{n}) = p(\mathbf{g} \mid \mathbf{f}, \mathbf{h}, \beta) = \det|\mathbf{W}| \left(\frac{\beta}{2\pi} \right)^{N/2} \exp \left[-\frac{1}{2}\beta \|\mathbf{g} - \mathbf{H}\mathbf{f}\|_{\mathbf{W}}^2 \right], \quad (1.8)$$

where N is the size of the image vector, $\mathbf{f}$, and the weighted norm used in the PDF is defined as $\|\mathbf{x}\|_{\mathbf{W}}^2 = \mathbf{x}^T \mathbf{W}^T \mathbf{W} \mathbf{x}$.

Alternative noise modeling, for instance Poisson noise arising in low-intensity imaging, is also assumed in certain BD problems. We will concentrate, however, on the Gaussian noise model presented above.

1.5.2   Parametric Prior Blur Models

The following analytical models are frequently used in Equation (1.2) to represent the LSI image degradation operator, i.e., the PSF. In this case the prior distribution for the blur is parameterized directly by the unknowns defining the parametric model.

Traditionally, when the parametric form of the PSF is assumed known, $p(\mathbf{h})$ is usually a uniform distribution. Then the unknown parameters may be estimated using, for example, Maximum Likelihood (ML) methods (see [28, 29]). Alternatively the unknown quantities defining the parametric function may be estimated *a priori* if the real underlying image is known. For instance, in astronomy the parameters of the atmospheric turbulence blur are estimated using observed known point sources (stars); similarly, the parameters describing the HST PSF model could also be estimated.

Note in the case when such a parametric model is not used, the values of the PSF coefficients directly parameterize the prior distribution for the blur. Ideally this prior should embody the physical constraints arising from an imaging system: positivity of the coefficients and energy preservation (the PSF coefficients should sum to one). Other prior assumptions often used for the PSF coefficients are smoothness or piecewise smoothness, symmetry, and finite support size. Due to similarities to the priors used for images, these priors will be discussed in Section 1.5.3.

1.5.2.1   Linear Motion Blur

In general, relative motion of the camera and scene to be imaged results in a PSF representing temporal integration along this motion path. If the camera movement or object motion is fast relative to the exposure period, we may approximate this as a linear motion blur. This is represented as the 1-D local averaging of neighboring pixels. An example of a horizontal motion blur model is given by (L is an even integer):

$$h(\mathbf{x}) = \begin{cases} \dfrac{1}{L+1}, & -\dfrac{L}{2} \leq x_1 \leq \dfrac{L}{2}, \\ & x_2 = 0 \\ 0 & , \text{ otherwise.} \end{cases} \tag{1.9}$$

1.5.2.2   Atmospheric Turbulence Blur

This type of blur is common in remote sensing and aerial imaging applications. For long-term exposure through the atmosphere a Gaussian PSF model is used:

$$h\left(\mathbf{x}\right) = K\, e^{-\frac{|\mathbf{x}|^2}{2\sigma^2}}, \tag{1.10}$$

where K is a normalizing constant ensuring that the blur has a unit volume, and σ^2 is the variance that determines the severity of the blur. Alternative atmospheric blur models have been suggested in [30, 31]. In these works the PSF is approximated by the function

$$h(\mathbf{x}) \propto (1 + \frac{|\mathbf{x}|^2}{R^2})^{-\delta}, \tag{1.11}$$

where δ and R are unknown parameters.

1.5.2.3   Out-of-Focus Blur

Photographical defocusing is another common type of blurring, primarily due to the finite size of the camera aperture. A complete model of the camera's image formation system depends on many parameters. These include the focal length, the camera aperture size and shape, the distance between object and camera, the wavelength of the incoming light, and effects due to diffraction (see [32] for further details). Furthermore, poor-quality optics introduce aberrations of their own. Accurate knowledge of all of these parameters is usually not available after the picture was taken. When the blur due to defocusing is large, the uniform circular PSF model is used as an approximation to these effects:

$$h\left(\mathbf{x}\right) = \begin{cases} \dfrac{1}{\pi r^2}, & |\mathbf{x}| \leq r \\ 0, & \text{otherwise.} \end{cases} \tag{1.12}$$

The uniform 2-D blur is sometimes used as a cruder approximation to an out-of-focus blur; it is also used as a model for sensor pixel integration in superresolution restoration. This model is defined (with L an even integer) as

$$h\left(\mathbf{x}\right) = \begin{cases} \dfrac{1}{(L+1)^2}, & -\dfrac{L}{2} \leq (x_1, x_2) \leq \dfrac{L}{2} \\ 0, & \text{otherwise.} \end{cases} \tag{1.13}$$

1.5.3   Prior Image and Blur Models

The prior distributions $p(\mathbf{f}|\Omega)$ and $p(\mathbf{h}|\Omega)$ should reflect our beliefs about the nature of $\mathbf{f}$ and $\mathbf{h}$ and constrain the space of possible solutions for them to the most probable ones. This is necessary due to the ill-posed nature of the problem. Abstract descriptions of natural images have been made: smooth, piecewise-smooth, or textured, for instance (of course some applications may have other specific constraints). We can attempt to model these descriptions in a stochastic sense using the priors. Typically this is done by specifying probabilistic relations between neighboring image pixels or their derivatives. Similar procedures may be followed for the PSF.

We will consider a general exponential model of the form

$$p\left(\mathbf{f} \,|\, \Omega\right) = \frac{1}{Z_f(\Omega)} \exp\left[-U_f(\mathbf{f}, \Omega)\right] \tag{1.14a}$$

$$p\left(\mathbf{h} \,|\, \Omega\right) = \frac{1}{Z_h(\Omega)} \exp\left[-U_h(\mathbf{h}, \Omega)\right] \tag{1.14b}$$

to represent the image and blur priors. The normalizing terms Z_f and Z_h depend on the hyperparameters for each distribution. They may be treated as constants if we assume the hyperparameters to be known; otherwise they must be calculated as $\int \exp\left[-U_f(\mathbf{f}, \Omega)\right] d\mathbf{f}$ and $\int \exp\left[-U_h(\mathbf{h}, \Omega)\right] d\mathbf{h}$, respectively, which may cause difficulties in inference unless we assume a special form for $U(\cdot)$. Note that $U(\cdot)$ is sometimes termed the *energy function*.

Many different image and blur models in the literature can be put in the form of Equation (1.14); particular cases will now be considered.

1.5.3.1 Stationary Gaussian Models

The most common model is the class of Gaussian models provided by $U_f = \frac{1}{2}\alpha\|\mathbf{L}\mathbf{f}\|^2$. Then, if $\det|\mathbf{L}| \neq 0$, the term Z_f in Equation (1.14) becomes simply $(2\pi)^{\frac{N}{2}} \alpha^{-\frac{N}{2}} \det|\mathbf{L}|^{-1}$, which if we use a fixed stationary form for $\mathbf{L}$ is simple to calculate. These models are often termed Simultaneous Autoregression (SAR) or Conditional Autoregression (CAR) models [33].

In the most basic case, we can use $\mathbf{L} = \mathbf{I}$, the identity. This imposes constraints on the magnitude of the intensity distribution of $\mathbf{f}$. A more common usage is $\mathbf{L} = \mathbf{C}$, the discrete Laplacian operator, which instead constrains the derivative of the image. For instance, Molina et al. [27] used this model for both image and blur, giving

$$p(\mathbf{f}|\alpha_{\mathrm{im}}) \propto \alpha_{\mathrm{im}}^{N/2} \exp\left[-\frac{1}{2}\alpha_{\mathrm{im}} \parallel \mathbf{C}\mathbf{f} \parallel^2\right] \tag{1.15a}$$

$$p(\mathbf{h}|\alpha_{\mathrm{bl}}) \propto \alpha_{\mathrm{bl}}^{M/2} \exp\left[-\frac{1}{2}\alpha_{\mathrm{bl}} \parallel \mathbf{C}\mathbf{h} \parallel^2\right]. \tag{1.15b}$$

Note that in these two equations N and M should in practice be replaced by $N-1$ and $M-1$, respectively, because $\mathbf{C}^T\mathbf{C}$ is singular.

This SAR model is suitable for $\mathbf{f}$ and $\mathbf{h}$ if it is assumed that the luminosity distribution is smooth on the image domain, and that the PSF is a partially smooth function.

In [34] the SAR model was used for the image prior, and a Gaussian PDF for the PSF; that is,

$$p(\mathbf{h}|\mu_{\mathrm{bl}}, \alpha_{\mathrm{bl}}) = \mathcal{N}(\mu_{\mathrm{bl}}, \alpha_{\mathrm{bl}}^{-1}\mathbf{I}). \tag{1.16}$$

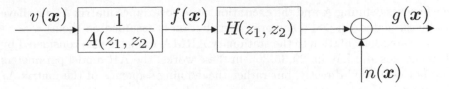

FIGURE 1.2: ARMA degradation model.

The components of the blur are assumed statistically independent so that the covariance matrix is diagonal. Clearly, a Gaussian PDF with unknown mean vector and fully populated covariance matrix might be used to model image and blur priors. However, this requires the simultaneous estimation of a very large number of hyperparameters, thus making the approach highly impractical unless we use additional hyperpriors (see Section 1.5.4) to impart *a priori* knowledge of the hyperparameters' values [27].

1.5.3.2 Autoregressive Models

A class of blind image deconvolution algorithms (see, e.g., [28, 29]) model the observed image **g** as an Autoregressive Moving Average (ARMA) process, as shown in Figure 1.2.

The observation Equation (1.2) forms the Moving Average (MA) part of the model. The original image is modeled as a 2-D Autoregressive (AR) process:

$$f(\mathbf{x}) = \sum_{\mathbf{s} \in S_a : \mathbf{s} \neq \mathbf{0}} a(\mathbf{s}) f(\dot{\mathbf{x}} - \mathbf{s}) + v(\mathbf{x}), \qquad (1.17)$$

or in matrix-vector form:

$$\mathbf{f} = \mathbf{A}\mathbf{f} + \mathbf{v}, \qquad (1.18a)$$

or equivalently,

$$\mathbf{f} = \mathbf{F}\mathbf{a} + \mathbf{v}, \qquad (1.18b)$$

where $S_a \subset \mathbb{R}^2$ is the support of the 2-D AR coefficients $a(\mathbf{x})$, and **A** has a BTTB form. The *excitation* noise signal, or modeling error, $v(\mathbf{x})$ is a zero-mean WGN process with diagonal covariance matrix Λ_v, that is independent of $f(\mathbf{x})$. Since $\mathbf{v} = (\mathbf{I} - \mathbf{A})\mathbf{f}$, the PDF of **f** is obtained via a probability transformation of **v**, as a Gaussian: $p(\mathbf{f} \,|\, \mathbf{a}, \Lambda_v) = \mathcal{N}(\mathbf{f} \,|\, 0, \Lambda_f)$. The image covariance matrix is defined as

$$\Lambda_f = (\mathbf{I} - \mathbf{A})^{-1} \Lambda_v (\mathbf{I} - \mathbf{A})^{-T}. \qquad (1.19)$$

It should be clear that $p(\mathbf{f})$ is in the form of Equation (1.14), with $U_f = \frac{1}{2} \|(\mathbf{I} - \mathbf{A})\mathbf{f}\|_{\Lambda_v}^2$ and $Z_f = (2\pi)^{\frac{N}{2}} \det|\Lambda_v|^{\frac{1}{2}} \det|\mathbf{I} - \mathbf{A}|^{-1}$. Unlike the SAR model above, however, where a deterministic form is used for the matrix **L**, the AR coefficients defining $\mathbf{A}$ and the excitation noise covariance matrix Λ_v also have to be estimated (in [28] a flat prior distribution is assumed on them).

A related formulation to the stationary ARMA model is also considered by Katsaggelos and Lay in [29, 35, 36]. In these works, the AR model parameters are not estimated directly, but rather the defining sequence of the matrix Λ_f is found in the discrete frequency domain, along with the other parameters, under the assumption that the image model is stationary. Observe that this approach does not assume a known model support size S_a or blur support S_h.

The AR image model is good at representing textured images; however, like any of the other stationary Gaussian models in this section, it is not such a good model for an original image that has prominent edges as part of the scene. Therefore it is possible to consider nonstationary extensions to the AR model.

This was done, for example, in [26], where the image is partitioned into blocks each assumed to be homogeneous regions with their own AR coefficients and excitation variance. The equations above remain the same, although the matrix $\mathbf{A}$ is no longer BTTB and the sizes of $\mathbf{F}$ and $\mathbf{a}$ increase. A local mean may be assumed for each region to further better model a real image. In general, a segmentation of the image could be assumed such that the blocks in the model coincide with the natural regions in the image. The use of a nonstationary image model in conjunction with a stationary blur can also aid blur identification [26].

1.5.3.3　Markov Random Field Models

A class of models encountered extensively in image segmentation [37], classical image restoration [38], and also in superresolution restoration [39] and BD [40, 41] are the Markov Random Field (MRF) models [42]. They are usually derived using local spatial dependencies; however, we may see they are closely related to the other models in this section.

Defining $U = \sum_{c \in \mathcal{C}} V_c(\mathbf{f})$ in Equation (1.14), we have the definition of a *Gibbs distribution*. In this context, Z is termed the partition function. $V_c(\mathbf{f})$ is a *potential function* defined over *cliques*, c in the image [42]. Briefly speaking, this gives a simple way of specifying interactions over local neighborhoods in the image field. If we use quadratic potentials, $V_c(\mathbf{f}) = \left(\mathbf{d}_c^T \mathbf{f}\right)^2$, we may rewrite the Gibbs distribution as a Gaussian:

$$p(\mathbf{f}) = \frac{1}{Z} \exp\left[-\mathbf{f}^T \mathbf{B} \mathbf{f}\right] = \frac{1}{Z} \exp\left[-\sum_{c \in \mathcal{C}} \mathbf{f}^T \mathbf{B}_c \mathbf{f}\right] \qquad (1.20)$$

where $\mathbf{B}_c$ is obtained from $\mathbf{d}_c$ and satisfies $[\mathbf{B}_c]_{\mathbf{x},\mathbf{s}}$ are only nonzero when pixels $\mathbf{x}$ and $\mathbf{s}$ are neighbors. Typically the vectors $\mathbf{d}_c$ represent finite difference operators. The partition function is now equal to $(2\pi)^{\frac{N}{2}} \det|\mathbf{B}|^{-\frac{1}{2}}$. This model is also then termed a Gaussian Markov Random Field (GMRF) [43] or CAR [33].

We may also use Generalized Gaussian MRFs (GGMRFs) with arbitrary nonquadratic potentials of a similar functional form: $V_c(\mathbf{f}) = \rho(\mathbf{d}_c^T \mathbf{f})$, where ρ is some (usually convex) function, such as the *Huber function* [43] or p-norm (with $p \geq 1$) based function, $\rho(u) = |u|^p$. This is similar to the use of potential functions used in anisotropic diffusion methods, the motivation being edge preservation in the reconstructed image. Other extensions to the model consider hierarchical, or Compound GMRFs (CGMRFs), also with the goal of avoiding oversmoothing of edges [38, 44].

1.5.3.4 Anisotropic Diffusion and Total Variation Type Models

Non-quadratic image priors have been investigated using *variational integrals* in the *anisotropic diffusion* [45] or *Total Variation* (TV) [46] regularization frameworks, with the aim of preserving edges by not over-penalizing discontinuities, i.e. outliers in the image gradient distribution, see [47, 48] for a unifying view of the probabilistic and variational approaches. The main difference to the GGMRFs models mentioned above is that these usually begin with a formulation in the continuous image domain resulting in Partial Differential Equations (PDEs) that must be solved. However, eventual discretization is eventually necessary, and hence the constraints may be reformulated as non-Gaussian priors, or Gaussians with a nonstationary covariance matrix [20]. Alternatively, other methods propose formulating the TV norm directly in the discrete domain [49, 50].

The generalized regularization approach using anisotropic diffusion has been proposed by You and Kaveh [45]. In this formulation, convex functions $\kappa(\cdot)$ and $\upsilon(\cdot)$ of the image gradient $|\nabla f(x)|$ and the PSF gradient $|\nabla h(x)|$ respectively are used in defining regularization functionals:

$$\mathcal{E}(f) = \int_{S_f} \kappa\left(|\nabla f(\mathbf{x})|\right) \, \mathrm{d}\mathbf{x} \tag{1.21a}$$

$$\mathcal{E}(h) = \int_{S_h} \upsilon\left(|\nabla h(\mathbf{s})|\right) \, \mathrm{d}\mathbf{s}. \tag{1.21b}$$

This is in analogy with standard regularization procedures. However as the functionals are continuous, variational calculus is used to perform the differentiation needed for minimization [51]. This results in a PDE which must be solved for each variable. Consider for instance minimization of Equation (1.21a) for f; the solution must satisfy

$$\nabla_f \mathcal{E}(f) = \nabla \cdot \left(\frac{\kappa'(|\nabla f|)}{|\nabla f|}\nabla f\right) = 0, \tag{1.22}$$

with appropriate boundary conditions. One method of solution is imposing an artificial time evolution variable t, and using a steepest descent method, i.e., for f,

$$\frac{\partial \hat{f}}{\partial t} = -\nabla_f \mathcal{E}(\hat{f}). \tag{1.23}$$

This may be interpreted as representing a physical anisotropic diffusion process [45, 51, 52]. That is, as the time variable t progresses, directional smoothing occurs depending on the local image gradient. The strength and type of smoothing depends on the *diffusion coefficient* or *flux variable, c*, which is related to the potential function by

$$c(|\nabla f|) = \frac{\kappa'(|\nabla f|)}{|\nabla f|} \tag{1.24}$$

We may consider $c(|\nabla f|)$ to be the amount of smoothing perpendicular to the edges. Appropriate choice of c or equivalently κ can result in spatially adaptive edge-preserving restoration.

Consider two cases of the potential function κ and related diffusion coefficient c. In the first case, $\kappa(x) = \frac{1}{2}x^2$ and hence $c(|\nabla f|) = 1$, and $\nabla_f \mathcal{E}(f) = \nabla^2(f)$, i.e., a Laplacian operator [53]. This corresponds to standard spatially invariant isotropic regularization, or a CAR model with the discrete Laplacian when discretized.

Another choice proposed for the BD problem by Chan and Wong [46] is given by the *Total Variation* (TV) norm. In this case, $\kappa(x) = x$ and hence $c(|\nabla f|) = \frac{1}{|\nabla f|}$. The result is that smoothing in the direction orthogonal to the edges is completely suppressed and is only applied parallel to the edge directions. This is demonstrated in [45] by decomposing the anisotropic diffusion equation, Equation (1.23) into components parallel and perpendicular to the edges. A very efficient way to solve the resulting optimization problem is shown in [46] in this particular case.

These two choices lead us to consider the following discrete prior image models:

$$p(\mathbf{f}) \propto \exp\left[-\alpha_{\text{im}} \sum_i ((\Delta_i^h \mathbf{f})_i^2 + (\Delta_i^v \mathbf{f})_i^2)\right] \tag{1.25}$$

for the Laplacian; and

$$p(\mathbf{f}) \propto \exp\left[-\alpha_{\text{im}} \sum_i \sqrt{(\Delta_i^h \mathbf{f})_i^2 + (\Delta_i^v \mathbf{f})_i^2}\right] \tag{1.26}$$

for the TV norm, where Δ_i^h and Δ_i^v are linear operators corresponding to horizontal and vertical first-order differences, at pixel i, respectively.

A combination of the two choices for c is considered in [45], resulting in a spatially adaptive diffusion coefficient. The smoothing strength is increased using the Laplacian in areas with low gradient magnitude, and decreased using the TV norm in areas where large intensity transitions occur in order to preserve edges while still removing noise. This is analogous to the use of the Huber function in Section 1.5.3.3. Many other diffusion coefficients are proposed in the literature, including very complex structural operators (see [52] for a review).

Šroubek and Flusser [20] use a similar scheme to those already mentioned, but (using the half-quadratic approach [54, 55]) demonstrate how the anisotropic diffusion model may be written in the form of Equation (1.14) by discretization of the functional in Equation (1.21a):

$$p\left(\mathbf{f}, c(\mathbf{f})\right) = \frac{1}{Z_f} \exp\left[-\frac{1}{2}\mathbf{f}^T \mathbf{B}\left(c\right)\mathbf{f}\right] \qquad (1.27)$$

They equate the diffusion, or flux variable, to the hidden line process often used in CGMRFs, that is it represents the edge strength between two pixels in the image. Therefore it is possible to build a spatially varying weights matrix $\mathbf{B}$ from the local image gradients. Note that as the flux variable is a function of $\mathbf{f}$, so is the covariance, so this is not strictly a Gaussian distribution unless $\mathbf{B}$ is assumed fixed. In practice, using an iterative scheme, $\mathbf{B}$ may be updated at each iteration.

A similar motivation was used in [19] to obtain a spatially varying weights matrix based on the local image variance, as was previously suggested in [56, 57]. The difference here is that the regularization is isotropic; better performance can be expected with the anisotropic schemes.

1.5.4   Hyperprior Models

So far we have studied the distributions $p(\mathbf{f}, \mathbf{h}|\Omega)$, $p(\mathbf{g}|\Omega, \mathbf{f}, \mathbf{h})$ that appear in the Bayesian modeling of the BD problem in Equation (1.4). We complete this modeling by studying now the distribution $p(\Omega)$.

An important problem is the estimation of the vector of parameters Ω when they are unknown. To deal with this estimation problem, the hierarchical Bayesian paradigm introduces a second stage (the first stage consisting again of the formulation of $p(\mathbf{f}|\Omega)$, $p(\mathbf{h}|\Omega)$, and $p(\mathbf{g}|\mathbf{f}, \mathbf{h}, \Omega)$). In this stage the hyperprior $p(\Omega)$ is also formulated.

A large part of the Bayesian literature is devoted to finding hyperprior distributions $p(\Omega)$ for which $p(\Omega, \mathbf{f}, \mathbf{h}|\mathbf{g})$ can be calculated in a straightforward way or be approximated. These are the so-called conjugate priors [58], which were developed extensively in Raiffa and Schlaifer [59].

Besides providing for easy calculation or approximations of $p(\Omega, \mathbf{f}, \mathbf{h}|\mathbf{g})$, conjugate priors have, as we will see later, the intuitive feature of allowing one to begin with a certain functional form for the prior and end up with a posterior of the same functional form, but with the parameters updated by the sample information.

Taking the above considerations about conjugate priors into account, the literature in BD uses different *a priori* models for the parameters depending on the type of unknown parameters. For parameters, ω, corresponding to inverses of variances, the gamma distribution is used. This is defined by:

$$p(\omega) = \Gamma(\omega|a_\omega, b_\omega) = \frac{(b_\omega)^{a_\omega}}{\Gamma(a_\omega)} \omega^{a_\omega - 1} \exp[-b_\omega\,\omega], \qquad (1.28)$$

where $\omega > 0$ denotes a hyperparameter, $b_\omega > 0$ is the scale parameter, and $a_\omega > 0$ is the shape parameter. These parameters are assumed known. The gamma distribution has the following mean, variance, and mode:

$$E[\omega] = \frac{a_\omega}{b_\omega}\,, \qquad Var[\omega] = \frac{a_\omega}{(b_\omega)^2}\,, \qquad \text{Mode}[\omega] = \frac{a_\omega - 1}{b_\omega}. \qquad (1.29)$$

Note that the mode does not exist when $a_\omega^o \leq 1$ and that mean and mode do not coincide.

For components of mean vectors the corresponding conjugate prior is a normal distribution. Additionally, for covariance matrices the hyperprior is given by an inverse Wishart distribution (see [60]).

We observe, however, that in general most of the methods proposed in the literature use the *uninformative* prior model

$$p(\Omega) = constant. \qquad (1.30)$$

1.6 Bayesian Inference Methods in Blind Image Deconvolution

There are a number of different ways that we may proceed to estimate the image and blur using Equation (1.5). Depending on the prior models chosen, finding analytic solutions may be difficult, so approximations are often needed. Many methods in the literature seek point estimates of the parameters $\mathbf{f}$ and $\mathbf{h}$. Typically, this reduces the problem to one of optimization. However, the Bayesian framework provides other methodologies for estimating the *distributions* of the parameters [60–62], which deal better with uncertainty; approximating or simulating the posterior distribution are two options. These different inference strategies and examples of their use will now be presented, proceeding from the simplest to the more complex.

1.6.1 Maximum *a Posteriori* and Maximum Likelihood

One possible point estimate is provided by the *Maximum A Posteriori* (MAP) solution, which are the values of $\mathbf{f}$, $\mathbf{h}$, and Ω that maximize the posterior probability density:

$$\{\hat{\mathbf{f}}, \hat{\mathbf{h}}, \hat{\Omega}\}_{\text{MAP}} = \underset{\mathbf{f},\mathbf{h},\Omega}{\text{argmax}}\; p\left(\mathbf{g}\,|\,\mathbf{f}, \mathbf{h}, \Omega\right) p\left(\mathbf{f}\,|\,\Omega\right) p\left(\mathbf{h}\,|\,\Omega\right) p\left(\Omega\right). \qquad (1.31)$$

The *Maximum Likelihood* (ML) solution attempts instead to maximize the likelihood $p(\mathbf{g} \mid \mathbf{f}, \mathbf{h}, \Omega)$ with respect to the parameters:

$$\{\hat{\mathbf{f}}, \hat{\mathbf{h}}, \hat{\Omega}\}_{\text{ML}} = \underset{\mathbf{f}, \mathbf{h}, \Omega}{\text{argmax}} \; p(\mathbf{g} \mid \mathbf{f}, \mathbf{h}, \Omega). \tag{1.32}$$

Note, however, that in this case we can only estimate the parameters in Ω that are present in the conditional distribution $p(\mathbf{g} \mid \mathbf{f}, \mathbf{h}, \Omega)$ but none of those present only in $p(\mathbf{f}, \mathbf{h} \mid \Omega)$.

The above maximization of the likelihood is typically seen as a non-Bayesian method, although it is identical to the MAP solution with uninformative (flat) prior distributions. Some approaches may use flat priors for some parameters but not others. Assuming known values for the parameters is equivalent to using degenerate distributions (delta functions) for priors. For instance, a degenerate distribution on Ω is defined as

$$p(\Omega) = \delta(\Omega, \Omega_0) = \begin{cases} 1, & \text{if } \Omega = \Omega_0 \\ 0, & \text{otherwise.} \end{cases} \tag{1.33}$$

Then, the MAP and ML solutions become, respectively,

$$\{\hat{\mathbf{f}}, \hat{\mathbf{h}}\}_{\text{MAP}} = \underset{\mathbf{f}, \mathbf{h}}{\text{argmax}} \; p(\mathbf{g} \mid \mathbf{f}, \mathbf{h}, \Omega_0) \, p(\mathbf{f} \mid \Omega_0) \, p(\mathbf{h} \mid \Omega_0) \tag{1.34}$$

$$\{\hat{\mathbf{f}}, \hat{\mathbf{h}}\}_{\text{ML}} = \underset{\mathbf{f}, \mathbf{h}}{\text{argmax}} \; p(\mathbf{g} \mid \mathbf{f}, \mathbf{h}, \Omega_0). \tag{1.35}$$

Many deconvolution methods can fit into this Bayesian formulation. The main differences among these algorithms come from the form of the likelihood, the particular choice of priors on the image, blur, and the hyperparameters, and the optimization methods used to find the solutions.

Observe that the regularization-based approaches using the L_2 norm frequently found in the literature also fall into this category. In these approaches the blind deconvolution problem is stated as a constrained minimization problem, where a cost function is minimized with a number of regularization constraint terms.

In regularization approaches the cost function is chosen as the error function $\|\mathbf{g} - \mathbf{Hf}\|_W^2$, which ensures fidelity to the data. The regularization terms are used to impose additional constraints on the optimization problem. Generally, these constraints ensure smoothness of the image and the blur, that is, the high-frequency energy of the image and the blur is minimized. The effect of the regularization terms is controlled by the regularization parameters, which basically represent the trade-off between fidelity to the data and desirable properties (smoothness) of the solutions.

For example, in [19], the classical regularized image restoration formulation used in [56, 57, 63] was extended to the BD case by adding a constraint for the blur. The problem is stated, in a relaxed minimization form, as

$$\hat{\mathbf{f}}, \hat{\mathbf{h}} = \underset{\mathbf{f}, \mathbf{h}}{\text{argmin}} \; \left[\|\mathbf{g} - \mathbf{Hf}\|_W^2 + \lambda_1 \|\mathbf{L}_f \mathbf{f}\|^2 + \lambda_2 \|\mathbf{L}_h \mathbf{h}\|^2 \right], \tag{1.36}$$

where λ_1 and λ_2 are the Lagrange multipliers for each constraint, and $\mathbf{L}_f$ and $\mathbf{L}_h$ are the regularization operators. In [19] each $\mathbf{L}$ is the Laplacian multiplied by a spatially varying weights term, calculated as in [57, 63–65] from the local image variance in order to provide some spatial adaptivity to avoid oversmoothing edges.

1.6.1.1 Iterated Conditional Modes

Let us consider again the solution of Equation (1.34). A major problem in the optimization is the simultaneous estimation of the variables $\mathbf{f}$ and $\mathbf{h}$. A widely used approach is that of *Alternating Minimization* (AM) of Equation (1.36) (or its continuous equivalent for PDE formulations), which follows the steepest descent with respect to one unknown while holding the other unknown constant. The advantage of this algorithm is its simplicity due to the linearization of the objective function. This optimization procedure corresponds to the Iterated Conditional Modes (ICM) proposed by Besag [66].

This estimation procedure has been applied to standard regularization approaches [19, 67], and to the anisotropic diffusion and TV type models described in Section 1.5.3.4, where the objective functional becomes

$$
\int_{S_f} \left(g(\mathbf{x}) - \int_{S_h} h(\mathbf{s} - \mathbf{x}) * f(\mathbf{x}) d\mathbf{s} \right)^2 d\mathbf{x} + \lambda_1 \int_{S_f} \kappa \left(|\nabla f(\mathbf{x})| \right) d\mathbf{x} + \lambda_2 \int_{S_h} \upsilon \left(|\nabla h(\mathbf{s})| \right) d\mathbf{s}.
$$

(1.37)

Partial derivatives with respect to f and h are taken to give the two PDEs for AM.

There are various numerical methods to solve the associated PDEs. These include the classical Euler, Newton, or Runge–Kutta methods; or recently developed approaches, such as time-marching [68], primal-dual methods [69], lagged diffusivity fixed point schemes [70], and half-quadratic regularization [54] (similar to the discrete schemes in [55, 71]). All of these methods employ techniques to discretize and linearize the PDEs to approximate the solution. The selection of a particular method depends on the computational limitations and speed requirements, since different techniques have different simplicity, stability, and convergence speed properties.

1.6.2 Minimum Mean Squared Error

The MAP estimate does not take into account the whole posterior PDF. If the posterior is sharply peaked about the maximum then this does not matter; however, in the case of high observation noise or a broad (heavy-tailed) posterior this estimate is likely to be unreliable. As mentioned in [72], for a Gaussian in high dimensions most of the probability *mass* is concentrated away from the probability *density* peak.

The Minimum Mean Squared Error (MMSE) estimate attempts to find the

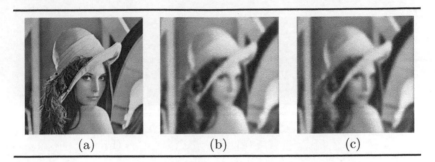

(a)   (b)   (c)

FIGURE 1.3: (a) Original Lena image; degraded images with Gaussian shaped PSF of variance 9, with: (b) BSNR=40 dB; (c) BSNR=20 dB.

optimal parameter values that minimize the expected mean squared error between the estimates and the true values. In other words we aim at calculating the mean value of $p(\mathbf{f}, \mathbf{h}, \Omega | \mathbf{g})$. In practice, finding MMSE estimates analytically is generally difficult, though it is possible with sampling-based methods (Section 1.6.5) and can be approximated using variational Bayesian methods (Section 1.6.4).

1.6.3   Marginalizing Hidden Variables

In the discussion so far none of the three unknowns, $\mathbf{f}$, $\mathbf{h}$, and Ω, have been marginalized out to perform inference on only a subset of $\mathbf{f}$, $\mathbf{h}$, and Ω.

We can, however, approach the BD inference problem by first calculating

$$\hat{\mathbf{h}}, \hat{\Omega} = \underset{\mathbf{h}, \Omega}{\operatorname{argmax}} \int_{\mathbf{f}} p(\Omega) p(\mathbf{f}, \mathbf{h}|\Omega) p(\mathbf{g}|\Omega, \mathbf{f}, \mathbf{h}) d\mathbf{f} \qquad (1.38)$$

and then selecting as restoration the image

$$\left. \hat{\mathbf{f}} \right|_{\hat{\mathbf{h}}, \hat{\Omega}} = \underset{\mathbf{f}}{\operatorname{argmax}} \, p(\mathbf{f}|\hat{\Omega}) p(\mathbf{g}|\hat{\Omega}, \mathbf{f}, \hat{\mathbf{h}}). \qquad (1.39)$$

We can also marginalize $\mathbf{h}$ and Ω to obtain

$$\hat{\mathbf{f}} = \underset{\mathbf{f}}{\operatorname{argmax}} \int_{\mathbf{h}, \Omega} p(\Omega) p(\mathbf{f}, \mathbf{h}|\Omega) p(\mathbf{g}|\Omega, \mathbf{f}, \mathbf{h}) d\mathbf{h} \cdot d\Omega \qquad (1.40)$$

The two above inference models are named Evidence- and Empirical-based analysis [73], respectively. The marginalized variables are called hidden variables.

The Expectation Maximization (EM) algorithm, first described in [74], is an incredibly popular technique in signal processing for iteratively solving

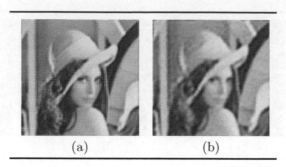

(a)   (b)

FIGURE 1.4: Blind deconvolution with the method in [35]. (a) 40 dB BSNR case; (b) 20 dB BSNR case.

ML and MAP problems that can be regarded as having *hidden data*. Its properties are well studied: convergence to a *local* maximum of the likelihood or the posterior distribution is guaranteed. It is particularly suited to inverse problems in image restoration and BD as it is obvious that the unobserved image, $\mathbf{f}$, represents a natural choice for the hidden data and in consequence for solving Equation (1.38).

The EM algorithm has been used in BD, for example, in [35, 36] in a general frequency domain formulation and in [28] using the ARMA model (Section 1.5.3.2). Some examples of the restorations possible using the method in [36] will now be demonstrated. Consider the original *Lena* image shown in Figure (1.3). This is synthetically degraded by blurring the original image by a Gaussian-shaped PSF with variance 9, and adding WGN with two different variances. This gives the images also in Figures (1.3b) and (1.3c), which have 40 dB and 20 dB Blurred-image SNR (BSNR) respectively. These test images will also be used in later sections. The restored images using the EM method [36] are shown in Figure (1.4).

Note that the Evidence-based analysis can also be used to marginalize the image $\mathbf{f}$ as well as the unknown parameters Ω to obtain $\mathrm{p}(\mathbf{h}|\mathbf{g})$, as in Equation (1.6), then the mode of this posterior distribution can be calculated. In order to estimate the original image $\mathbf{f}$ we can then use only the observation model (see [26] for details). An example of the results obtained by this method is shown in Figure (1.5). The original 256×256 pixel Cameraman image is blurred with a causal blur, and noise is added at 35 dB BSNR. The blind deconvolved image is shown in Figure (1.5).

It is rarely possible to calculate in closed form the integrals involved in the Evidence- and Empirical-based Bayesian inference. To solve this problem we can use approximations of the integrands. Let us consider the integral in

| | | |
| (a) | (b) | (c) |

FIGURE 1.5: (a) Original Cameraman image; (b) degraded image; (c) reconstruction with the method in [26].

Equation (1.38), then for each value of $\mathbf{h}$ and Ω we can calculate

$$\left.\hat{\mathbf{f}}\right|_{\mathbf{h},\Omega} = \underset{\mathbf{f}}{\mathrm{argmax}}\, p(\mathbf{f}|\Omega)p(\mathbf{g}|\mathbf{f},\mathbf{h},\Omega) \qquad (1.41)$$

and perform the second-order Taylor's expansion of $\log p(\mathbf{g}|\mathbf{f},\mathbf{h},\Omega)$ around $\hat{\mathbf{f}}$. As a consequence of the approximation, the integral in Equation (1.38) is performed over a distribution on $\mathbf{f}$ that is Gaussian and usually easy to calculate. This methodology is called Laplace distribution approximation [75, 76] and has been applied, for instance, by Galatsanos *et al.* [77,78] to partially known blur deconvolution problems.

1.6.4 Variational Bayesian Approach

Variational Bayesian methods are generalizations of the EM algorithm to compute ML or MAP estimates. The EM algorithm has proven to be very useful in a wide range of applications; however, in many problems its application is not possible because the posterior distribution cannot be specified. The variational methods overcome this shortcoming by approximating $p(\mathbf{f},\mathbf{h},\Omega|\mathbf{g})$ by a simpler distribution $q(\mathbf{f},\mathbf{h},\Omega)$ obtained by minimizing the Kullback–Leibler (KL) divergence between the variational approximation and the exact distribution. Additionally to providing approximations to the estimates based on $p(\mathbf{f},\mathbf{h},\Omega|\mathbf{g})$, the study of the distribution $q(\mathbf{f},\mathbf{h},\Omega)$ allows us to examine the quality of these estimates.

Note that the Laplace approximation [77,78] can be considered as an intermediate step between inference based on the true posterior and the one based on a variational distribution approximation.

The variational approximation applied to BD aims at approximating the intractable posterior distribution $p(\mathbf{f},\mathbf{h},\Omega|\mathbf{g})$ by a tractable one denoted by $q(\mathbf{f},\mathbf{h},\Omega)$. For an arbitrary PDF $q(\mathbf{f},\mathbf{h},\Omega)$, the goal is to minimize the KL divergence, given by: [79]

$$KL(q(\mathbf{f}, \mathbf{h}, \Omega) \parallel p(\mathbf{f}, \mathbf{h}, \Omega | \mathbf{g})) = \int q(\mathbf{f}, \mathbf{h}, \Omega) \log \left(\frac{q(\mathbf{f}, \mathbf{h}, \Omega)}{p(\mathbf{f}, \mathbf{h}, \Omega | \mathbf{g})} \right) d\mathbf{f} \cdot d\mathbf{h} \cdot d\Omega$$

$$= \int q(\mathbf{f}, \mathbf{h}, \Omega) \log \left(\frac{q(\mathbf{f}, \mathbf{h}, \Omega)}{p(\mathbf{f}, \mathbf{h}, \Omega, \mathbf{g})} \right) d\mathbf{f} \cdot d\mathbf{h} \cdot d\Omega$$

$$+ \text{const},$$

(1.42)

which is always nonnegative and equal to zero only when $q(\mathbf{f}, \mathbf{h}, \Omega) = p(\mathbf{f}, \mathbf{h}, \Omega | \mathbf{g})$, which corresponds to the EM result.

To reduce computational complexity and enable the approximate parameter distributions to be found in an analytic form, the PDF $q(\mathbf{f}, \mathbf{h}, \Omega)$ is factorized using the mean field approximation, such that

$$q(\mathbf{f}, \mathbf{h}, \Omega) = q(\mathbf{f})q(\mathbf{h})q(\Omega).$$

(1.43)

For a vector parameter $\theta \in \Theta = \{\mathbf{f}, \mathbf{h}, \Omega\}$, we denote by Θ_θ the subset of Θ with θ removed; for example, for $\theta = \mathbf{f}$, $\Theta_\mathbf{f} = \{\mathbf{h}, \Omega\}$ and $q(\Theta_\mathbf{f}) = q(\Omega)q(\mathbf{h})$. An iterative procedure can be developed to estimate the distributions of the parameters $\{\mathbf{f}, \mathbf{h}, \Omega\}$. At each iteration, the distribution of the parameter θ is estimated using the current estimates of the distribution of Θ_θ:

$$q^k(\theta) = \arg\min_{q(\theta)} KL(q^k(\Theta_\theta)q(\theta) \parallel p(\Theta \mid \mathbf{g})).$$

(1.44)

AM strategies can be employed. For example, in [34], a cascaded EM algorithm is proposed similar to the AM algorithm, where at each iteration the distributions of the parameters $\mathbf{f}$ and $\mathbf{h}$ are calculated in an alternating fashion while assuming one parameter to be constant. This approach can also be interpreted as an EM algorithm where at each stage the current estimate of one parameter is assumed "known" in estimating the other parameter, as in the classical image restoration problems.

Some example restorations using the variational method are now presented. Using the *VAR1* method presented by Likas and Galatsanos [34] produces the results shown in Figure (1.6). The *BR* method in Molina *et al.* [27] produces the restored images in Figure (1.7).

1.6.5 Sampling Methods

The most general approach to performing inference for the BD problem is to simulate the posterior distribution in Equation (1.5). This in theory allows us to perform inference on arbitrarily complex models in high-dimensional spaces, where no analytic solution is available. Markov Chain Monte Carlo (MCMC) methods (see, e.g., [61,80,81]) attempt to approximate the posterior distribution by the statistics of samples generated from a Markov Chain.

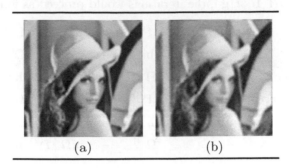

(a)   (b)

FIGURE 1.6: Restorations with the variational Bayesian method in [34]: (a) 40 dB BSNR case; (b) 20 dB BSNR case.

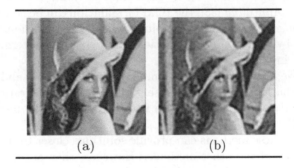

(a)   (b)

FIGURE 1.7: Restorations with variational Bayesian method in [27]: (a) 40 dB BSNR case; (b) 20 dB BSNR case.

The most simple example of MCMC is the Gibbs sampler which has been used in classical image restoration in conjunction with MRF image models [38]. If we can write down analytic expressions for the *conditional* distributions of all the parameters we wish to estimate, given the others, we simply draw samples from each of the distributions in turn, conditioned on the most recently sampled values of the other parameters. For example if we want to simulate $p(\mathbf{f}, \mathbf{h}, \Omega \,|\, \mathbf{g})$, the iterations would proceed as follows:

$$
\begin{aligned}
\text{First iteration:} \quad \mathbf{f}^{(1)} \;&\leftarrow\; p(\mathbf{f} \mid \mathbf{h}^{(0)}, \Omega^{(0)}, \mathbf{g}) \\
\mathbf{h}^{(1)} \;&\leftarrow\; p(\mathbf{h} \mid \mathbf{f}^{(1)}, \Omega^{(0)}, \mathbf{g}) \\
\Omega^{(1)} \;&\leftarrow\; p(\Omega \mid \mathbf{f}^{(1)}, \mathbf{h}^{(1)}, \mathbf{g}) \\
\text{Second iteration:} \quad \mathbf{f}^{(2)} \;&\leftarrow\; p(\mathbf{f} \mid \mathbf{h}^{(1)}, \Omega^{(1)}, \mathbf{g}) \\
\mathbf{h}^{(2)} \;&\leftarrow\; p(\mathbf{h} \mid \mathbf{f}^{(2)}, \Omega^{(1)}, \mathbf{g}) \\
\Omega^{(2)} \;&\leftarrow\; p(\Omega \mid \mathbf{f}^{(2)}, \mathbf{h}^{(2)}, \mathbf{g}) \\
&\;\;\vdots \\
t^{\text{th}} \text{ iteration:} \quad \mathbf{f}^{(t)} \;&\leftarrow\; p(\mathbf{f} \mid \mathbf{h}^{(t-1)}, \Omega^{(t-1)}, \mathbf{g}) \\
&\;\;\vdots
\end{aligned}
$$

where the symbol $\leftarrow$ means the value is drawn from the distribution on the right. Notice the similarity to the iterative procedure for the Variational Bayesian approach in Equation (1.44), where instead of drawing samples we are taking expectations of the same distributions. Similarly, with the use of *Simulated Annealing* [38], the ICM formulation can be considered a deterministic approximation of the sampling, where the conditional distributions are replaced by degenerate distributions at their modes (termed "instantaneous freezing" in [66]).

Once we have accumulated samples, point estimates and other statistics of the distribution may be found using Monte Carlo integration; for example, to find the MMSE estimate of the $\mathbf{f}$ we simply take the mean of the samples, $\frac{1}{n}\sum_{t=1}^{n} \mathbf{f}^{(t)}$.

Clearly, these methods can provide solutions closer to the optimal one than AM or any of the other methods. However, they are very computationally intensive by comparison, and although in theory convergence to the posterior is guaranteed, in practice it can be hard to tell when this has occurred; it may take a long time to explore the parameter space. Sampling methods could, for instance, be of use for the boundary condition model proposed in [82] where, because of the blur prior used, both direct inference as well as approximations by variational methods are difficult to perform.

1.7 Non–Bayesian Blind Image Deconvolution Models

In this section other blind deconvolution models, which have appeared in the literature and cannot be obtained from the Bayesian formulation described in the previous sections, are briefly reviewed.

1.7.1 Spectral and Cepstral Zero Methods

These algorithms fall into the *a priori* class of approaches described in Section 1.3, according to which the PSF of the blurring system is estimated separately from the image. They were the first examples of BD to be developed for images (see Stockham *et al.* [3] and Cannon [4]).

These blur identification algorithms are well suited to the problem when the frequency response of the blurring system has a known parametric form that is completely characterized by its frequency domain zeros: this is the case for linear motion blur and circular defocusing blur (Section 1.5.2). Let us rewrite Equation (1.2) in the frequency domain, while ignoring the noise term, i.e.,

$$G(\omega) = F(\omega) H(\omega). \tag{1.45}$$

Due to the multiplication of the spectra, the zeros of $G(\omega)$ are the zeros of $F(\omega)$ and $H(\omega)$ combined. Therefore the problem reduces to identifying which of the zeros of $G(\omega)$ belong to $H(\omega)$. The use of the parametric model makes this possible: the Fourier Transform (FT) of the linear and circular blurs are Sinc and Bessel functions that have periodic patterns of zeros in their spectra. The spacing between the zeros depends on the parameter L in Equation (1.9) or r in Equation (1.12). Hence if this pattern can be detected in $G(\omega)$, the parameters can be identified.

In practice this type of identification often fails due to the presence of noise masking the periodic patterns. The *homomorphic* [3] or *Cepstral* [4] methods attempt to exploit the effect of the nonstationary image and stationary blur to alleviate this difficulty. The homomorphic procedure begins by partitioning the image into blocks f_i, of size larger than the PSF. Each blurred block g_i is equal to the blur convolved with the unblurred block:

$$g_i(\mathbf{x}) = f_i(\mathbf{x}) * h_i(\mathbf{x}) + n_i(\mathbf{x}). \tag{1.46}$$

This expression holds apart from at the block boundaries, due to contamination from neighbouring blocks (in practice the blocks are windowed to reduce these edge effects). The log operator may then be applied to the FT of the blurred image, which has the effect of converting the original convolution to an addition. The average of these blocks may then be calculated (assuming N blocks):

$$\frac{1}{N} \sum_{i=1}^{N} \log(G_i(\omega)) \approx \frac{1}{N} \sum_{i=1}^{N} \log(F_i(\omega)) + \log(H_i(\omega)). \tag{1.47}$$

This summation now consists of the average of the contributions from the blocks $H_i(\omega)$, which are assumed to be equal, and $F_i(\omega)$, which are not, since the image blocks have varying spectral content. Therefore the blur component should tend to dominate in the summation. It may be possible to remove this average image component, subtracting the average of a collection of representative unblurred images.

As an alternative to Equation (1.47), the *Power Cepstrum* may be used, which involves taking the FT of the log power spectra of the signals. Combined with the block-based method just described, the result is that a large spike will occur in the Cepstral domain wherever there was a periodic pattern of zeros in the original Fourier domain; the distance of this spike from the origin represents the spacing between the zeros and hence may be used to identify the parameters of the blurs. The large contribution from the blur term repeated in each block will dominate, and the high-frequency noise and spectrally averaged image content tend to be separated from the blur spike in the Cepstral domain. These methods have been extended to use the Power Bispectrum instead in [83] which shows improved performance in low SNR conditions. However, these methods are all limited to the parametric PSF models.

1.7.2 Zero Sheet Separation Algorithms

Lane and Bates [6] have shown that any signal $g(\mathbf{x})$, formed by multiple convolutions, in theory, is *automatically deconvolvable*, provided its dimension is greater than one. This argument rests on the analytical properties of the Z-Transform (ZT) of N-dimensional signals with finite support, which is necessarily zero on $(2N - 2)$-dimensional hypersurfaces in a $2N$-dimensional space.

The method assumes that there is no additive noise, i.e., the relation $G(z_1, z_2) = H(z_1, z_2)F(z_1, z_2)$ holds, where $F(z_1, z_2)$ and $H(z_1, z_2)$ are the 2-D ZTs of the original image and blur. Then the task is to separate the 2-D ZT of the blurred image, $G(z_1, z_2)$, into the two convolutive factors. As $G(z_1, z_2)$ is a polynomial in z_1, z_2, the solution is equivalent to factorizing the polynomial, i.e., identifying the zeros that belong to each component.

In order to do this, several assumptions are made: the convolutive factors (image and blur) should have compact (finite) support; the ZT of each factor should be zero on a single continuous surface, its *zero sheet*; that these zero sheets do not coalesce — they only intersect at discrete points (which it is suggested often holds in practice).

Zero sheets appear conceptually useful in analyzing the BD problem; however, in practice their applicability is limited. The main problem of this method is its high computational complexity and sensitivity to noise.

1.7.3 ARMA Parameter Estimation Algorithms

The estimation of the ARMA parameters in the model described in Section 1.5.3.2 can be done in many different ways; in addition to the Bayesian approaches already described, there also exist Generalized Cross-Validation (GCV) and neural network-based algorithms, which are based on second-order statistics. Also, Higher-Order Statistics (HOS) approaches can be used for estimating the ARMA parameters for non-Gaussian models (it should be noted that with the second-order statistics algorithms the phase of the PSF can not be recovered, unless it is assumed that the PSF is minimum phase). The GCV algorithm will now be described.

Cross-validation is a parameter estimation method that partitions the observed data into two sets. Given some parameter values, the estimation set is used to form a prediction, and then the validation set is used to test the validity of this prediction. If the parameter values used were close to the correct ones, the validation criterion will be small. To make full use of the data, the criterion may be averaged across all the choices of estimation sets, using only one data element for each validation set. Hence the procedure is also referred to as the "leave one out" algorithm. Prior to its use in BD, GCV was used in the estimation of the regularization parameter for classical restoration [84,85].

When GCV is used to solve the BD problem [86], the original image $\hat{\mathbf{f}}$ is estimated using all but one of the pixels, $\mathbf{x}$, from the degraded image $\mathbf{g}$, for a particular value of the parameter set $\Theta = \{\mathbf{h}, \Omega\}$ under test. The validation criterion is then to test the difference between the pixel $\mathbf{x}$ of $\mathbf{g}$ that was left out, and the corresponding reblurred pixel in $\mathbf{H}\hat{\mathbf{f}}$. This is averaged over all choices of $\mathbf{x}$. With the GCV criterion defined, a numerical search technique may be used to search for the parameter set that minimizes the criterion.

1.7.4 Nonparametric Deterministic Constraints Algorithms

This class of algorithms differs from the other joint blur and image identification methods in that they do not explicitly model the original image or the PSF with a stochastic or deterministic model. Instead, they typically use an iterative formulation to impose deterministic constraints on the unknowns at each step. These deterministic constraints may include nonnegativity, finite support, and energy bounds on the image, the blur, or both. These constraints are incorporated into an optimality criterion which is minimized with numerical iterative techniques.

Examples in the literature based on deterministic constraint principles include the Iterative Blind Deconvolution (IBD) algorithm [7,87,88], McCallum's simulated annealing algorithm [89], the Nonnegativity And Support constraints with Recursive Image Filtering (NAS-RIF) algorithm [90], and the blind superresolution algorithm [91], among others. Some of these methods will now be briefly described.

1.7.4.1 The Iterative Blind Deconvolution Algorithms

One of the early IBD algorithms is the one proposed by Ayers and Dainty [7]. In addition to nonnegativity and finite support, it uses Wiener-like constraints to estimate image and blur in the Fourier domain at each iteration. Beginning with an initial random PSF estimate $\widehat{h}_0(\mathbf{x})$ and image estimate $\widehat{f}_0(\mathbf{x})$, the following sequence defines the algorithm at iteration i:

1. Find $\widehat{F}_i(\mathbf{k})$, the DFT of $\widehat{f}_i(\mathbf{x})$.

2. Impose blur constraints in the Fourier domain (where $(.)^*$ denotes complex conjugation):

$$\widetilde{H}_i\left(\mathbf{k}\right) = \frac{G(\mathbf{k})\widehat{F}_i^*\left(\mathbf{k}\right)}{\left|\widehat{F}_i\left(\mathbf{k}\right)\right|^2 + \alpha/\left|\widehat{H}_{i-1}\left(\mathbf{k}\right)\right|^2} \tag{1.48}$$

3. Find $\widetilde{h}_i(\mathbf{x})$, the IDFT of $\widetilde{H}_i\left(\mathbf{k}\right)$.

4. Impose spatial domain positivity and finite support constraints on $\widetilde{h}_i(\mathbf{x})$ to give $\widehat{h}_i(\mathbf{x})$

5. Find $\widehat{H}_i\left(\mathbf{k}\right)$, the DFT of $\widehat{h}_i(\mathbf{x})$.

6. Impose image constraints in the Fourier domain:

$$\widetilde{F}_i\left(\mathbf{k}\right) = \frac{G(\mathbf{k})\widehat{H}_i^*\left(\mathbf{k}\right)}{\left|\widehat{H}_i\left(\mathbf{k}\right)\right|^2 + \alpha/\left|\widehat{F}_i\left(\mathbf{k}\right)\right|^2}, \tag{1.49}$$

7. Find $\widetilde{f}_i(\mathbf{x})$, the IDFT of $\widetilde{F}_i\left(\mathbf{k}\right)$,

8. Impose spatial domain positivity and finite support constraints on $\widetilde{f}_i(\mathbf{x})$ to give $\widehat{f}_{i+1}(\mathbf{x})$.

9. Next iteration: set $i = i + 1$; go to step 1.

The real constant α represents the energy of the additive noise, and it has to be carefully chosen in order to obtain a reliable restoration. While this method is intuitively appealing, its convergence properties are undefined and tend to be highly sensitive to the initial guess [10]. Notice the similarity to the EM algorithm in the Fourier domain [29, 35]; however, the IBD algorithm is heuristically derived, and does not include estimation of the image and noise model parameters.

Further IBD-type algorithms using set-theoretic projection have been proposed by Yang *et al.* [92] and Lane [93], for the special case of astronomical speckle imaging.

1.7.4.2   The NAS-RIF Algorithms

The NAS-RIF method by Kundur and Hatzinakos [10, 90] is a similar method to IBD. It seeks to minimize a cost function at each step, by updating an FIR restoration filter, which is convolved with $\mathbf{g}$ to give an estimate $\widehat{\mathbf{f}}$. The cost function is based on the same constraints as those in IBD, apart from that no assumptions are made on the PSF other than it having an inverse, both of which must be absolutely summable. The constraints are applied via the method of Projection Onto Convex Sets (POCS). NAS-RIF seems to

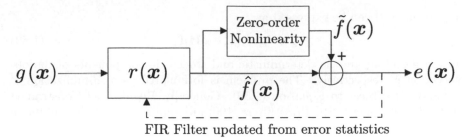

FIGURE 1.8: The blind image deconvolution based on the higher order statistics.

have been fairly successful in its goals, with good convergence properties and reasonable quality restorations. However, it is only applicable to the class of images with finite support, i.e., it is entirely contained in the image frame on a uniform black background. This may include applications in medical imaging and astronomy, but prevents its widespread use with other natural images. Extensions to this method have appeared in [94, 95].

1.7.5 Nonparametric Algorithms Based on Higher-Order Statistics

The principle of the HOS algorithms [96–99] is to use a nonlinear or non-Gaussian representation of the original image, allowing higher -order moments of the signal to be represented. These models have been typically applied when the image or its edges are modeled as sparse "spike-like" signals (for example, star fields).

In order to exploit HOS, an adaptive filter combined with a nonlinearity is used to restore the blurred image. The filter is updated to give a restored image that best fits the model. The adaptive filter structure for the HOS algorithms is shown in Figure 1.8.

Note that $g(\mathbf{x})$ represents the degraded image, $\widehat{f}(\mathbf{x})$ represents the original image estimate, $\widetilde{f}(\mathbf{x})$ represents the output of the zero-order nonlinearity, $r(\mathbf{x})$ represents the restoration filter used to obtain the original image estimate, and $e(\mathbf{x})$ represents an error sample. The restoration filter $r(\mathbf{x})$ is optimized in order to minimize some cost function J that involves a sequence of error samples $e(\mathbf{x})$.

1.7.6 Total Least Squares (TLS)

Total Least Squares approaches are extensions of the standard least squares methods. The PSF is assumed to be the sum of a deterministic and a stochastic component, that is, $\mathbf{H} = \overline{\mathbf{H}} + \delta\mathbf{H}$. Using this, the degradation model in

Equation (1.3) can be expressed as

$$\bar{\mathbf{g}} + \delta\mathbf{g} = (\bar{\mathbf{H}} + \delta\mathbf{H})\mathbf{f}, \tag{1.50}$$

where $\bar{\mathbf{g}}$ and $\delta\mathbf{g}$ are the deterministic and stochastic components of the observation $\mathbf{g}$, respectively. The problem is formulated as a minimization of $\delta\mathbf{g}$ and $\delta\mathbf{H}$ subject to Equation (1.50). Generally, Regularized Constrained Total Least Squares (RCTLS) filters [100] are applied to find the minimum values, where the matrices are assumed to have a special form, such as BTTB. In addition to the TLS solutions already presented that use the hierarchical Bayesian framework [78], linear algebra-based solutions are applied in [101].

1.7.7 Learning-Based Algorithms

Learning-based algorithms for image restoration and blind image restoration have been recently proposed [102, 103]. The basic idea with such an approach is that the prior knowledge required for solving various (inverse) problems can be learned from training data, i.e., a set of prototype images belonging to the same (statistical) class of images to the ones processed.

Original images and their degraded versions by the known degradation operator (restoration problem) are used for designing Vector Quantizer (VQ) codebooks. The codevectors are designed using the blurred images. For each such vector, the high frequency information obtained from the original images is also available. During restoration, the high-frequency information of a given degraded image is estimated from its low-frequency information based on the codebooks. For the BD problem, a number of codebooks are designed corresponding to various versions of the blurring function. Given a noisy and blurred image, one of the codebooks is chosen based on a similarity measure, therefore providing the identification of the blur. To make the restoration process computationally efficient, Principal Component Analysis (PCA) and VQ-Nearest Neighborhood approaches are utilized in [103].

1.7.8 Methods for Spatially Varying Degradation

Blind deconvolution in case of Spatially Varying (SV) degradation is a more difficult problem than the spatially invariant case. The blur is generally assumed to be varying smoothly or piecewise-smoothly, and the variation to be slow in the spatial domain. The standard EM procedure has been extended to use sectioned methods where the image is divided into blocks [104, 105]. A hierarchical sliding window approach with the local Fourier transform is employed in [106]. In [107], SV PSF identification for a known image is considered using an MRF model for the parameterization of the SV blur. Some of the spatially invariant methods described in previous sections are also extended to the SV blur case, for example, a parameterized piecewise-smooth degradation model is used to extend the anisotropic regularization-based restoration method in [45].

1.7.9 Multichannel Methods

Multichannel images are typically acquired using an imaging system with multiple sensors, multiple time instants, or multiple frequency bands. Examples of multichannel images include multispectral images, where different channels present different frequency bands, wave radiometric images, and image sequences, such as video. Reviews of classical and blind multichannel restoration methods are presented in [108, 109]. Multichannel methods can be classified into two approaches, Single-Input-Multiple Output (SIMO) and Multiple-Input-Multiple-Output (MIMO).

The EM approach in [35] is combined with the MIMO restoration method in [110] to obtain a blind multichannel method in [111], where cross-channel blurs are also taken into account. The SIMO multichannel restoration problem is addressed in [112] by an ARMA model, where the likelihood function maximization is performed using a steepest descent algorithm.

Another class of algorithms have been developed using Greatest Common Divisor (GCD) methods. Under the *relative coprimeness* condition — the channel PSFs share no common factor other than a scalar — the GCD of the outputs of the channels will be the original image. By exploiting the commutativity of the convolution operator, a matrix equation is formed involving channel outputs and PSFs. Then, as an extension of 1-D blind equalization methods, eigenstructure properties of the matrices may be used to directly estimate the PSFs and the original image. Perfect restoration algorithms for noise-free degradation with FIR PSFs have been proposed in [113] and [114]. A similar approach with a direct estimation method has been proposed in [115]. The noise is a major problem for these approaches. To deal with noise amplification, direct vector-space methods have been proposed in [116] and [117], and sufficiency conditions are derived for exact restoration in [118]. Rav-Acha and Peleg [119] have proposed a multichannel method with an *a priori* blur identification method via an exhaustive search, and a coprimeness condition is imposed on the channel model.

Recently an extension of [113] and [114] has been proposed in [120], that exploits anisotropic regularization priors mentioned in Section 1.5.3.4. As well as the standard TV form, a more advanced Mumford-Shah regularization term is also used. Very good restoration results are achieved even in low SNR conditions. This method is extended to deal with unknown PSF support and global translational motion in [20], where a MAP formulation is utilized.

Observe that there is an inherent advantage with multichannel methods in the amount of information available to aid both blur identification and image restoration. The coprimeness condition ensures that the problem becomes less ill-posed; therefore we should expect better results than in single-channel methods.

1.8 Conclusions

In this chapter we have analyzed the methods proposed in the literature to tackle BD problems from the point of view of Bayesian modeling and inference. We have shown that most of the proposed methods can be considered as particular selections of probability distributions and inference models within the Bayesian framework. The study of inference models that go from single-point estimates to distribution simulations makes possible the introduction of image and blur models encapsulating information that goes beyond simple prior constraints.

References

[1] A. K. Katsaggelos, ed., *Digital Image Restoration*. Springer–Verlag, 1991.

[2] M. R. Banham and A. K. Katsaggelos, "Digital image restoration," *IEEE Signal Processing Magazine*, vol. 14, no. 2, pp. 24–41, 1997.

[3] T. G. Stockham, Jr., T. M. Cannon, and R. B. Ingebretsen, "Blind deconvolution through digital signal processing," *Proceedings IEEE*, vol. 63, no. 4, pp. 678–692, 1975.

[4] M. Cannon, "Blind deconvolution of spatially invariant image blurs with phase," *IEEE Transactions on Acoustic, Speech, and Signal Processing*, vol. 24, no. 1, pp. 58–63, 1976.

[5] A. M. Tekalp, H. Kaufman, and J. W. Woods, "Identification of image and blur parameters for the restoration of noncausal blurs," *IEEE Transactions on Acoustic, Speech, and Signal Processing*, vol. 34, pp. 963–972, August 1986.

[6] R. G. Lane and R. H. T. Bates, "Automatic multidimensional deconvolution," *Journal of the Optical Society of America-A*, vol. 4, no. 1, pp. 180–188, 1987.

[7] G. R. Ayers and J. C. Dainty, "Iterative blind deconvolution method and its applications," *Optics Letters*, vol. 13, no. 7, pp. 547–549, 1988.

[8] K. T. Lay and A. K. Katsaggelos, "Simultaneous identification and restoration of images using maximum likelihood estimation and the EM algorithm," in *Proc. 26th Annual Allerton Conf. on Commun., Control and Computing* (Monticello, IL), pp. 661–662, September 1988.

[9] R. L. Lagendijk, J. Biemond, and D. E. Boekee, "Blur identification using the expectation-maximization algorithm," *ICASSP, IEEE International Conference on Acoustics, Speech and Signal Processing — Proceedings*, vol. 3, pp. 1397–1400, 1989.

[10] D. Kundur and D. Hatzinakos, "Blind image deconvolution," *IEEE Signal Processing Magazine*, vol. 13, no. 3, pp. 43–64, 1996.

[11] D. Kundur and D. Hatzinakos, "Blind image deconvolution revisited," *IEEE Signal Processing Magazine*, vol. 13, no. 6, pp. 61–63, 1996.

[12] J. Krist, "Simulation of HST PSFs using Tiny Tim," in *Astronomical Data Analysis Software and Systems IV* (R. A. Shaw, H. E. Payne, and J. J. E. Hayes, eds.), (San Francisco, USA), pp. 349–353, Astronomical Society of the Pacific, 1995.

[13] T. J. Schultz, "Multiframe blind deconvolution of astronomical images," *Journal of the Optical Society of America-A*, vol. 10, pp. 1064–1073, 1993.

[14] T. Bretschneider, P. Bones, S. McNeill, and D. Pairman, "Image-based quality assessment of SPOT data," in *Proceedings of the American Society for Photogrammetry & Remote Sensing*, 2001. Unpaginated CD-ROM.

[15] F. S. Gibson and F. Lanni, "Experimental test of an analytical model of aberration in an oil-immersion objective lens used in three-dimensional light microscopy," *Journal of the Optical Society of America-A*, vol. 8, pp. 1601–1613, 1991.

[16] O. Michailovich and D. Adam, "A novel approach to the 2-D blind deconvolution problem in medical ultrasound," *IEEE Transactions on Medical Imaging*, vol. 24, no. 1, pp. 86–104, 2005.

[17] M. Roggemann, "Limited degree-of-freedom adaptive optics and image reconstruction," *Applied Optics*, vol. 30, pp. 4227–4233, 1991.

[18] P. Nisenson and R. Barakat, "Partial atmospheric correction with adaptive optics," *Journal of the Optical Society of America-A*, vol. 4, pp. 2249–2253, 1991.

[19] Y. L. You and M. Kaveh, "A regularization approach to joint blur and image restoration," *IEEE Transactions on Image Processing*, vol. 5, no. 3, pp. 416–428, 1996.

[20] F. Šroubek and J. Flusser, "Multichannel blind deconvolution of spatially misaligned images," *IEEE Transactions on Image Processing*, vol. 7, pp. 45–53, July 2005.

[21] C. A. Segall, R. Molina, and A. K. Katsaggelos, "High-resolution images from low-resolution compressed video," *IEEE Signal Processing Magazine*, vol. 20, no. 3, pp. 37–48, 2003.

[22] S. Dai, M. Yang, Y. Wu, and A. K. Katsaggelos, "Tracking motion-blurred targets in video," in *International Conference on Image Processing (ICIP'06)*, Atlanta, GA, October 2006.

[23] K. Faulkner, C. J. Kotre, and M. Louka, "Veiling glare deconvolution of images produced by X-ray image intensifiers," *Third Int. Conf. on Image Proc. and Its Applications*, pp. 669–673, 1989.

[24] A. K. Jain, *Fundamentals of Digital Image Processing*. New Jersey: Prentice Hall, 1 ed., 1989.

[25] M. Bertero and P. Boccacci, *Introduction to Inverse Problems in Imaging*. Institute of Physics Publishing, 1 ed., 1998.

[26] T. E. Bishop and J. R. Hopgood, "Blind image restoration using a block-stationary signal model," in *ICASSP, IEEE International Conference on Acoustics, Speech and Signal Processing — Proceedings*, May 2006.

[27] R. Molina, J. Mateos, and A. Katsaggelos, "Blind deconvolution using a variational approach to parameter, image, and blur estimation," *IEEE Transactions on Image Processing*, vol. 15, no. 12, pp. 3715–3727, December 2006. .

[28] R. L. Lagendijk, J. Biemond, and D. E. Boekee, "Identification and restoration of noisy blurred images using the expectation-maximization algorithm," *IEEE Transactions on Acoustic, Speech, and Signal Processing*, vol. 38, pp. 1180-1191, July 1990.

[29] A. K. Katsaggelos and K. T. Lay, "Maximum likelihood identification and restoration of images using the expectation-maximization algorithm," in *Digital Image Restoration* (A. K. Katsaggelos, ed.), Springer-Verlag, 1991.

[30] A. F. J. Moffat, "A theoretical investigation of focal stellar images in the photographic emulsion and application to photographic photometry," *Astronomy and Astrophysics*, vol. 3, pp. 455–461, 1969.

[31] R. Molina and B. D. Ripley, "Using spatial models as priors in astronomical image analysis," *Journal of Applied Statistics*, vol. 16, pp. 193–206, 1989.

[32] H.-C. Lee, "Review of image-blur models in a photographic system using the principles of optics," *Optical Engineering*, vol. 29, pp. 405–421, May 1990.

[33] B. D. Ripley, *Spatial Statistics*, pp. 88–90, John Wiley, 1981.

[34] A. C. Likas and N. P. Galatsanos, "A variational approach for Bayesian blind image deconvolution," *IEEE Transactions on Signal Processing*, vol. 52, no. 8, pp. 2222–2233, 2004.

[35] K. T. Lay and A. K. Katsaggelos, "Image identification and image restoration based on the expectation-maximization algorithm," *Optical Engineering*, vol. 29, pp. 436–445, May 1990.

[36] A. K. Katsaggelos and K. T. Lay, "Maximum likelihood blur identification and image restoration using the EM algorithm," *IEEE Transactions on Signal Processing*, vol. 39, no. 3, pp. 729–733, 1991.

[37] H. Derin and H. Elliott, "Modelling and segmentation of noisy and textured images using Gibbs random fields," *IEEE Transactions on Pattern Analysis and Machine Intelligence*, vol. PAMI-9, pp. 39–55, January 1987.

[38] S. Geman and D. Geman, "Stochastic relaxation, Gibbs distributions, and the Bayesian restoration of images," *IEEE Transactions on Pattern Analysis and Machine Intelligence*, vol. PAMI-6, no. 6, pp. 721–741, 1984.

[39] R. R. Schultz and R. L. Stevenson, "Extraction of high-resolution frames from video sequences," *IEEE Transactions on Image Processing*, vol. 5, no. 6, pp. 996–1011, 1996.

[40] J. Zhang, "The mean field theory in EM procedures for blind Markov random field image restoration," *IEEE Transactions on Image Processing*, vol. 2, no. 1, pp. 27–40, 1993.

[41] B. A. Chipman and B. D. Jeffs, "Blind multiframe point source image restoration using MAP estimation," *Conference Record of the Asilomar Conference on Signals, Systems and Computers*, vol. 2, pp. 1267–1271, 1999.

[42] C. S. Won and R. M. Gray, *Stochastic Image Processing*. Information Technology: Transmission, Processing, and Storage, Kluwer Academic/Plenum Publishers, 2004.

[43] C. A. Bouman and K. Sauer, "Generalized Gaussian image model for edge-preserving MAP estimation," *IEEE Transactions on Image Processing*, vol. 2, no. 3, pp. 296–310, 1993.

[44] F.-C. Jeng and J. W. Woods, "Compound Gauss-Markov random fields for image estimation," *IEEE Transactions on Signal Processing*, vol. 39, no. 3, pp. 683–697, 1991.

[45] Y. L. You and M. Kaveh, "Blind image restoration by anisotropic regularization," *IEEE Transactions on Image Processing*, vol. 8, no. 3, pp. 396–407, 1999.

[46] T. F. Chan and C.-K. Wong, "Total variation blind deconvolution," *IEEE Transactions on Image Processing*, vol. 7, no. 3, pp. 370–375, 1998.

[47] A. Hamza, H. Krim, and G. Unal, "Unifying probabilistic and variational estimation," *IEEE Signal Processing Magazine*, vol. 19, no. 5, pp. 37–47, 2002.

[48] T. F. Chan and J. Shen, *Image Processing and Analysis: Variational, Pde, Wavelet, and Stochastic Methods.* SIAM, 2005.

[49] T. F. Chan, S. Osher, and J. Shen, "The digital TV filter and nonlinear denoising," *IEEE Transactions on Image Processing*, vol. 10, no. 2, pp. 231–241, 2001.

[50] J. M. Bioucas-Dias, M. A. T. Figueiredo, and J. P. Oliveira, "Total variation-based image deconvolution: a majorization-minimization approach," in *ICASSP, IEEE International Conference on Acoustics, Speech and Signal Processing — Proceedings*, May 2006.

[51] Y. You, W. Xu, A. Tannenbaum, and M. Kaveh, "Behavioral analysis of anisotropic diffusion in image processing," *IEEE Transactions on Image Processing*, vol. 5, no. 11, pp. 1539–1553, 1996.

[52] J. Weickert, "A review of nonlinear diffusion filtering," in *SCALE-SPACE '97: Proceedings of the First International Conference on Scale-Space Theory in Computer Vision*, pp. 3–28, London, UK, Springer-Verlag, 1997.

[53] Y.-L. You and M. Kaveh, "Ringing reduction in image restoration by orientation-selective regularization," *IEEE Signal Processing Letters*, vol. 3, no. 2, pp. 29–31, 1996.

[54] A. Chambolle and P.-L. Lions, "Image recovery via total variation minimization and related problems," *Numerische Mathematik*, vol. 76, no. 2, pp. 167–188, 1997.

[55] D. Geman and G. Reynolds, "Constrained restoration and the recovery of discontinuities," *IEEE Transactions on Pattern Analysis and Machine Intelligence*, vol. 14, no. 3, pp. 367–383, 1992.

[56] R. L. Lagendijk, J. Biemond, and D. E. Boekee, "Regularized iterative image restoration with ringing reduction," *IEEE Transactions on Acoustic, Speech, and Signal Processing*, vol. 36, pp. 1874–1888, December 1988.

[57] A. K. Katsaggelos, J. Biemond, R. W. Schafer, and R. M. Mersereau, "A regularized iterative image restoration algorithm," *IEEE Transactions on Signal Processing*, vol. 39, pp. 914–929, April 1991.

[58] J. O. Berger, *Statistical Decision Theory and Bayesian Analysis*, ch. 3 and 4. New York, Springer-Verlag, 1985.

[59] H. Raiffa and R. Schlaifer, *Applied Statistical Decision Theory.* Division of Research, Graduate School of Business, Administration, Harvard University, Boston, 1961.

[60] A. Gelman, J. B. Carlin, H. S. Stern, and D. R. Rubin, *Bayesian Data Analysis*, Chapman & Hall, 2003.

[61] R. M. Neal, "Probabilistic inference using Markov chain Monte Carlo methods," Tech. Rep. CRG-TR-93-1, Dept. of Computer Science, University of Toronto, 1993. available online at http://www.cs.toronto.edu/~radford/res-mcmc.html.

[62] M. I. Jordan, Z. Ghahramani, T. S. Jaakola, and L. K. Saul, "An introduction to variational methods for graphical models," in *Learning in Graphical Models*, pp. 105–162, MIT Press, 1998.

[63] A. K. Katsaggelos, *Iterative Image Restoration Algorithms*, PhD thesis, Georgia Institute of Technology, School of Electrical Engineering, August 1985.

[64] S. N. Efstratiadis and A. K. Katsaggelos, "Adaptive iterative image restoration with reduced computational load," *Optical Engineering*, vol. 29, pp. 1458–1468, 1990.

[65] M. G. Kang and A. K. Katsaggelos, "General choice of the regularization functional in regularized image restoration," *IEEE Transactions on Image Processing*, vol. 4, no. 5, pp. 594–602, 1995.

[66] J. Besag, "On the statistical analysis of dirty pictures," *Journal of the Royal Statistical Society. Series B (Methodological)*, vol. 48, no. 3, pp. 259–302, 1986.

[67] L. Chen and K.-H. Yap, "A soft double regularization approach to parametric blind image deconvolution," *IEEE Transactions on Image Processing*, vol. 14, no. 5, pp. 624–633, 2005.

[68] L. I. Rudin, S. Osher, and E. Fatemi, "Nonlinear total variation based noise removal algorithms," in *Proceedings of the eleventh annual international conference of the Center for Nonlinear Studies on Experimental mathematics: computational issues in nonlinear science*, Amsterdam, The Netherlands, pp. 259–268, Elsevier North-Holland, Inc., 1992.

[69] T. F. Chan, G. H. Golub, and P. Mulet, "A nonlinear primal-dual method for total variation-based image restoration," *SIAM Journal on Scientific Computing*, vol. 20, pp. 1964–1977, November 1999.

[70] C. R. Vogel and M. E. Oman, "Iterative methods for total variation denoising," *SIAM Journal on Scientific Computing*, vol. 17, no. 1, pp. 227–238, 1996.

[71] D. Geman and C. Yang, "Nonlinear image recovery with half-quadratic regularization," *IEEE Transactions on Image Processing*, vol. 4, no. 7, pp. 932–946, 1995.

[72] R. Molina, A. K. Katsaggelos, and J. Mateos, "Bayesian and regularization methods for hyperparameter estimation in image restoration," *IEEE Transactions on Image Processing*, vol. 8, no. 2, pp. 231–246, 1999.

[73] R. Molina, "On the hierarchical Bayesian approach to image restoration. Applications to Astronomical images," *IEEE Transactions on Pattern Analysis and Machine Intell.*, vol. 16, no. 11, pp. 1122–1128, 1994.

[74] A. D. Dempster, N. M. Laird, and D. B. Rubin, "Maximum likelihood from incomplete data via the E-M algorithm," *Journal of the Royal Statistical Society: Series B*, vol. 39, pp. 1–37, 1977.

[75] R. Kass and A. E. Raftery, "Bayes factors," *Journal of the American Statistical Association*, vol. 90, pp. 773–795, 1995.

[76] D. J. C. MacKay, "Probable networks and plausible predictions: a review of practical Bayesian methods for supervised neural networks," *Network: Computation in Neural Systems*, no. 6, pp. 469-505, 1995.

[77] N. P. Galatsanos, V. Z. Mesarovic, R. Molina, A. K. Katsaggelos, and J. Mateos, "Hyperparameter estimation in image restoration problems with partially-known blurs," *Optical Engineering*, vol. 41, no. 8, pp. 1845–1854, 2002.

[78] N. P. Galatsanos, V. Z. Mesarovic, R. Molina, and A. K. Katsaggelos, "Hierarchical Bayesian image restoration for partially-known blur," *IEEE Transactions on Image Processing*, vol. 9, no. 10, pp. 1784–1797, 2000.

[79] S. Kullback, *Information Theory and Statistics*, New York, Dover Publications, 1959.

[80] C. Andrieu, N. de Freitras, A. Doucet, and M. Jordan, "An introduction to MCMC for machine learning," *Machine Learning*, vol. 50, pp. 5–43, 2003.

[81] J. J. K. O'Ruanaidh and W. Fitzgerald, *Numerical Bayesian Methods Applied to Signal Processing*. Springer Series in Statistics and Computing, New York, Springer, 1 ed., 1996.

[82] R. Molina, A. K. Katsaggelos, J. Abad, and J. Mateos, "A Bayesian approach to blind deconvolution based on Dirichlet distributions," in *1997 International Conference on Acoustics, Speech and Signal Processing (ICASSP'97)*, vol. IV, Munich, Germany, pp. 2809–2812, 1997.

[83] M. M. Chang, A. M. Tekalp, and A. T. Erdem, "Blur identification using the bispectrum," *IEEE Transactions on Signal Processing*, vol. 39, no. 10, pp. 2323–2325, 1991.

[84] N. P. Galatsanos and A. K. Katsaggelos, "Methods for choosing the regularization parameter and estimating the noise variance in image restoration and their relation," *IEEE Transactions on Image Processing*, vol. 1, pp. 322–336, 1992.

[85] S. J. Reeves and R. M. Mersereau, "Optimal estimation of the regularization parameters and stabilizing functional for regularized image restoration," *Optical Optical Engineering*, vol. 29, no. 5, pp. 446–454, 1990.

[86] S. Reeves and R. Mersereau, "Blur identification by the method of generalized cross-validation," *IEEE Transactions on Image Processing*, vol. 1, pp. 301–311, July 1992.

[87] N. Miura and N. Baba, "Extended-object reconstruction with sequential use of the iterative blind deconvolution method," *Optics Communications*, vol. 89, pp. 375–379, 1992.

[88] N. M. F. Tsumuraya and N. Baba, "Iterative blind deconvolution method using Lucy's algorithm," *Astronomy and Astrophysics*, vol. 282, no. 2, pp. 699–708, 1994.

[89] B. C. McCallum, "Blind deconvolution by simulated annealing," *Optics Communications*, vol. 75, no. 2, pp. 101–105, 1990.

[90] D. Kundur and D. Hatzinakos, "A novel blind deconvolution scheme for image restoration using recursive filtering," *IEEE Transactions on Signal Processing*, vol. 46, no. 2, pp. 375–390, 1998.

[91] K. Nishi and S. Ando, "Blind superresolving image recovery from blur-invariant edges," *IEEE Transactions on Acoustic Speech, and Signal Processing*, vol. 5, pp. 85–88, 1994.

[92] Y. Yang, N. P. Galatsanos, and H. Stark, "Projection based blind deconvolution," *Journal of the Optical Society of America-A*, vol. 11, no. 9, pp. 2401–2409, 1994.

[93] R. G. Lane, "Blind deconvolution of speckle images," *Journal of the Optical Society of America-A*, vol. 9, pp. 1508–1514, September 1992.

[94] M. Ng, R. Plemmons, and S. Qiao, "Regularization of RIF blind image deconvolution," *IEEE Transactions on Image Processing*, vol. 9, no. 6, pp. 1130–1134, 2000.

[95] C. A. Ong and J. Chambers, "An enhanced NAS-RIF algorithm for blind image deconvolution," *IEEE Transactions on Image Processing*, vol. 8, no. 7, pp. 988–992, 1999.

[96] P. Campisi and G. Scarano, "A multiresolution approach for texture synthesis using the circular harmonic functions," *IEEE Transactions on Image Processing*, vol. 11, pp. 37–51, January 2002.

[97] G. Panci, P. Campisi, C. Colonnese, and G. Scarano, "Multichannel blind image deconvolution using the Bussgang algorithm: spatial and multiresolution approaches," *IEEE Transactions on Image Processing*, vol. 12, pp. 1324–1337, November 2003.

[98] H. S. Wu, "Minimum entropy deconvolution for restoration of blurred two-tone images," *Electronics Letters*, vol. 26, no. 15, pp. 1183–1184, 1990.

[99] R. A. Wiggins, "Minimum entropy deconvolution," *Geoexploration*, vol. 16, pp. 21–35, 1978.

[100] V. Z. Mesarovic, N. P. Galatsanos, and A. K. Katsaggelos, "Regularized constrained total least–squares image restoration," *IEEE Transactions on Image Processing*, vol. 4, pp. 1096–1108, August 1995.

[101] N. Mastronardi, P. Lemmerling, S. V. Huffel, A. Kalsi, and D. O'Leary, "Implementation of regularized structured total least squares algorithms for blind image blurring," *Linear Algebra and Its Applications*, vol. 391, no. 1–3, pp. 203–221, 2004.

[102] K. Panchapakesan, D. G. Sheppard, M. W. Marcellin, and B. R. Hunt, "Blur identification from vector quantizer encoder distortion," *IEEE Transactions on Image Processing*, vol. 10, pp. 465–470, March 2001.

[103] R. Nakagaki and A. K. Katsaggelos, "A VQ-based blind image restoration algorithm," *IEEE Transactions on Image Processing*, vol. 12, pp. 1044–1053, September 2003.

[104] R. L. Lagendijk and J. Biemond, "Block–adaptive image identification and restoration," *Proceedings ICASSP, IEEE International Conference on Acoustics, Speech, and Signal Processing*, vol. 4, pp. 2497–2500, 1991.

[105] Y. P. Guo, H. P. Lee, and C. L. Teo, "Blind restoration of images degraded by space-variant blurs using iterative algorithms for both blur identification and image restoration," *Image and Vision Computing*, vol. 15, pp. 399–410, May 1997.

[106] M. K. Ozkan, A. Tekalp, and M. Sezan, "Identification of a class of space-variant image blurs," *Proceedings of SPIE – The International Society for Optical Engineering*, vol. 1452, pp. 146–156, 1991.

[107] A. Rajagopalan and S. Chaudhuri, "MRF model-based identification of shift-variant point spread function for a class of imaging systems," *Signal Processing*, vol. 76, no. 3, pp. 285–299, 1999.

[108] F. Šroubek and J. Flusser, "An overview of multichannel image restoration techniques," in *Week of Doctoral Students* (J. Safrnkov, ed.), Prague, pp. 580–585, Matfyzpress, 1999.

[109] N. P. Galatsanos, M. Wernick, and A. K. Katsaggelos, "Multi-channel image recovery," in *Handbook of Image and Video Processing* (A. Bovik, ed.), ch. 3.7, pp. 161–174, Academic Press, 2000.

[110] A. K. Katsaggelos, K. T. Lay, and N. Galatsanos, "A general framework for frequency domain multichannel signal processing," *IEEE Transactions on Image Processing*, vol. 2, no. 3, pp. 417–420, 1993.

[111] B. C. Tom, K. Lay, and A. K. Katsaggelos, "Multichannel image identification and restoration using the expectation-maximization algorithm," *Optical Engineering*, vol. 35, no. 1, pp. 241–254, 1996.

[112] A. Rajagopalan and S. Chaudhuri, "A recursive algorithm for maximum likelihood-based identification of blur from multiple observations," *IEEE Transactions on Image Processing*, vol. 7, no. 7, pp. 1075–1079, 1998.

[113] G. Harikumar and Y. Bresler, "Perfect blind restoration of images blurred by multiple filters: theory and efficient algorithms," *IEEE Transactions on Image Processing*, vol. 8, pp. 202–219, February 1999.

[114] G. Harikumar and Y. Bresler, "Exact image deconvolution from multiple FIR blurs," *IEEE Transactions on Image Processing*, vol. 8, no. 6, pp. 846–862, 1999.

[115] G. Giannakis and R. J. Heath, "Blind identification of multichannel FIR blurs and perfect image restoration," *IEEE Transactions on Image Processing*, vol. 9, no. 11, pp. 1877–1896, 2000.

[116] H. Pai and A. C. Bovik, "Exact multichannel blind image restoration," *IEEE Signal Processing Letters*, vol. 4, no. 8, pp. 217–220, 1997.

[117] H. Pai and A. C. Bovik, "On eigenstructure-based direct multichannel blind image restoration," *IEEE Transactions on Image Processing*, vol. 10, no. 10, pp. 1434–1446, 2001.

[118] H. Pai, J. Havlicek, and A. C. Bovik, "Generically sufficient conditions for exact multichannel blind image restoration," *Proceedings of the 1998 IEEE International Conference on Acoustics, Speech, and Signal Processing*, vol. 5, pp. 2861–2864, 1998.

[119] A. Rav-Acha and S. Peleg, "Two motion-blurred images are better than one," *Pattern Recognition Letters*, vol. 26, pp. 311–317, 2005.

[120] F. Šroubek and J. Flusser, "Multichannel blind iterative image restoration," *IEEE Transactions on Image Processing*, vol. 12, no. 9, pp. 1094–1106, 2003.

2

Blind Image Deconvolution Using Bussgang Techniques: Applications to Image Deblurring and Texture Synthesis

Patrizio Campisi, Alessandro Neri

Dipartimento di Elettronica Applicata, Università degli Studi Roma Tre,
Via della Vasca Navale 84, I-00146 Roma, Italy
e-mail: (campisi, neri)@uniroma3.it

Stefania Colonnese, Gianpiero Panci, Gaetano Scarano

Dipartimento INFOCOM, Università "La Sapienza" di Roma
via Eudossiana 18, I-00184 Roma, Italy
e-mail: gaetano@infocom.ing.uniroma1.it

Abstract

In this chapter, Bussgang blind deconvolution techniques are reviewed in the general Bayesian framework of minimum mean square error (MMSE) estimation, and some recent activities of the authors on both single-channel and multichannel blind image deconvolution, under the general framework of Bussgang deconvolution, are described.

Applications of the single-channel Bussgang blind deconvolution approach to perform unsupervised texture synthesis is detailed. Moreover, the Bussgang blind deconvolution method is generalized to the multichannel case with application to image deblurring problems. Specifically, we address the restoration problem of poorly spatially correlated images as well as strongly correlated (natural) images and experimental results pertaining to restoration of motion blurred text images, out-of-focus spiky images, and blurred natural images are given. A theoretical analysis to show the local convergence properties of the Bussgang deconvolution algorithm is also reported.

2.1 Introduction

Image restoration [1, 2] has been widely studied in the literature because of its theoretical as well as practical importance in many application fields. Its goal consists in recovering the original image from a single or multiple degradated observations. The different proposed restoration techniques depend on both the image model formation and the occurred degradation.

A linear degradation model is usually employed in many applications. According to this model the observed image is obtained by *convolving* the original image by means of a linear filter (the blurring filter) in the presence of additive independent noise. In order to restore the original image, the *deconvolution* of the degradated image by means of an estimation of the blur filter is necessary.

Many approaches to perform image restoration using deconvolution approaches in fields such as astronomical imaging [3–5], remote sensing [6, 7], and medical imaging [8, 9], to cite only a few, have been proposed.

In some application cases the blur is assumed known, and well known deconvolution methods, such as Wiener filtering, recursive Kalman filtering, and constrained iterative deconvolution methods, are fruitfully employed for restoration.

However, in many practical situations, the blur is partially known [10] or unknown, because an exact knowledge about the mechanism of the image degradation process is not available. Therefore, the blurring process needs to be characterized on the basis of the available blurred data and blind image restoration techniques have to be devised for restoration. These techniques aim at the retrieval of the original image, observed through a nonideal channel whose characteristics are unknown or partially known in the restoration stage.

Many blind restoration algorithms have been proposed in the past and an extended survey for the single-channel formulation can be found in [11, 12]. A maximum likelihood method to the blur identification is presented in [13]. The resulting algorithm is an iterative approach which simultaneously identifies and restores noisy blurred images. In [14] the generalized cross-validation criterion, introduced in the context of image restoration, has been successfully extended to the blur identification problem. The problem of restoring an image distorted by additive noise and by a linear space-invariant point spread function, which is not exactly known, is formulated in [15] as the solution of a perturbed set of linear equations by means of the regularized constrained total least squares method. In [16] a blind deconvolution approach based on the total variation (TV) minimization method [17] is proposed. The minimization of the TV norm has been proven to be extremely effective especially for recovering edges of images as well as motion blur and out-of-focus blur. In [18] the authors make use of the Fourier phase for image restoration [19], applying appropriate constraints in the Radon domain. In [20] an algorithm for the blind restoration of images with unknown, spatially varying blur is presented. The proposed method consists in partitioning the image into tiles whose blur functions can be approximated as correlated and approximately spatially invariant. Each blur function is then estimated by using phase diversity [21]. In [22] the blind image deconvolution problem is addressed in the Bayesian framework by using an approach which can be seen as a generalization of the expectation maximization (EM) algorithm. In [23] a modified maximum likelihood (ML) approach, namely the quasi-ML (QML) approach, has been used for blind deconvolution of images. In this approach, an approximation of the probability density function is used since the exact source distribution, necessary for the ML approach, is often unknown.

In some applications the observation system is able to give multiple observations of the original image. In electron microscopy, for example, many differently focused versions of the same image are acquired during a single experiment, due to an intrinsic trade-off between the bandwidth of the imaging system and the contrast of the resulting image. In other applications, such as telesurveillance, multiple observed images can be acquired in order to better counteract, in the restoration phase, possible degradations due to motion, defocus, or noise. In remote sensing applications, by employing sensor diversity, the same scene can be acquired at different times through the atmosphere that can be modeled as a time-variant channel.

Different approaches have been proposed in the recent past on multichannel methods. In [24] an algorithm for the deconvolution of an unknown, possibly colored, Gaussian stationary or nonstationary signal, observed through two or more unknown channels, is described. The proposed EigenVector-based Algorithm for Multichannel (EVAM) blind deconvolution is based on eigenvalue decomposition of a sample correlation matrix. In [25] the problem of restoring an image from its noisy convolution in a multichannel framework is presented. Under some mild assumptions, both the filters and the image can be exactly determined from noise-free observations. In [26] blind order determination, blind blur identification, and blind image restoration are approached from a novel deterministic framework that allows the image to be nonstationary and have unknown color or distribution. Both in [25] and in [26] the channel estimation phase precedes the restoration phase. Once the channel has been estimated, image restoration is performed either by subspace-based and likelihood-based algorithms [25], or by a bank of finite impulse response (FIR) filters optimized with respect to a deterministic criterion [26]. In [27] an algorithm for direct multichannel blind image restoration is proposed. In the noise-free case the original image can be exactly restored but for a scalar ambiguity factor. In [28] a novel iterative algorithm based on anisotropic denoising techniques of total variation and a Mumford–Shah functional with the EVAM condition included. However, this approach is not able to cope with channel misregistration and therefore would lead to strong artifacts in the restored image.

In [29] a multichannel method, which does not require perfect alignment of the individual channels and the knowledge of the blur size, is introduced. Preliminary results on blind multichannel image deconvolution using the intersection of the confidence interval (ICI) are presented in [30].

In this chapter some recent activities of the authors on both single-channel and multichannel blind image deconvolution are presented in the unifying framework of the Bussgang deconvolution method. Roughly speaking, the Bussgang restoration algorithm is based on iteratively filtering the measurements, updating the filter coefficients using a suitable nonlinear estimate of the original image. Specifically, the filter coefficients are updated by solving a linear system whose coefficients take into account the cross-correlation between the nonlinear estimate of the original image and the measurements. Moreover, the nonlinearity design is possibly obtained resorting to a Bayesian criterion. Applications of the single-channel blind Bussgang deconvolution approach to the texture synthesis are described. Also the generalization to the multichannel case of the Bussgang deconvolution is proposed and its application to the restoration of binary text images, spiky images, and natural images is presented.

The chapter is organized as follows. In Section 2.2 Bussgang processes are defined and their basic properties are outlined. An iterative deconvolution approach making use of the Bussgang deconvolution approach is described in Section 2.3 for the single-channel case. Also some theoretical considerations on single-channel Bussgang deconvolution convergence are outlined and some applications oriented to texture synthesis-by-analysis are given. In Section 2.4 a generalization to the multichannel case of the single-channel Bussgang deconvolution algorithm is theorized. Then, applications to the deblurring of both synthetic images, like binary text and spiky images, and natural images are given. Specifically, it is assumed that the observation system is able to give multiple observations of the original image. Eventually, conclusions are drawn in Section 2.5.

2.2 Bussgang Processes

In the early 1950s Bussgang [31] demonstrated that, for Gaussian processes, the cross-correlation between the input and the output of a zero memory nonlinearity (ZNL) is proportional to the autocorrelation of the input.

However, it is possible not to restrict this property only to Gaussian processes. In fact, by indicating with $E\{\cdot\}$ the expected value operator, we can state that:

Definition 1 *A stochastic process $x(t)$ is said to be a "Bussgang" process if for any ZNL $\Gamma(\cdot)$:*

$$E\{x(t_1)\Gamma(x(t_2))\} = \text{const} \cdot E\{x(t_1)x(t_2)\}. \tag{2.1}$$

A theorem that gives the necessary and sufficient condition for a generic process to belong to the class of the Bussgang processes follows:

Theorem 1 *A second-order, stationary stochastic process, $x(t)$, is Bussgang if and only if:*

$$E\{x(t+\tau)|x(t)\} = \frac{R_x(\tau)}{R_x(0)} \cdot x(t), \tag{2.2}$$

where $R_x(\tau) = E\{x(t)x(t-\tau)\}$ is the autocorrelation function of the stationary stochastic process $x(t)$.

The proof of this theorem, reported also in [32], follows for the sake of completeness.

Proof: From the Bussgang property given in (2.1) it follows that

$$E\{x_1\Gamma(x_2)\} = k \cdot E\{x_1 x_2\}, \tag{2.3}$$

with k a constant, $x_1 = x(t+\tau)$, and $x_2 = x(t)$. The left-hand side of (2.3) can be written as

$$
\begin{aligned}
E\{x_1\Gamma(x_2)\} &= \int\int x_1\Gamma(x_2)p_{X_1,X_2}(x_1,x_2)dx_1 dx_2 = \\
&= \int \Gamma(x_2)\left(\int x_1 p_{X_1|X_2}(x_1|x_2)dx_1\right)dx_2 = \\
&= \int \Gamma(x_2)E\{x_1|x_2\}p_{X_2}(x_2)dx_2.
\end{aligned}
\tag{2.4}
$$

By equating the right-hand side of (2.3) to (2.4) it follows that

$$k = \frac{\int \Gamma(x_2)E\{x_1|x_2\}p_{X_2}(x_2)dx_2}{E\{x_1 x_2\}}. \tag{2.5}$$

It is evident that k is independent of τ if and only if

$$E\{x_1|x_2\} = h(x_2)E\{x_1 x_2\} = h(x_2)R_x(\tau), \tag{2.6}$$

$h(x_2)$ being a function not yet specified.

For $\tau = 0$, (2.6) gives

$$h(x_2) = \frac{x_2}{R_x(0)} \tag{2.7}$$

By substituting (2.7) into (2.6) the theorem is demonstrated.

In the next section, it is demonstrated that any iterative deconvolution algorithm utilizing a ZNL converges to a Bussgang process.

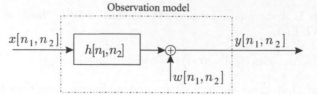

FIGURE 2.1:   Image formation model.

2.3   Single-Channel Bussgang Deconvolution

Let us consider the image formation model depicted in Figure 2.1 where $x[n_1, n_2]$ represents the original image, $h[n_1, n_2]$ the linear observation filter, and $w[n_1, n_2]$ a realization of white additive Gaussian random field:

$$y[n_1, n_2] = (x * h)[n_1, n_2] + w[n_1, n_2]. \tag{2.8}$$

Restoration performed by means of the Bussgang deconvolution approach was originally proposed by Godfrey and Rocca in [32] for the deconvolution of seismic traces. Bussgang deconvolution belongs to the class of iterative deconvolution algorithms. Roughly speaking, it is based on iteratively filtering the observation by means of an inverse linear filter, then a suitable nonlinear estimate of the original image is obtained, and eventually the inverse linear filter coefficients are updated using the nonlinear estimate of the original data and the observation. The observation is deconvolved with the so obtained inverse filter and the previous operations are iterated until convergence is reached.

More specifically, in the restoration stage, depicted in Figure 2.2, the linear estimate $\hat{x}^{(k)}[n_1, n_2]$ of the original image $x[n_1, n_2]$ at the k-th iteration is accomplished using a FIR restoration filter $f^{(k-1)}[n_1, n_2]$, with finite support of size $(2P + 1) \times (2P + 1)$, thus obtaining the restored image:

$$\hat{x}^{(k)}[n_1, n_2] = \sum_{t_1, t_2 = -P}^{P} f^{(k-1)}[t_1, t_2] \, y[n_1 - t_1, n_2 - t_2]. \tag{2.9}$$

The updated version of the filter $f^{(k-1)}[n_1, n_2]$, namely $f^{(k)}[n_1, n_2]$, is estimated by minimizing the Mean Square Error (MSE) between $\hat{x}^{(k+1)}[n_1, n_2]$ and the original image $x[n_1, n_2]$:

$$\text{MSE} \stackrel{\text{def}}{=} \text{E}\left\{ \left| \hat{x}^{(k+1)}[n_1, n_2] - x[n_1, n_2] \right|^2 \right\}. \tag{2.10}$$

The application of the orthogonality principle to the minimization of (2.10) yields the following linear set of $(2P + 1)^2$ *normal equations* for the determi-

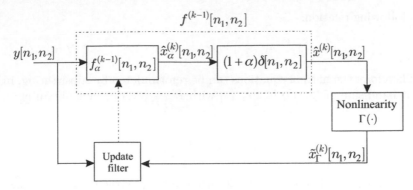

FIGURE 2.2: Bussgang restoration stage.

nation of the $(2P + 1)^2$ unknowns, i.e., the coefficients of the Wiener filter,

$$\sum_{t_1,t_2=-P}^{P} R_{yy}[r_1 - t_1, r_2 - t_2] f^{(k)}[t_1, t_2] = R_{xy}[r_1, r_2] \qquad (2.11)$$

where

$$R_{yy}[r_1, r_2] \stackrel{\text{def}}{=} \mathrm{E}\{y[n_1, n_2]\, y[n_1 - r_1, n_2 - r_2]\},$$

and

$$R_{xy}[r_1, r_2] \stackrel{\text{def}}{=} \mathrm{E}\{x[n_1, n_2]\, y[n_1 - r_1, n_2 - r_2]\}$$

with $r_1, r_2 = -P, \cdots, P$.

The solution of (2.11) requires the knowledge of the cross-correlation $R_{xy}[r_1, r_2]$ between the original image $x[n_1, n_2]$ and the observation $y[n_1, n_2]$, in which case it provides the optimal linear estimate of the original image in the Minimum MSE (MMSE) sense and no iterations are needed. However, since such cross-correlation is not available in practice, a suitable nonlinear estimate of $x[n_1, n_2]$ can be used.

Specifically, as for the nonlinear estimator, a sensible and robust choice is the MMSE estimate of $x[n_1, n_2]$ *given* the observed image $y[n_1, n_2]$, namely the conditional *a posteriori* mean

$$\tilde{x}[n_1, n_2] \stackrel{\text{def}}{=} \mathrm{E}\{x[n_1, n_2]|y\}. \qquad (2.12)$$

However, as pointed out in [33] $\hat{x}^{(k)}[n_1, n_2]$ is a sufficient statistic for the estimation of $x[n_1, n_2]$ given $y[n_1, n_2]$, since it is obtained from $y[n_1, n_2]$ through the invertible filter $f^{(k-1)}[n_1, n_2]$ (see Figure 2.2). Therefore, (2.12) can be rewritten as

$$\mathrm{E}\{x[n_1, n_2]|y\} = \mathrm{E}\{x[n_1, n_2]|\hat{x}^{(k)}\} = \Gamma(\hat{x}^{(k)}) \stackrel{\text{def}}{=} \tilde{x}_{\Gamma}^{(k)}[n_1, n_2]. \qquad (2.13)$$

Since the minimum mean square estimation error $e^{(k)}[n_1, n_2] = x[n_1, n_2] - \tilde{x}_{\Gamma}^{(k)}[n_1, n_2]$ is statistically orthogonal to the observed image $y[n_1, n_2]$ we have the following relation:

$$\mathrm{E}\left\{ \left(x[n_1, n_2] - \tilde{x}_\Gamma^{(k)}[n_1, n_2] \right) \cdot y[n_1 - r_1, n_2 - r_2] \right\} = 0 \;\; \forall \, (n_1, n_2).$$

Therefore, the normal equations can be reformulated by substituting, in the right-hand side of (2.11), $R_{xy}[r_1, r_2]$ with $R_{\tilde{x}_\Gamma^{(k)} y}[r_1, r_2]$, thus obtaining:

$$\sum_{t_1, t_2 = -P}^{P} R_{yy}[r_1 - t_1, r_2 - t_2] f^{(k)}[t_1, t_2] = R_{\tilde{x}_\Gamma^{(k)} y}[r_1, r_2]. \tag{2.14}$$

The Bussgang deconvolution algorithm, also illustrated in Figure 2.2, can be summarized as follows:

1. **Linear estimation step:** the deconvolved image $\hat{x}^{(k)}[n_1, n_2]$ is computed by filtering the observation $y[n_1, n_2]$ through a previous estimation of the Wiener filter $f^{(k-1)}[n_1, n_2]$, i.e.,

$$\hat{x}^{(k)}[n_1, n_2] = (f^{(k-1)} * y)[n_1, n_2].$$

Since the original image can be retrieved except for an amplitude scale factor, at each iteration the Wiener filter is normalized to yield a unitary energy output.

2. **Nonlinear estimation step:** according to a suitable stochastic model of the image $x[n_1, n_2]$, the nonlinear MMSE estimate $\tilde{x}_\Gamma^{(k)}[n_1, n_2]$ is computed through the conditional *a posteriori* expectation

$$\tilde{x}_\Gamma^{(k)}[n_1, n_2] = \Gamma(\hat{x}^{(k)}) \stackrel{\text{def}}{=} E\{x[n_1, n_2] | \hat{x}^{(k)}\}.$$

In general, the nonlinear estimator $\Gamma(\cdot)$ exhibits nonzero spatial memory.

3. **Wiener filter coefficients update step:** the Wiener filter $f^{(k)}[n_1, n_2]$ is recomputed by solving the normal equations (2.14) where $R_{yy}[r_1, r_2]$ and $R_{\tilde{x}_\Gamma^{(k)} y}[r_1, r_2]$ are substituted by their sample estimates $\widehat{R}_{yy}[r_1, r_2]$ and $\widehat{R}_{\tilde{x}_\Gamma^{(k)} y}[r_1, r_2]$ [1], respectively.

4. **Convergence test step:** convergence is tested by a suitable criterion.

The algorithm is initialized, for $k = 1$, by an initial guess $f^{(0)}[n_1, n_2]$ of the Wiener filter.

As outlined in [32, 34], the iterative algorithm reaches an equilibrium point, namely $f^{(k-1)}[n_1, n_2] = \text{const} \cdot f^{(k)}[n_1, n_2]$, when the cross-correlation between

---

[1]The cross-correlation estimate is here obtained as

$$\widehat{R}_{xy}[k_1, k_2] = \frac{1}{N_1 N_2} \sum_{n_1=0}^{N_1 - |k_1| - 1} \sum_{n_2=0}^{N_2 - |k_2| - 1} x[n_1, n_2] y[n_1 - k_1, n_2 - k_2].$$

the deconvolved image $\hat{x}^{(k)}[n_1, n_2]$ and the measured images $y[n_1, n_2]$ become proportional to the cross-correlation between the estimate $\tilde{x}_\Gamma^{(k)}[n_1, n_2]$ and the observations, i.e.

$$R_{\hat{x}^{(k)}y}[r_1, r_2] = \text{const} \cdot R_{\tilde{x}_\Gamma^{(k)}y}[r_1, r_2], \qquad (2.15)$$

which is equivalent to (2.1) as it is straightforward to verify. Since the equilibrium point is characterized by the invariance between cross-correlation (2.15), also known as the "Bussgang" property of stationary processes under nonlinear transformations, the deconvolution algorithm is commonly referred to as the "Bussgang" deconvolution algorithm [32,34,35]. However, it is well known that if we deal with a realization of a stationary Gaussian random field, the restoration process becomes inherently ambiguous, being the Bussgang property (2.15) satisfied irrespective of the filter $f[n_1, n_2]$.

A discussion on the convergence of the algorithm was proposed by Campisi and Scarano in [36] where the convergence of the Bussgang algorithm is analyzed in the sense of error energy reduction for the single-channel case. The analysis, reported in Section 2.3.1, properly takes into account the fact that the deconvolved image is amplitude scaled at each iteration, and it shows that the convergence of the algorithm is assured when the error between the original image and the (amplitude-scaled) nonlinear MMSE estimate is scarcely correlated with the original image itself.

2.3.1 Convergency Issue

In this section we demonstrate the error energy reduction property of the Bussgang deconvolution algorithm, i.e., the nonincreasing of the mean square error done in estimating the original image $x[n_1, n_2]$ from one iteration to the next.

The iterative Bussgang deconvolution algorithm is depicted in Figure 2.2 in its flow-graph form.

Specifically, with reference to Figure 2.2 the error at the (k)-th iteration is defined as

$$e^{(k)}[n_1, n_2] = x[n_1, n_2] - \tilde{x}_\Gamma^{(k)}[n_1, n_2]$$

and the error energy reduction property is stated as follows:

$$\text{Ms}\left\{e^{(k-1)}[n_1, n_2]\right\} \geq \text{Ms}\left\{e^{(k)}[n_1, n_2]\right\} \qquad (2.16)$$

being $\text{Ms}\{(\cdot)\} = \text{E}\{(\cdot)^2\}$ the mean square value operator.

In order to prove (2.16), let us assume the following hypothesis:

H1. The output $\tilde{x}_\Gamma^{(k)}[n_1, n_2]$ of the nonlinear, with memory, estimator $\Gamma(\cdot)$ satisfies the statistical orthogonality property:

$$\text{E}\left\{\left(x[n_1, n_2] - \tilde{x}_\Gamma^{(k)}[n_1, n_2]\right)\hat{x}^{(k)}[n_1, n_2]\right\} = 0. \qquad (2.17)$$

This hypothesis surely holds true when a MMSE estimator is employed. This MMSE nonlinear estimator needs to be approximated in concrete applications.

Moreover, the following facts will be invoked in the following:

F1. In Figure 2.2, the filter $f_\alpha^{(k-1)}[n_1, n_2]$ is the MMSE (Wiener) filter that, at the (k)-th iteration, linearly estimates $\tilde{x}_r^{(k-1)}[n_1, n_2]$ from its input $y[n_1, n_2]$. Being the output of a Wiener filter, the random field $\hat{x}_\alpha^{(k)}[n_1, n_2]$ satisfies the following orthogonality property:

$$\mathrm{E}\left\{\left(\tilde{x}_r^{(k-1)}[n_1, n_2] - \hat{x}_\alpha^{(k)}[n_1, n_2]\right)\hat{x}_\alpha^{(k)}[n_1, n_2]\right\} = 0. \qquad (2.18)$$

F2. The filter $f_\alpha^{(k-1)}[n_1, n_2]$ is followed by a normalizing scalar transformation that sets the mean square value of its output $\hat{x}^{(k)}[n_1, n_2] = (1 + \alpha)\hat{x}_\alpha^{(k)}[n_1, n_2]$ to be equal to the mean square value of $x[n_1, n_2]$, i.e.,

$$\mathrm{Ms}\left\{\hat{x}^{(k)}[n_1, n_2]\right\} = (1 + \alpha)^2 \mathrm{Ms}\left\{\hat{x}_\alpha^{(k)}[n_1, n_2]\right\}$$
$$= \mathrm{Ms}\left\{x[n_1, n_2]\right\}. \qquad (2.19)$$

Note that (2.18) also implies

$$\mathrm{E}\left\{\left(\tilde{x}_r^{(k-1)}[n_1, n_2] - \hat{x}_\alpha^{(k)}[n_1, n_2]\right)\hat{x}^{(k)}[n_1, n_2]\right\} = 0. \qquad (2.20)$$

F3. The hypothesis **H1**, as expressed by (2.17), also implies

$$\mathrm{E}\left\{\left(x[n_1, n_2] - \tilde{x}_r^{(k)}[n_1, n_2]\right)\hat{x}_\alpha^{(k)}[n_1, n_2]\right\} = 0. \qquad (2.21)$$

F4. Note that it always results

$$\mathrm{Ms}\left\{x[n_1, n_2]\right\} \geq \mathrm{Ms}\left\{\tilde{x}_r^{(k)}[n_1, n_2]\right\}$$

and that normalization with $\alpha \geq 0$ is needed to restore the correct power level.

For the sake of notation compactness, the indices $[n_1, n_2]$ will be omitted in the following formulas.

By indicating with $\pm(\cdot)$ addition and subtraction of the same quantity, let us write the error at the $(k-1)$-th iteration can be expressed as follows:

$$e^{(k-1)} = x - \tilde{x}_r^{(k-1)} \pm \tilde{x}_r^{(k)} \pm \hat{x}^{(k)}$$
$$= e^{(k)} + \left(\tilde{x}_r^{(k)} - \hat{x}^{(k)}\right) - \left(\tilde{x}_r^{(k-1)} - \hat{x}_\alpha^{(k)}\right) + \alpha\,\hat{x}_\alpha^{(k)}.$$

Its mean square value assumes the following expression:

$$\text{Ms}\left\{e^{(k-1)}\right\} = \text{Ms}\left\{e^{(k)}\right\} + \text{Ms}\left\{\tilde{x}_{\text{r}}^{(k)} - \hat{x}^{(k)}\right\}$$
$$+ \text{Ms}\left\{\tilde{x}_{\text{r}}^{(k-1)} - \hat{x}_{\alpha}^{(k)}\right\} + \alpha^2 \text{Ms}\left\{\hat{x}_{\alpha}^{(k)}\right\}$$
$$+ 2\text{E}\left\{e^{(k)}\left(\tilde{x}_{\text{r}}^{(k)} - \hat{x}^{(k)}\right)\right\} - 2\text{E}\left\{e^{(k)}\left(\tilde{x}_{\text{r}}^{(k-1)} - \hat{x}_{\alpha}^{(k)}\right)\right\}$$
$$+ 2\alpha\text{E}\left\{e^{(k)}\hat{x}_{\alpha}^{(k)}\right\} - 2\text{E}\left\{\left(\tilde{x}_{\text{r}}^{(k)} - \hat{x}^{(k)}\right)\left(\tilde{x}_{\text{r}}^{(k-1)} - \hat{x}_{\alpha}^{(k)}\right)\right\}$$
$$+ 2\alpha\text{E}\left\{\left(\tilde{x}_{\text{r}}^{(k)} - \hat{x}^{(k)}\right)\hat{x}_{\alpha}^{(k)}\right\} - 2\alpha\text{E}\left\{\left(\tilde{x}_{\text{r}}^{(k-1)} - \hat{x}_{\alpha}^{(k)}\right)\hat{x}_{\alpha}^{(k)}\right\}.$$

Recalling (2.19) and eliminating the null contributions due to (2.17), (2.18), (2.20), and (2.21), we obtain :

$$\text{Ms}\left\{e^{(k-1)}\right\} = \text{Ms}\left\{e^{(k)}\right\} + \text{Ms}\left\{\tilde{x}_{\text{r}}^{(k)} - \hat{x}^{(k)}\right\}$$
$$+ \text{Ms}\left\{\tilde{x}_{\text{r}}^{(k-1)} - \hat{x}_{\alpha}^{(k)}\right\} + \alpha^2 \text{Ms}\left\{\hat{x}_{\alpha}^{(k)}\right\}$$
$$- 2\text{E}\left\{e^{(k)}\tilde{x}_{\text{r}}^{(k-1)}\right\} - 2\text{E}\left\{\tilde{x}_{\text{r}}^{(k)}\left(\tilde{x}_{\text{r}}^{(k-1)} - \hat{x}_{\alpha}^{(k)}\right)\right\}$$
$$+ 2\alpha\text{E}\left\{\tilde{x}_{\text{r}}^{(k)}\hat{x}_{\alpha}^{(k)}\right\} - 2\alpha(1+\alpha)\text{Ms}\left\{\hat{x}_{\alpha}^{(k)}\right\}$$
$$= \text{Ms}\left\{e^{(k)}\right\} + \text{Ms}\left\{\tilde{x}_{\text{r}}^{(k)} - \hat{x}^{(k)}\right\}$$
$$+ \text{Ms}\left\{\tilde{x}_{\text{r}}^{(k-1)} - \hat{x}_{\alpha}^{(k)}\right\} - 2\text{E}\left\{x\tilde{x}_{\text{r}}^{(k-1)}\right\}$$
$$+ \underbrace{2(1+\alpha)\text{E}\left\{\tilde{x}_{\text{r}}^{(k)}\hat{x}_{\alpha}^{(k)}\right\}}_{2\text{E}\left\{\tilde{x}_{\text{r}}^{(k)}\hat{x}^{(k)}\right\}=2\text{E}\left\{x\hat{x}^{(k)}\right\}} \quad \underbrace{-\alpha(2+\alpha)\text{Ms}\left\{\hat{x}_{\alpha}^{(k)}\right\}}_{-\frac{\alpha(2+\alpha)}{(1+\alpha)^2}\text{Ms}\{x\}} \pm 2\text{Ms}\left\{x\right\}$$
$$= \text{Ms}\left\{e^{(k)}\right\} + \text{Ms}\left\{\tilde{x}_{\text{r}}^{(k)} - \hat{x}^{(k)}\right\} + \text{Ms}\left\{\tilde{x}_{\text{r}}^{(k-1)} - \hat{x}_{\alpha}^{(k)}\right\}$$
$$+ 2\left(\text{E}\left\{x\hat{x}^{(k)}\right\} - \text{Ms}\left\{x\right\}\right) + \frac{2+2\alpha+\alpha^2}{(1+\alpha)^2}\text{Ms}\left\{x\right\}$$
$$- 2\text{E}\left\{x\,\tilde{x}_{\text{r}}^{(k-1)}\right\}.$$

$$(2.22)$$

Error energy reduction (2.16) occurs whenever the last three terms in (2.22) are negligible or result in a nonnegative quantity. We shall see that this occurs when the scalar normalization $(1+\alpha)$ has almost adjusted the overall gain to have

$$x[n_1, n_2] = (1+\alpha)\,\tilde{x}_{\text{r}}^{(k)}[n_1, n_2] + \epsilon^{(k)}[n_1, n_2]$$
$$\text{with} \qquad\qquad (2.23)$$
$$\text{E}\left\{x[n_1, n_2]\,\epsilon^{(k-1)}[n_1, n_2]\right\} \simeq 0.$$

Using (2.23) we can write

$$\mathrm{E}\left\{x[n_1, n_2]\,\tilde{x}_{\mathrm{r}}^{(k-1)}[n_1, n_2]\right\} = \frac{1}{1+\alpha}\mathrm{Ms}\left\{x[n_1, n_2]\right\}. \tag{2.24}$$

The substitution of (2.24) in the last two terms of (2.22) yields

$$\frac{2 + 2\alpha + \alpha^2}{(1+\alpha)^2}\,\mathrm{Ms}\left\{x\right\} - 2\mathrm{E}\left\{x\,\tilde{x}_{\mathrm{r}}^{(k-1)}\right\} = \frac{\alpha^2}{(1+\alpha)^2}\,\mathrm{Ms}\left\{x\right\} \geq 0. \tag{2.25}$$

As far as the first of the last three terms of (2.22) is concerned, we have

$$\mathrm{E}\left\{x\hat{x}^{(k)}\right\} - \mathrm{Ms}\left\{x\right\} =$$

$$= \frac{2}{4\pi^2}\iint\limits_{-\pi}^{\pi} P_{xx}(e^{j\omega_1}, e^{j\omega_2})\cdot\left(G^{(k-1)}(e^{j\omega_1}, e^{j\omega_2}) - 1\right)d\omega_1 d\omega_2$$

where $P_{xx}(e^{j\omega_1}, e^{j\omega_2})$ is the power spectral density of the random field $x[n_1, n_2]$, $H(e^{j\omega_1}, e^{j\omega_2})$ and $F(e^{j\omega_1}, e^{j\omega_2})$ the Fourier transform of $h[n_1, n_2]$ and $f[n_1, n_2]$ respectively. The term

$$G^{(k-1)}(e^{j\omega_1}, e^{j\omega_2}) - 1 = H(e^{j\omega_1}, e^{j\omega_2})\cdot F^{(k-1)}(e^{j\omega_1}, e^{j\omega_2}) - 1$$

measures how effective is the deconvolution filter $f^{(k-1)}[n_1, n_2]$ in inverting the filter $h[n_1, n_2]$ represented in Figure 2.2. In digital transmissions it constitutes the Fourier spectrum of the so-called *Inter-Symbol Interference* (ISI). In particular, by setting $g^{(k-1)}[n_1, n_2] = h[n_1, n_2] * f^{(k-1)}[n_1, n_2]$, we can write

$$\frac{1}{4\pi^2}\iint\limits_{-\pi}^{\pi}\left(G^{(k-1)}(e^{j\omega_1}, e^{j\omega_2}) - 1\right)d\omega_1 d\omega_2 = g^{(k-1)}[0, 0] - 1.$$

This quantity, namely the ISI evaluated in the origin, tends to be close to zero when power normalization has almost achieved gain equalization, i.e., when (2.23) is fulfilled. In our case, $x[n_1, n_2]$ cannot be assumed spatially white, but, since its power spectral density $P_{xx}(e^{j\omega_1}, e^{j\omega_2})$ is always nonnegative, we can still argue that the term $2\left(\mathrm{E}\left\{x\hat{x}^{(k)}\right\} - \mathrm{Ms}\left\{x\right\}\right)$ tends to be small and comparable with the nonnegative quantity (2.25).

In summary, under condition (2.23), (2.22) can be written as follows:

$$\mathrm{Ms}\left\{e^{(k-1)}\right\} = \mathrm{Ms}\left\{e^{(k)}\right\} + \mathrm{Ms}\left\{\tilde{x}_{\mathrm{r}}^{(k)} - \hat{x}^{(k)}\right\} + \mathrm{Ms}\left\{\tilde{x}_{\mathrm{r}}^{(k-1)} - \hat{x}_{\alpha}^{(k)}\right\}$$

$$+ \underbrace{2\left(\mathrm{E}\left\{x\hat{x}^{(k)}\right\} - \mathrm{Ms}\left\{x\right\}\right)}_{\text{negligible under (2.23)}} + \underbrace{\frac{\alpha^2}{(1+\alpha)^2}\mathrm{Ms}\left\{x\right\}}_{\geq 0}$$

$$\geq \mathrm{Ms}\left\{e^{(k)}\right\}$$

thus satisfying the error energy reduction condition.

2.3.2 Application to Texture Synthesis

In this section the Bussgang deconvolution algorithm is used in a synthesis-by-analysis approach for texture synthesis proposed by the authors.

In the last decades the texture synthesis problem has been extensively investigated since it has both interesting theoretical implications and several applications. To cite only a few, in the area of computer graphics, synthetic textures are projected onto surfaces to give them a realistic appearance. Moreover, in very low bit-rate compression techniques, synthetic textures obtained using reduced dimensionality parameters space may replace backgrounds in natural scenes, thus leading to a dramatic bit saving.

However, the design of a texture reproduction method allowing an exact match between the synthesized texture and the given prototype is a difficult task to accomplish, since it is difficult to capture with a unique mathematical model the unlimited variety of structures combined with different illumination conditions that can be encountered when dealing with textures. Moreover, the design of a synthesis algorithm cannot be done without taking into account the characteristics of the human visual system (HVS). Psychophysical studies [37,38] have outlined that the HVS does not indulge in a detailed examination of the textured regions of an image but quickly preattentively discriminates among them [39], while devoting more attention to the objects present in the scene. This implies that an acceptable result of the synthesis process is a synthetic texture that, although different from the prototype, gives its same visual impression.

Simple schemes to characterize textures are based on linear models such as FIR or infinite impulse response (IIR) filters excited by independent, identically distributed (i.i.d.) random fields [40–42]. Identification of filters is done using deconvolution techniques (see, for instance, [43]). In general, these models are simple to identify, but they are substantially limited by the inherent linear stochastic dependence between pixels. However, these models are often unable to preserve the underlying structure of the given prototype, so that, in order to achieve a better texture quality, a human-supervised step is often required for the definition of a more structured excitation [42].

More complex models employing morphological operators, fractals [44,45], and Markov random fields [46,47] have also been used. However, their identification remains difficult, especially when samples of moderate size are available.

In the recent past some alternative approaches dealing with some fundamental characteristics of textures, namely "repetitiveness," "directionality," and "granularity and complexity," have been proposed in [48–50]. The texture is preliminarily decomposed into its deterministic (periodic and directional) and indeterministic (unpredictable) components, according to a 2-D extension of Wold decomposition paradigm.

In [51] texture modeling is carried out in a complex wavelet multiresolution framework by exploiting the local autocorrelation of the coefficients in each subband along with the local autocorrelation and cross–correlation of coefficient magnitudes at other orientations and spatial scales. The method gives good results for both random and directional textures. However, the synthesis process is iterative and this makes the textures' reproduction algorithm expensive in terms of both computational complexity and processing time.

In [36], [52], the authors replace the i.i.d. excitations usually employed in linear texture models with non-i.i.d. binary sources, thus leading to a texture model constituted by the cascade of a filter followed by a hard-limiter, which yields the binary excitation, and a final filter (synthesis filter). The binary field, driving the synthesis filter, is devised to retain the morphological structure of the given prototype. The binary excitation and the synthesis filter are optimized in the sense that the first- and second-order distributions of the original texture match their counterpart of the synthesized texture in a Bussgang deconvolution framework. This is in accordance with the so-called "Julesz's conjecture" [54], which assumes that in a preattentive stage of vision the HVS is unable to distinguish between textures having the same first- and second-order distributions but different higher-order distributions, at least in a preattentive stage of vision. As a consequence, a method to synthesize textures perceptually close to a given prototype can rely on the mimic of the first- and second-order distributions of the texture sample. In Section 2.3.2.1 the proposed texture model is illustrated. The texture parameters identification procedure using the Bussgang deconvolution approach is given in Section 2.3.2.2. Some experimental results are given in Section 2.3.2.3.

2.3.2.1 Texture Model

A given texture $t[n_1, n_2]$ is modeled according to the following factorization:

$$t[n_1, n_2] = t_{synt}[n_1, n_2] + \varepsilon[n_1, n_2]$$
$$t_{synt}[n_1, n_2] = \eta_{\boldsymbol{\theta}}\big(x[n_1, n_2]\big) \qquad (2.26)$$
$$x[n_1, n_2] = (s * v)[n_1, n_2]$$

as shown in Figure 2.3, where

- $v[n_1, n_2]$ is a random binary excitation, which is designed in order to capture the basic morphology of the texture, by retaining the zero-crossing locations of the texture prototype;

- $s[n_1, n_2]$ is a shaping filter, with inverse denoted by $f[n_1, n_2]$, which adds gray-scale details to the binary image $v[n_1, n_2]$;

- $x[n_1, n_2]$ is the reproduced texture before the final histogram modification;

- $\eta_{\boldsymbol{\theta}}(\cdot)$ is an invertible zero-memory nonlinearity, with inverse denoted by $\eta_{\boldsymbol{\theta}}^{-1}(\cdot)$. It performs a histogram matching between the original texture $t[n_1, n_2]$ and the reproduced one, $x[n_1, n_2]$.

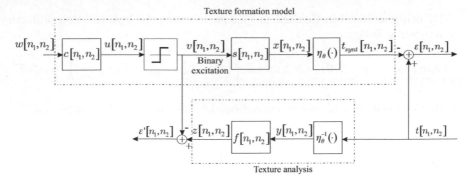

FIGURE 2.3: Texture model.

In essence, the model replaces the usual independent and identically distributed (i.i.d.) sources employed in linear texture models (see, for example, [40], [42]) with a non-i.i.d. binary source. This choice is motivated by the simplicity of controlling the second-order distribution of binary random fields, which, under weak symmetry constraints, is completely described by its autocorrelation function (ACF). Therefore the binary excitation is designed in order to have the same ACF of the binarized version of the texture sample under analysis. Specifically, the binary excitation $v[n_1, n_2]$ is obtained by filtering a realization $w[n_1, n_2]$ of a white Gaussian random field by means of a linear filter $c[n_1, n_2]$ and then by hard-limiting the output. As described in detail in Section 2.3.2.3, the filter $c[n_1, n_2]$ is designed by means of the *arcsin law* [53] in order to generate a binary excitation $v[n_1, n_2]$ having the same spatial ACF of the corresponding binarized texture obtained in the analysis stage. The synthetic texture is then obtained by passing the synthetic binary excitation through the shaping filter $s[n_1, n_2]$, followed by the histogram matcher. Since most of the visual morphology of a texture is contained in the binary excitation, this scheme is in accordance with the Julesz's conjecture.

The effectiveness of this model for texture reproduction has been discussed in [36] and [52]. The model has been successfully extended for the reproduction of color textures in [55]. It has been also employed to design a very low bit-rate texture-oriented compression technique [56]. Moreover, in [57] and [58] the model has been successfully used for texture classification.

2.3.2.2 Texture Parameters Identification Procedure

With reference to Figure 2.3, the texture model parameters ($v[n_1, n_2]$, $s[n_1, n_2]$, $\boldsymbol{\theta}$) are obtained by minimizing a suitable cost function of the synthesis residual $\varepsilon[n_1, n_2] = t[n_1, n_2] - t_{synt}[n_1, n_2]$. The synthesized texture is obtained by filtering a *binary* random field $v[n_1, n_2]$ followed by a final nonlinear stage $\eta_{\boldsymbol{\theta}}(\cdot)$, i.e.,

$$t_{synt}[n_1, n_2] = \eta_{\boldsymbol{\theta}}(s * v)[n_1, n_2].$$

By using as cost function the mean square error, the optimization problem is written as follows:

$$(v[n_1,n_2], s[n_1,n_2], \boldsymbol{\theta}) = \arg \min_{\tilde{v},\tilde{s},\tilde{\theta}} \mathrm{E}\left\{ \left(t[n_1,n] - \eta_{\tilde{\theta}}(\tilde{s} * \tilde{v})[n_1,n_2] \right)^2 \right\} \quad (2.27)$$

with the constraint of binary excitation, i.e., given the following probability density function of the stationary random field $v[n_1, n_2]$:

$$p_V(v) = \frac{1}{2}\Big(\delta(v-1) + \delta(v+1)\Big).$$

By considering both invertible nonlinearities $\eta_{\boldsymbol{\theta}}(\cdot)$ and invertible filters $s[n_1, n_2]$, the optimization problem given by (2.27) is equivalent to the minimization of the mean square error of the residual error (see also Figure 2.3):

$$\varepsilon'[n_1,n_2] = z[n_1,n_2] - v[n_1,n_2] = \left(\eta_{\boldsymbol{\theta}}^{-1}(t) * f \right)[n_1,n_2] - v[n_1,n_2]$$

being $f[n_1, n_2]$ the inverse filter of $s[n_1, n_2]$.

The minimization of the mean square error of $\varepsilon'[n_1,n_2]$, conditioned to the knowledge of $\boldsymbol{\theta}$, i.e., given the transformed texture $\eta_{\boldsymbol{\theta}}^{-1}(t[n_1,n_2])$, can be iteratively conducted using the following Bussgang deconvolution algorithm:

$$v^{(k)}[n_1,n_2] =$$

$$= \arg \min_{\tilde{v}} \mathrm{E}_{t|f^{(k-1)},\boldsymbol{\theta}^{(k-1)}} \left\{ \left((\eta_{\boldsymbol{\theta}^{(k-1)}}^{-1}(t) * f^{(k-1)})[n_1,n_2] - \tilde{v}[n_1,n_2] \right)^2 \right\},$$
$$(2.28)$$

$$f^{(k)}[n_1,n_2] =$$

$$= \arg \min_{\tilde{f}} \mathrm{E}_{t|v^{(k)},\boldsymbol{\theta}^{(k-1)}} \left\{ \left((\eta_{\boldsymbol{\theta}^{(k-1)}}^{-1}(t) * \tilde{f})[n_1,n_2] - v^{(k)}[n_1,n_2] \right)^2 \right\}. \quad (2.29)$$

As previously seen, (2.28) turns out to be the nonlinear estimation step of the Bussgang deconvolution algorithm while (2.29) turns out to be the Wiener filter design step.

The estimation of $\boldsymbol{\theta}$ is performed as follows:

$$\boldsymbol{\theta}^{(k)} = \arg \min_{\tilde{\boldsymbol{\theta}}} \mathrm{E}_{t|s^{(k-1)},v^{(k)}} \left\{ \left(t[n_1,n_2] - \eta_{\tilde{\theta}}\left(s^{(k-1)} * v^{(k)} \right)[n_1,n_2] \right)^2 \right\}.$$
$$(2.30)$$

Near convergence and making the approximation that the nonlinearity is a zero-memory one, (2.28) is very close to a hard-limiter acting on the deconvolved texture $z[n_1, n_2] = (\eta_{\boldsymbol{\theta}}^{-1}(t) * f)[n_1,n_2]$. We observed that no significant convergence speed is lost when it is kept fixed to a hard-limiter during all the iterations, allowing for a more simple overall algorithm implementation. The nonlinearity $\eta_{\boldsymbol{\theta}}(\cdot)$ obtained from the parameters $\boldsymbol{\theta}$ estimated in (2.30) can be reasonably approximated by means of a histogram matcher, also taking into account that the texture $t[n_1, n_2]$ is measured with a finite number of bits per pixels, typically 8 bits per pixel.

In summary, referring to Figure 2.4, the texture parameters identification algorithm can be summarized as follows:

1. **Linear estimation step:** the deconvolved texture $z^{(k)}[n_1, n_2]$ is computed from the histogram equalized texture $y^{(k)}[n_1, n_2] = \eta_{\boldsymbol{\theta}^{(k-1)}}^{-1}(t[n_1, n_2])$ by filtering it through the previous estimate of the Wiener filter $f^{(k-1)}[n_1, n_2]$,

$$z^{(k)}[n_1, n_2] = \left(f^{(k-1)} * y^{(k)} \right)[n_1, n_2].$$

Since the original image can be retrieved except for an amplitude scale factor, at each iteration the Wiener filter is normalized to yield a unitary energy output.

2. **Nonlinear estimation step:** the binary random field $v^{(k)}[n_1, n_2]$ is computed through hard-limiting the deconvolved texture $z^{(k)}[n_1, n_2]$.

3. **Wiener filter coefficients update step:** the Wiener filter $f^{(k)}[n_1, n_2]$ is recomputed by solving the normal equations

$$\sum_{q_1, q_2 = -P}^{P} R_{y^{(k)} y^{(k)}}[r_1 - q_1, r_2 - q_2] \cdot f^{(k)}[q_1, q_2] = R_{v^{(k)} y^{(k)}}[r_1, r_2] \quad (2.31)$$

after substituting statistical averages with sample averages.

4. **Histogram matcher update step:** the updated nonlinearity $\eta_{\boldsymbol{\theta}^{(k)}}$ is determined performing the histogram matching between the observed texture $t[n_1, n_2]$ and the linear reproduced texture $(v^{(k)} * s^{(k-1)})[n_1, n_2]$.

5. **Convergence test step:** convergence is tested by a suitable criterion.

The algorithm is initialized, for $k = 1$, by a suitable initial guess $f^{(0)}[n_1, n_2]$, of the Wiener filter and of the histogram matcher $\eta_{\boldsymbol{\theta}^{(0)}}$, usually set to the identity operators.

2.3.2.3   Texture Synthesis and Experimental Results

In summary, the synthesis burden is shared between the generation of the binary excitation $v[n_1, n_2]$ and the filter $s[n_1, n_2]$. The first allows sketching the structure of the target texture; in order to achieve this task, $v[n_1, n_2]$ is forced to have the same second-order statistics of the given prototype. This task is accomplished by observing that the ACF $R_{uu}[k_1, k_2]$ of a realization of a Gaussian random field $z[n_1, n_2]$ and the ACF $R_{vv}[k_1, k_2]$ of its binarized version $v[n_1, n_2] = \text{sign}(u[n_1, n_2])$ are related by the *arcsin law* [53]:

$$\frac{R_{vv}[k_1, k_2]}{\sigma_v^2} = \frac{2}{\pi} \cdot \arcsin\left[\frac{R_{uu}[k_1, k_2]}{\sigma_u^2} \right]. \quad (2.32)$$

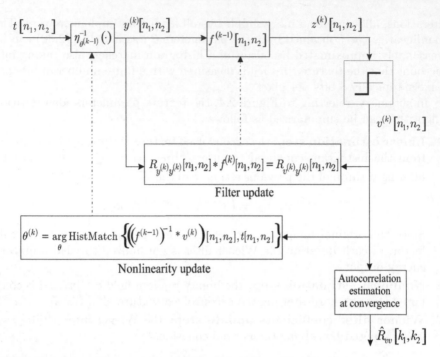

FIGURE 2.4:   Texture's parameter identification procedure.

Thus, a binary field possessing the desired second-order distribution is directly obtained by hard-limiting a Gaussian random field having the following ACF

$$\widehat{R}_{uu}[k_1, k_2] = \sigma_u^2 \cdot \sin\left[\frac{\pi}{2}\frac{\widehat{R}_{vv}[k_1, k_2]}{\sigma_v^2}\right] \tag{2.33}$$

where $\widehat{R}_{vv}[k_1, k_2]$ is an estimate of the ACF $R_{vv}[k_1, k_2]$, drawn from the binarized texture $v[n_1, n_2]$ obtained in analysis. Therefore, a Gaussian random field characterized by the desired ACF $\widehat{R}_{uu}[k_1, k_2]$ is obtained by filtering a white Gaussian random field through a "generating" filter $c[n_1, n_2]$ such that

$$\left|\mathcal{F}\{c[n_1, n_2]\}\right|^2 \propto \mathcal{F}\left\{\widehat{R}_{uu}[k_1, k_2]\right\}. \tag{2.34}$$

To summarize, according to the scheme of Figure 2.3 the synthesis is performed reversing the identification process. A realization of a white Gaussian random field is first passed through the coloring filter $c[n_1, n_2]$ (see (2.34)) and the hard-limiter, to generate a binary excitation $v[n_1, n_2]$ mimicking the underlying structure of the given texture. Then, the synthetic texture $t_{synt}[n_1, n_2]$ is obtained by filtering $v[n_1, n_2]$ by means of the estimated reconstruction filter $f[n_1, n_2]$ (see (2.31)) and by performing final histogram matcher.

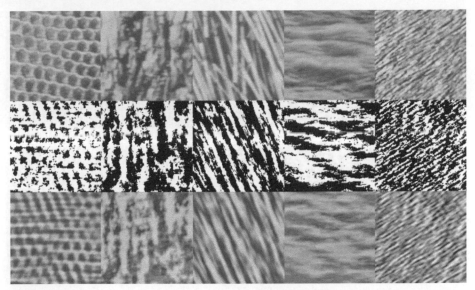

FIGURE 2.5: From left to right: D39 (Reptile Skin), D12 (Bark of Tree), D15 (Straw), D37 (Water), D93 (Fur). First row: original textures. Second row: synthetic binary excitations. Third row: synthetic textures.

Textures with different structural characteristics, like pseudo-periodic, directional, or random, are used for the quality assessment of the synthesis procedure. Referring to Figure 2.5 and Figure 2.6, some original textures extracted from the Brodatz [59] collection are displayed along with the synthetic binary excitations and the synthetic textures. Limiting ourselves to a preattentive vision stage, the results yield a quite good perceptual impression.

2.4 Multichannel Bussgang Deconvolution

In some applications the observation system is able to give multiple observations of the original image. In electron microscopy, for example, many differently focused versions of the same image are acquired during a single experiment, due to an intrinsic trade-off between the bandwidth of the imaging system and the contrast of the resulting image. In other applications, such as telesurveillance or remote sensing, multiple observed images can be acquired in order to better counteract, in the restoration phase, possible degradations due to motion, defocus, or noise.

In case of multiple measurements, the restoration algorithm can exploit the redundancy present in the observations and, in principle, it can achieve better

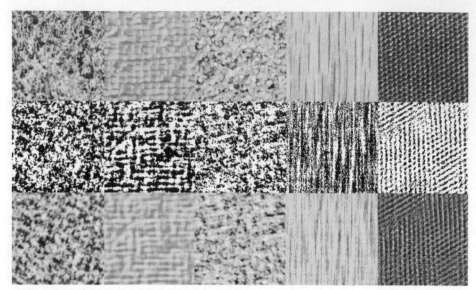

FIGURE 2.6: From left to right: D9 (Grass), D84 (Raffia), D29 (Sand), D68 (Wood), D77 (Cotton Canvas). First row: original textures. Second row: synthetic binary excitations. Third row: synthetic textures.

performance than the ones obtainable from a single measure.

In this section the Bussgang deconvolution algorithm is extended to the multichannel case with applications to image deconvolution problems. In its basic outline, the Bussgang restoration algorithm is based on iteratively filtering the measurements updating the filter coefficients using a suitable nonlinear estimate of the original image. Specifically, the filter coefficients are updated by solving a linear system (*normal equations*) whose coefficients' matrix takes into account the cross-correlation between the measurements and the nonlinear estimate of the original image. We must outline that the nonlinearity design, possibly obtained resorting to a Bayesian criterion, plays a key role in the overall algorithm convergence. In Section 2.4.1 we characterize both the observation and restoration models. In Section 2.4.2 we derive the multichannel Wiener filter, while in Section 2.4.3 the multichannel Bussgang algorithm method is outlined. In Sections 2.4.4 and 2.4.5 the Bussgang nonlinearities are explicitly derived for some reference cases, and some experimental results along with some performance evaluations are also given.

2.4.1 The Observation Model

The single-input multiple-output (SIMO) observation model of images is represented by M linear observation filters in presence of additive noise. This model, depicted in Figure 2.7, is given by:

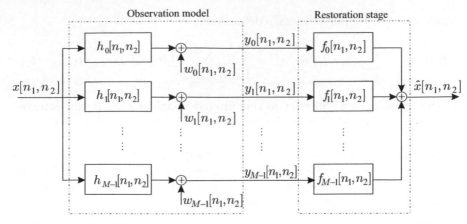

FIGURE 2.7: Blurred image generation model and multichannel Wiener restoration stage.

$$
\begin{cases}
y_0[n_1, n_2] = (x * h_0)[n_1, n_2] + w_0[n_1, n_2] \\
y_1[n_1, n_2] = (x * h_1)[n_1, n_2] + w_1[n_1, n_2] \\
\quad\vdots \\
y_{M-1}[n_1, n_2] = (x * h_{M-1})[n_1, n_2] + w_{M-1}[n_1, n_2].
\end{cases}
\tag{2.35}
$$

Moreover, $w_i[n_1, n_2]$, $i = 0, \cdots, M - 1$, are white Gaussian random fields, statistically independent and mutually uncorrelated:

$$
\mathrm{E}\{w_i[n_1, n_2]\, w_j[m - r_1, n - r_2]\} = \sigma_{w_i}^2\, \delta[r_1, r_2] \cdot \delta[i - j] =
\begin{cases}
\sigma_{w_i}^2 \cdot \delta[r_1, r_2] & \text{for } i = j \\
0 & \text{for } i \neq j.
\end{cases}
$$

2.4.2 Multichannel Wiener Filter

In this section, restoration by means of multichannel Wiener filtering is described. The results obtained hereafter will be used in Section 2.4.3. With reference to the observation model, depicted in Figure 2.7, multichannel Wiener filtering is performed by means of the filter bank $f_i[n_1, n_2]$, $i = 0, \cdots, M - 1$, with finite support of size $(2P + 1) \times (2P + 1)$, which minimize the Mean Square Error (MSE),

$$
\mathrm{MSE} \overset{\mathrm{def}}{=} \mathrm{E}\left\{ \left| \hat{x}[n_1, n_2] - x[n_1, n_2] \right|^2 \right\},
\tag{2.36}
$$

being $\hat{x}[n_1, n_2]$ the filtered image defined as

$$\hat{x}[n_1, n_2] = \sum_{i=0}^{M-1} (y_i * f_i)[n_1, n_2] = \sum_{i=0}^{M-1} \sum_{t_1, t_2 = -P}^{P} f_i[t_1, t_2] \, y_i[n_1 - t_1, n_2 - t_2].$$
(2.37)

In the following, we will refer to this filtered image as the "linear estimate" to clearly distinguish it from the "nonlinear estimate" that characterizes the Bussgang deconvolution algorithm.

The application of the orthogonality principle to the minimization of the above cost function yields the following linear set of $M \times (2P + 1)^2$ *normal equations* for the determination of the $M \times (2P + 1)^2$ unknowns, i.e. the coefficients of the multichannel Wiener filter

$$\sum_{j=0}^{M-1} \sum_{t_1, t_2 = -P}^{P} R_{y_j y_i}[r_1 - t_1, r_2 - t_2] f_j[t_1, t_2] = R_{x y_i}[r_1, r_2]$$
(2.38)

where

$$R_{y_j y_i}[r_1, r_2] \overset{\text{def}}{=} E\{y_j[n_1, n_2] \, y_i[n_1 - r_1, n_2 - r_2]\}$$

and

$$R_{x y_i}[r_1, r_2] \overset{\text{def}}{=} E\{x[n_1, n_2] \, y_i[n_1 - r_1, n_2 - r_2]\},$$

with $i = 0, \cdots, M - 1$ and $r_1, r_2 = -P, \cdots, P$. In principle, no assumption on the blur convolution kernels (support size, shape, and mutual coprimeness) is needed. However, the mutual coprimeness influences the numerical stability of the inversion involved in the solution of the normal equations. If the different observation channels are not coprime, the rank of the coefficients matrix formed by the samples $R_{y_j y_i}[r_1, r_2]$ reduces, and the normal equations (2.38) must be solved resorting to a pseudo-inversion.

Deconvolution by Wiener filtering requires, in principle, the knowledge of the cross-correlations between the original image and the measurements; it provides the optimal linear estimate of the original image in the Minimum MSE (MMSE) sense. If such cross-correlations are not available, one can resort to suitable estimates, as done in the Bussgang blind deconvolution algorithm described in Section 2.4.3.

2.4.3 Multichannel Bussgang Algorithm

Bussgang blind deconvolution [32, 34, 60] has been extended by the authors in [61] to the multichannel blind image restoration problem. The multichannel Bussgang blind deconvolution algorithm outlined in [61] is here reported. The algorithm stems from the following facts:

F1. Since the minimum mean square estimation error $x[n_1, n_2] - \tilde{x}_\Gamma[n_1, n_2]$ is statistically orthogonal to the observed images $y_0, \cdots, y_{M-1}$, the conditional *a posteriori* mean $\tilde{x}_\Gamma[n_1, n_2] \overset{\text{def}}{=} E\{x[n_1, n_2]|y_0, \cdots, y_{M-1}\}$, i.e. the

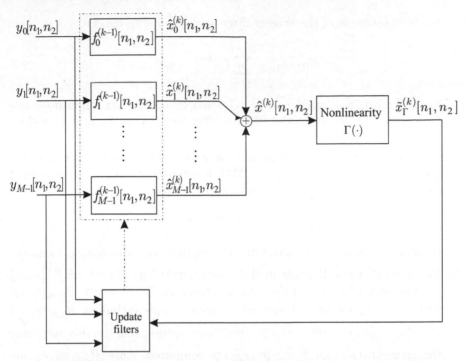

FIGURE 2.8: General form of the Bussgang deconvolution algorithm.

MMSE estimate of $x[n_1, n_2]$ *given* the observed samples of the sequences $y_0, \cdots, y_{M-1}$, satisfies:

$$E\{(x[n_1, n_2] - \tilde{x}_\Gamma[n_1, n_2]) \cdot y_i[n_1 - r_1, n_2 - r_2]\} = 0$$

for $i = 0, 1, \cdots, M-1$ and $\forall$ (n_1, n_2). Hence, the cross-correlations in the right-hand side of (2.38) possesses the following remarkable property:

$$R_{xy_i}[r_1, r_2] = R_{\tilde{x}_\Gamma y_i}[r_1, r_2].$$

F2. Thus, the multichannel Wiener filter $f_i[n_1, n_2]$ solves the normal equations (2.38) even if we substitute $R_{xy_i}[r_1, r_2] = R_{\tilde{x}_\Gamma y_i}[r_1, r_2]$, $i = 0, \cdots, M-1$, in the right-hand side of (2.38).

F3. The deconvolved image $\hat{x}[n_1, n_2]$ at the output of the multichannel Wiener filter constitutes a sufficient statistic for the estimation of $x[n_1, n_2]$ given $y_0, \cdots, y_{M-1}$; therefore we also have $\tilde{x}_\Gamma[n_1, n_2] = E\{x[n_1, n_2]|\hat{x}\}$.

Stemming from this analysis, we devise an iterative blind deconvolution algorithm, also illustrated in Figure 2.8, whose k-th iteration is here summarized:

1. **Linear estimation step:** the deconvolved image $\hat{x}^{(k)}[n_1, n_2]$ is computed by filtering the observations set $y_0[n_1, n_2], \cdots, y_{M-1}[n_1, n_2]$ by means of a

previous estimate of the Wiener filter bank $f_i^{(k-1)}[n_1, n_2]$, i.e.,

$$\hat{x}^{(k)}[n_1, n_2] = \sum_{i=0}^{M-1} (f_i^{(k-1)} * y_i)[n_1, n_2].$$

Since the original image can be retrieved except for an amplitude scale factor, at each iteration the Wiener filter bank is normalized to yield a unitary energy output.

2. **Nonlinear estimation step:** according to a suitable stochastic model of the image $x[n_1, n_2]$, the nonlinear MMSE estimate $\tilde{x}_\Gamma^{(k)}[n_1, n_2]$ is computed through the conditional *a posteriori* expectation

$$\tilde{x}_\Gamma^{(k)}[n_1, n_2] = \Gamma(\hat{x}^{(k)}) \overset{\text{def}}{=} E\{x[n_1, n_2]|\hat{x}^{(k)}\}.$$

In general, the nonlinear estimator $\Gamma(\cdot)$ exhibits nonzero spatial memory.

3. **Wiener filter coefficients update step:** the Wiener filter bank $f_i^{(k)}[n_1, n_2]$ is recomputed by solving the normal equations (2.38) where $R_{y_j y_i}[r_1, r_2]$ and $R_{\tilde{x}_\Gamma^{(k)} y_i}[r_1, r_2]$ are substituted by their sample estimates $\widehat{R}_{y_j y_i}[r_1, r_2]$ and $\widehat{R}_{\tilde{x}_\Gamma^{(k)} y_i}[r_1, r_2]$, respectively. We must outline that in this step only the cross-correlations $\widehat{R}_{\tilde{x}_\Gamma^{(k)} y_i}[r_1, r_2]$ are computed, since $\widehat{R}_{y_j y_i}[r_1, r_2]$ are computed only once before starting the iterations.

4. **Convergence test step:** convergence is tested by a suitable criterion.

The algorithm is initialized, for $k = 1$, by a suitable initial guess $f_i^{(0)}[n_1, n_2]$, with $i = 0, \cdots, M - 1$, of the Wiener filter bank. The algorithm is also illustrated in Figure 2.8, where we put in evidence the multiple-input single-output structure of the restoration system. The restoration filter bank, in fact, aims at yielding the (single) MMSE estimate of the original image given the measurements. The nonlinear MMSE estimate drives the restoration filter bank towards the MMSE (Wiener) filter bank. Since the MMSE filter bank preserves the information about the measurements pertaining to the original image, there is no loss in applying the nonlinear MMSE estimate directly to the single deconvolved image $\hat{x}[n_1, n_2]$ rather than to the different outputs of the restoration filters.

As outlined in [32, 34] and in Section 2.3.1 the iterative algorithm reaches an equilibrium point, namely $f_i^{(k-1)}[n_1, n_2] = \text{const} \cdot f_i^{(k)}[n_1, n_2]$, with $i = 0, \ldots, M - 1$, when the cross-correlations between the deconvolved image $\hat{x}^{(k)}[n_1, n_2]$ and the measured images $y_i[n_1, n_2]$, become proportional to the cross-correlations between the estimate $\tilde{x}_\Gamma^{(k)}[n_1, n_2]$ and the observations, i.e. :

$$R_{\hat{x}^{(k)} y_i}[r_1, r_2] = \text{const} \cdot R_{\tilde{x}_\Gamma^{(k)} y_i}[r_1, r_2], \tag{2.39}$$

with $i = 0, \ldots, M - 1$.

A discussion on the convergence of the algorithm is given in Section 2.3.1 for the single-channel case. The condition for the convergence, discussed in Section 2.3.1 for the single-channel case, straightforwardly extends to the multichannel case, since the variables involved in the analysis (original image, deconvolved image, and nonlinear estimate) do not change.

We must remark that the convergence of the algorithm is affected by the nonlinearity $\Gamma(\cdot)$, whose analytical determination can result quite difficult since it depends on the assumed probabilistic model of the image to be restored. In the next Sections we discuss two different image models that can be usefully employed in our restoration algorithm.

2.4.4 Application to Image Deblurring: Binary Text and Spiky Images

Let us consider the case of non-Gaussian images exhibiting negligible spatial correlation. Such images lead to rather simple restoration models. In particular, it can be assumed that at the generic k-th iteration of the algorithm, the deconvolution error $w^{(k)}[n_1, n_2] = \hat{x}^{(k)}[n_1, n_2] - x[n_1, n_2]$ is a normally distributed, white random field, statistically independent of the original image.

Then, the nonlinear estimator turns out to be a zero memory function of the deconvolved image $\hat{x}^{(k)}[n_1, n_2]$, whose form depends only on the marginal probability density function (PDF) of the original image $p_X(x)$ and on the deconvolution noise variance $\sigma^2_{w^{(k)}}$. The nonlinearity $\Gamma(\cdot)$ is thus given by:

$$\Gamma(\hat{x}^{(k)}[n_1, n_2]) \overset{\text{def}}{=} \mathrm{E}\left\{ x[n_1, n_2] \mid \hat{x}^{(k)}[n_1, n_2] \right\} = $$

$$= \frac{\displaystyle\int_{-\infty}^{+\infty} \xi \cdot p_X(\xi) \cdot \mathcal{N}\left(\hat{x}[n_1, n_2] - \xi; 0, \sigma^2_{w^{(k)}}\right) d\xi}{\displaystyle\int_{-\infty}^{+\infty} p_X(\xi) \cdot \mathcal{N}\left(\hat{x}[n_1, n_2] - \xi; 0, \sigma^2_{w^{(k)}}\right) d\xi}. \qquad (2.40)$$

The deconvolution noise variance $\sigma^2_{w^{(k)}}$ depends on both the blur and the observation noise and, when the algorithm begins to converge, it can be estimated as

$$\mathrm{E}\left\{ |w^{(k)}[n_1, n_2]|^2 \right\} \approx \mathrm{E}\left\{ |\tilde{x}_\Gamma^{(k)}[n_1, n_2] - \hat{x}[n_1, n_2]|^2 \right\}.$$

In practice, by varying this parameter we select the nonlinearity more adequate to represent any a priori information about the original image.

2.4.4.1   Binary Text Images

In the case of binary text images, $x[n_1, n_2]$ can be modeled as a realization of a white stationary random field with PDF

$$p_x(x) = p_0\, \delta(x) + (1 - p_0)\, \delta(x - a). \tag{2.41}$$

Hence, the nonlinear MMSE estimate $\tilde{x}_\Gamma^{(k)}[n_1, n_2]$ defined in (2.40) is evaluated as follows:

$$\tilde{x}_\Gamma^{(k)}[n_1, n_2] = \Gamma\left(\hat{x}^{(k)}[n_1, n_2]\right) = a\left[1 + \frac{p_0}{1 - p_0}\exp\left(-\frac{\hat{x}^{(k)}[n_1, n_2] - a/2}{\sigma_{w^{(k)}}^2/a}\right)\right]^{-1}. \tag{2.42}$$

By indicating with $\sigma_x^2 = a^2 p_0(1 - p_0)$ the signal power and with $\text{SNR}_d = \sigma_x^2/\sigma_{w^{(k)}}^2$ the signal to deconvolution noise ratio, (2.42) can be expressed as follows:

$$\frac{\tilde{x}_\Gamma^{(k)}[n_1, n_2]}{\sigma_x} = \Gamma\left(\frac{\hat{x}^{(k)}[n_1, n_2]}{\sigma_x}\right) =$$

$$= \frac{1}{\sqrt{p_0(1 - p_0)}}\left[1 + \frac{p_0}{1 - p_0}\exp\left(-\frac{\dfrac{\hat{x}^{(k)}[n_1, n_2]}{\sigma_x} - \dfrac{1}{2\sqrt{p_0(1 - p_0)}}}{\dfrac{\sqrt{p_0(1 - p_0)}}{\text{SNR}_d}}\right)\right]^{-1}. \tag{2.43}$$

In Figure 2.9, the nonlinear function given in (2.43) is shown for different values of the SNR_d having assumed for p_0 the value 0.5, whereas in Figure 2.10 the same function is shown for different values of the parameter p_0 and for a fixed value of $\text{SNR}_d = 20$ dB. It is worth pointing out that the nonlinearity takes a soft decision for the high value of the deconvolution noise, which occurs at the beginning of the iterative deconvolution algorithm, while it takes a hard decision for low-level noise, which occurs at the end of the iterative deconvolution algorithm.

A few experimental results are here shown, together with the results achievable by the ideal Wiener restoration FIR filter bank of equal support size, corresponding to the case of perfect knowledge of the right-hand side of the normal equations.

In Figure 2.11(a) the original binary text images before application of the blurring filters is shown. The text image is blurred using two motion blur filters having finite impulse responses $h_0[n_1, n_2]$ and $h_1[n_1, n_2]$. They are respectively given by:

$$h_0[n_1, n_2] = \begin{cases} \delta[n_1 - n_2] & \text{for } 0 \le n_1, n_2 \le 5 \\ 0 & \text{elsewhere}, \end{cases}$$

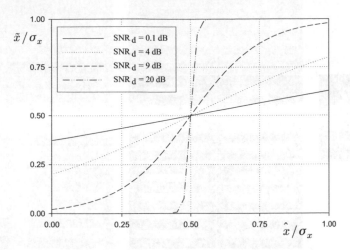

FIGURE 2.9: Nonlinearity $\Gamma(\cdot)$, given by (2.42), employed for binary text image deblurring, parameterized with respect to the deconvolution SNR_d for $p_0 = 0.5$.

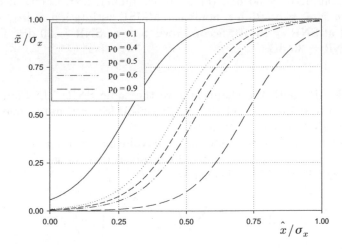

FIGURE 2.10: Nonlinearity $\Gamma(\cdot)$, given by (2.42), employed for binary text image deblurring, parameterized with respect to p_0 for $\mathrm{SNR}_d = 20$ dB.

$$h_1[n_1, n_2] = \begin{cases} \delta[n_1] & \text{for } 0 \leq n_2 \leq 6 \\ 0 & \text{elsewhere.} \end{cases}$$

In Figure 2.11(b) the blurred text images are shown for SNR values of 10,

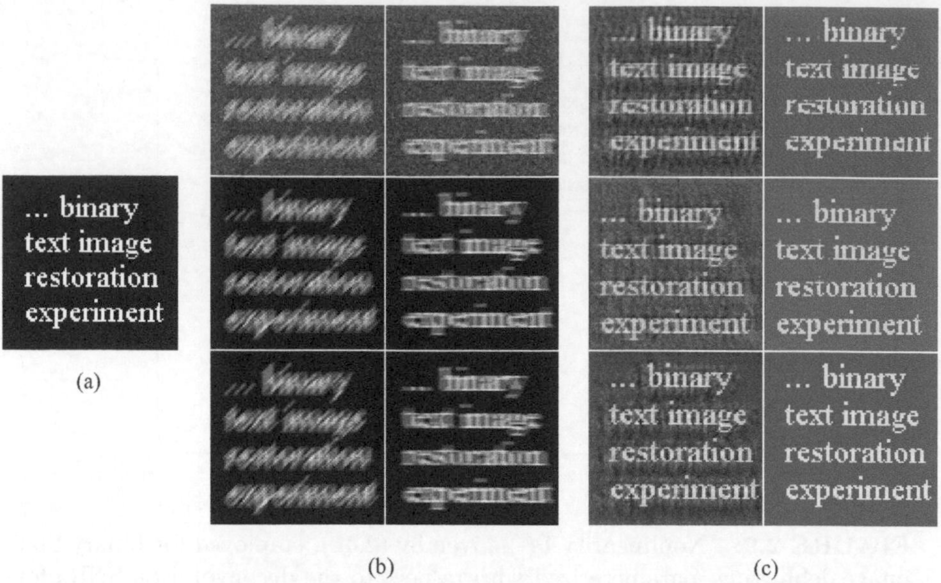

FIGURE 2.11: Binary text. (a) Original image. (b) Blurred binary text. First row: blur at SNR = 10 dB. Second row: blur at SNR = 20 dB. Third row: blur at SNR = 30 dB. (c) Restored binary text. Left: deconvolved estimation $\hat{x}[n_1, n_2]$. Right: nonlinear MMSE estimation $\tilde{x}_{\Gamma}[n_1, n_2]$. First row: blur at SNR = 10 dB. Second row: blur at SNR = 20 dB. Third row: blur at SNR = 30 dB.

20, and 30 dB (both the channels have the same SNR). In Figure 2.11(c) the deconvolved images and the nonlinear MMSE estimated ones, for the considered SNR values are depicted.

It is worth pointing out that the deconvolution technique here described allows obtaining clear and readable deconvolved text images.

In Figure 2.12 the MSE, defined as

$$\text{MSE} \stackrel{\text{def}}{=} \frac{1}{N^2} \sum_{i,j=0}^{N-1} \left(x[i,j] - x^{est}[i,j] \right)^2,$$

where $x^{est} \in \{\hat{x}, \tilde{x}_{\Gamma}\}$, is plotted vs. the iteration number at different SNR values for both the deconvolved estimate $\hat{x}[n_1, n_2]$ and the nonlinear MMSE estimate $\tilde{x}_{\Gamma}[n_1, n_2]$. From Figure 2.12 we observe that the iterative deconvolution converges after 40 to 45 iterations. During the iterations, the deconvolution noise decays, and thus the nonlinear estimator tends to behave like a hard (bilevel) detector. At convergence, the nonlinearly restored image not only shows sharper lines and edges, but also reaches a lower MSE than the deconvolved image.

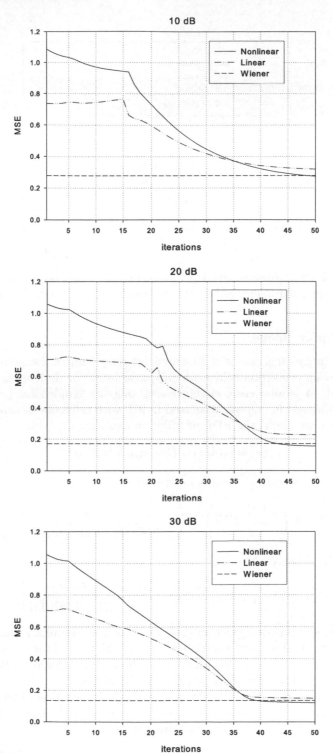

FIGURE 2.12: Binary text image deconvolution: MSE vs. iterations at SNR=10 dB, 20 dB, 30 dB.

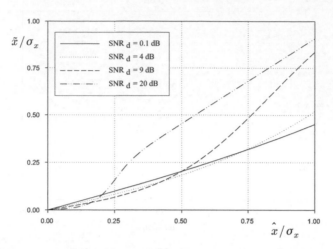

FIGURE 2.13: Nonlinearity $\Gamma(\cdot)$, given by (2.45), employed for spiky images deblurring, parameterized w.r.t. the deconvolution SNR_d for $p_0 = 0.5$.

2.4.4.2 Spiky Images

Another interesting case is represented by spiky images, i.e., images characterized by an excitation field made of sparse bright or dark spots in a gray background. A similar case, closely resembling the geophysical prospecting problem addressed in [32] was discussed in [62] with reference to astronomical images. Here we model the PDF of spiky images as:

$$p_x(x) = p_0\,\delta(x) + (1 - p_0)\,\mathcal{N}(x; 0, \sigma_x^2). \tag{2.44}$$

The nonlinear estimate $\tilde{x}_\Gamma^{(k)}$ turns out to be

$$\frac{\tilde{x}_\Gamma^{(k)}[n_1, n_2]}{\sigma_x} = \Gamma\left(\frac{\hat{x}^{(k)}[n_1, n_2]}{\sigma_x}\right) = \frac{1}{1 + 1/\mathrm{SNR}_d} \cdot \frac{\hat{x}^{(k)}[n_1, n_2]/\sigma_x}{g\left(\hat{x}^{(k)}[n_1, n_2]/\sigma_x\right)}, \tag{2.45}$$

being $g(\cdot)$ the function given by:

$$g(\cdot) = 1 + \frac{p_0}{1 - p_0}\,\sqrt{1 + \mathrm{SNR}_d}\,\exp\left[-\frac{(\cdot)^2}{2} \cdot \frac{\mathrm{SNR}_d}{(1 + 1/\mathrm{SNR}_d)}\right]. \tag{2.46}$$

The nonlinear function given in (2.45) is shown in Figure 2.13 for different values of the signal to deconvolution noise ratio SNR_d and for the value $p_0 = 0.5$. In Figure 2.14 the same function is shown for different values of the parameter p_0 and for a fixed value of $\mathrm{SNR}_d = 20$ dB.

In Figure 2.15 the original spiky image before application of the blurring filters is shown. The impulse responses of the employed blurring filters are Gaussian shaped, and take the following form:

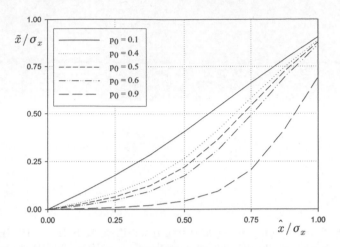

FIGURE 2.14: Nonlinearity $\Gamma(\cdot)$, given by (2.45), employed for spiky images deblurring, parameterized w.r.t. to p_0 for $\text{SNR}_d = 20\text{dB}$.

FIGURE 2.15: Original spiky image.

$$h_i[n_1, n_2] = e^{-(n_1-u_i)^2/2\sigma^2} \cos(\theta_i\, n_1) \cdot e^{-(n_2-v_i)^2/2\sigma^2} \cos(\phi_i\, n_2) \qquad (2.47)$$

with $\theta_i/\pi = 0.2, 0.7, 0.2, 0.9$, $\phi_i/\pi = 0.1, 0.2, 0.9, 0.8$, $u_i = 0, 1, -1, 0$, $v_i = 0, 1, 0, 1$, and $\sigma = 3$. In Figure 2.16 the blurred spiky images are shown for SNR values of 10, 20, and 30 dB (all the channels have the same SNR). The deconvolved images are also displayed in Figure 2.16. We see that the spikes are clearly distinguishable after deconvolution, even in those pairs that lie very close each other. In this case convergence is attained after five iterations, as illustrated in Figure 2.17 where the MSE is plotted vs. the iteration number for values of the SNR equal to 10, 20, and 30 dB respectively. Note also that the nonlinear estimate outperforms the linear Wiener estimate.

2.4.5 Application to Image Deblurring: Natural Images

The quality of the restored image obtained by means of the Bussgang algorithm strictly depends on how the adopted nonlinear processing is able to restore specific characteristics or properties of the original image. If the un-

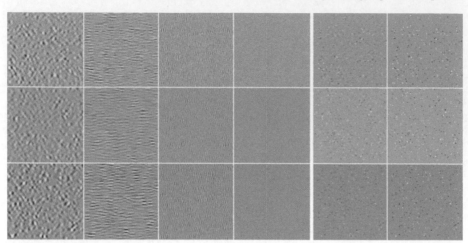

FIGURE 2.16: Spiky images. First row: blur at SNR = 10 dB. Second row: blur at SNR = 20 dB. Third row: blur at SNR = 30 dB. From column one to column four: four different blurred observations. Column five: restored spiky image (deconvolved estimation $\hat{x}[n_1, n_2]$). Column six: restored spiky image (nonlinear MMSE estimation $\tilde{x}_\Gamma[n_1, n_2]$).

known image is well characterized using a probabilistic description, as for text images, the nonlinearity $\Gamma(\cdot)$ can be designed on the basis of a Bayesian criterion, as the "best" estimate of $x[n_1, n_2]$ given $\hat{x}^{(k)}[n_1, n_2]$. Often, the Minimum Mean Square Error (MMSE) criterion is adopted.

However, for natural images, the image characterization and hence the nonlinearity design is much more complex. In [63], the authors design the nonlinearity $\Gamma(\cdot)$, after having represented the linear estimate[2] $\hat{x}[n_1, n_2]$ in a transform domain in which both the blur effect and the original image structural characteristics are more easily understood.

Specifically, in [63] the design of the nonlinear processing stage using the Radon Transform (RT) [64] of the image edges has been investigated. This choice is motivated by the fact that the RT of the image edges well describes the structural image features and the effect of blur, thus simplifying the non-linearity design.

As for the image representation in the edge domain, let us consider the decomposition of the linear estimate $\hat{x}[n_1, n_2]$ by means of a filter pair composed by the low-pass filter $\psi^{(0)}[n_1, n_2]$ and a band-pass filter $\psi^{(1)}[n_1, n_2]$ (see

---

[2]To simplify the notation, in the following we will drop the superscript (k) referring to the k-th iteration of the deconvolution algorithm.

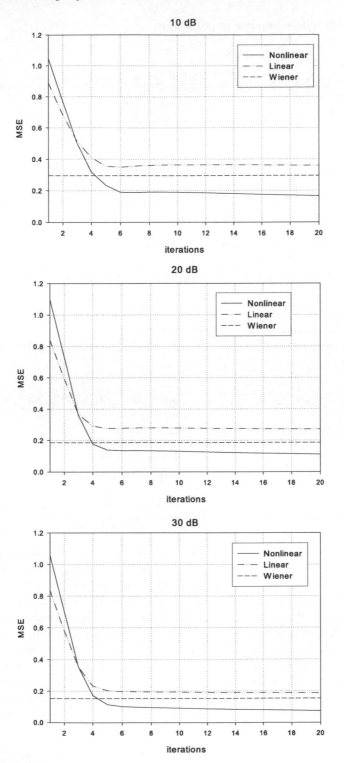

FIGURE 2.17: Spiky images deconvolution: MSE vs. iterations at SNR=10 dB, 20 dB, 30 dB.

Figure 2.18) whose impulse responses are

$$
\begin{cases}
\psi^{(0)}[n_1, n_2] = e^{-r^2[n_1,n_2]/\sigma_0^2} \\
\psi^{(1)}[n_1, n_2] = \dfrac{r[n_1, n_2]}{\sigma_1} e^{-r^2[n_1,n_2]/\sigma_1^2} e^{-j\theta[n_1,n_2]}
\end{cases}
\tag{2.48}
$$

where $r[n_1, n_2] \stackrel{\text{def}}{=} \sqrt{n_1^2 + n_2^2}$ and $\theta[n_1, n_2] \stackrel{\text{def}}{=} \arctan n_2/n_1$ are discrete polar pixel coordinates. These impulse responses belong to the class of the Circular Harmonic Functions (CHF), whose detailed analysis can be found in [65, 66], and possess the interesting characteristic of being invertible by a suitable filter pair $\phi^{(0)}[n_1, n_2], \phi^{(1)}[n_1, n_2]$.

For the values of the form factors σ_0 and σ_1 of interest, the corresponding transfer functions can be well approximated as follows:

$$
\begin{cases}
\Psi^{(0)}(e^{j\omega_1}, e^{j\omega_2}) \simeq \pi\sigma_0^2 e^{-\rho^2(\omega_1,\omega_2)\sigma_0^2/4} \\
\Psi^{(1)}(e^{j\omega_1}, e^{j\omega_2}) \simeq \dfrac{-j\pi\sigma_1^3}{2} \rho(\omega_1,\omega_2) e^{-\rho^2(\omega_1,\omega_2)\sigma_1^2/4} e^{-j\gamma(\omega_1,\omega_2)}
\end{cases}
\tag{2.49}
$$

being $\rho(\omega_1, \omega_2) \stackrel{\text{def}}{=} \sqrt{\omega_1^2 + \omega_2^2}$ and $\gamma(\omega_1, \omega_2) \stackrel{\text{def}}{=} \arctan \omega_2/\omega_1$, the polar coordinates in the spatial radian frequency domain.

The reconstruction filters $\phi^{(0)}[n_1, n_2]$ and $\phi^{(1)}[n_1, n_2]$ satisfy the invertibility condition

$$
\Psi^{(0)}(e^{j\omega_1}, e^{j\omega_2}) \cdot \Phi^{(0)}(e^{j\omega_1}, e^{j\omega_2}) + \Psi^{(1)}(e^{j\omega_1}, e^{j\omega_2}) \cdot \Phi^{(1)}(e^{j\omega_1}, e^{j\omega_2}) = 1.
$$

In the performed experiments, we have chosen

$$
\Phi^{(0)}(e^{j\omega_1}, e^{j\omega_2}) = \frac{\overline{\Psi^{(0)}}(e^{j\omega_1}, e^{j\omega_2})}{\mid \Psi^{(0)}(e^{j\omega_1}, e^{j\omega_2}) \mid^2 + \mid \Psi^{(1)}(e^{j\omega_1}, e^{j\omega_2}) \mid^2}
\tag{2.50}
$$

$$
\Phi^{(1)}(e^{j\omega_1}, e^{j\omega_2}) = \frac{\overline{\Psi^{(1)}}(e^{j\omega_1}, e^{j\omega_2})}{\mid \Psi^{(0)}(e^{j\omega_1}, e^{j\omega_2}) \mid^2 + \mid \Psi^{(1)}(e^{j\omega_1}, e^{j\omega_2}) \mid^2}
\tag{2.51}
$$

to prevent amplification of spurious components occurring at those spatial frequencies where $\Psi^{(0)}(e^{j\omega_1}, e^{j\omega_2})$ and $\Psi^{(1)}(e^{j\omega_1}, e^{j\omega_2})$ are small in magnitude. The optimality of these reconstruction filters is discussed in [67].

The zero-order circular harmonic filter $\psi^{(0)}[n_1, n_2]$ extracts a low-pass version $\hat{x}_0[n_1, n_2]$ of the input image; the form factor σ_0 is chosen so to retain only very low spatial frequencies, so obtaining a low-pass component exhibiting high spatial correlation. The first-order circular harmonic filter $\psi^{(1)}[n_1, n_2]$ is a band-pass filter, with frequency selectivity set by properly choosing the form factor σ_1. The output of this filter is a complex image $\hat{x}_1[n_1, n_2]$, which will be referred to in the following as the "edge image," whose magnitude is related to the presence of edges and whose phase is proportional to their orientation.

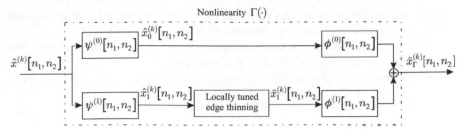

FIGURE 2.18: Multichannel nonlinear estimator $\Gamma(\cdot)$.

Coarsely speaking, the edge image $\hat{x}_1[n_1, n_2]$ is composed by curves, representing edges occurring in $x[n_1, n_2]$, whose width is controlled by the form factor σ_1, and by low-magnitude values representing the interior of uniform or textured regions occurring in $x[n_1, n_2]$. Strong intensity curves in $\hat{x}_1[n_1, n_2]$ are well analyzed by the local application of the bidimensional RT. This transform maps a straight line into a point in the transformed domain, and therefore it yields a compact and meaningful representation of the image's edges. However, since most images' edges are curves, the analysis must be performed locally by partitioning the image into regions small enough such that in each block may occur only straight lines. Specifically, after having chosen the region dimensions, the value of the filter parameter σ_1 is set such that the width of the observed curve is a small fraction of its length. In more detail, the evaluation of the edge image is performed by the CH filter of order one $\psi^{(1)}[n_1, n_2]$ that can be seen as the cascade of a derivative filter followed by a Gaussian smoothing filter. The response to an abrupt edge of the original image is a line in $\hat{x}_1[n_1, n_2]$. The line is centered in correspondence of the edge, whose energy is concentrated in an interval of $\pm\sigma_1^{-1}$ pixels and that slowly decays to zero in an interval of $\pm 3\sigma_1^{-1}$ pixels. Therefore by partitioning the image into blocks of 8×8 pixels, the choice of $\sigma_1 \approx 1$ yields edge structures that are well readable in the partitions of the edge image.

Then each region is classified either as a "strong edge" region or as a "weak edge" and "textured" region. The proposed enhancement procedures for the different kind of regions are described in detail in Section 2.4.5.2.

It is worth pointing out that our approach shares the local RT as a common tool with a family of recently proposed image transforms — the curvelet transforms [68–70] — that represent a significant alternative to wavelet representation of natural images. In fact, the curvelet transform yields a sparse representation of both smooth image and edges, either straight or curved.

2.4.5.1   Local Radon Transform of the Edge Image: Original Image and Blur Characterization

Once the edge image has been obtained, it is analyzed by partitioning it into areas of proper size. A number of regions where the edge curves locally appear as straight elements can be found. On such regions, a local application of the RT yields a compact representation of the edge itself.

Let us here discuss the Radon transform representation in more detail. For a continuous image $\xi(t_1, t_2)$ the RT [64] is defined as

$$p_\beta^\xi(s) \stackrel{\text{def}}{=} \int_{-\infty}^{\infty} \xi(\cos \beta \cdot s - \sin \beta \cdot u, \ \sin \beta \cdot s + \cos \beta \cdot u) \, du, \tag{2.52}$$

$$-\infty < s < \infty, \ \beta \in [0, \pi)$$

that represents the summation of $\xi(t_1, t_2)$ along a ray at distance s and angle β.

It is well known [64] that it can be inverted by

$$\xi(t_1, t_2) = \frac{1}{4\pi^2} \int_0^\pi \int_{-\infty}^{\infty} P_\beta^\xi(j\sigma) e^{j\sigma(\sigma \cos \beta \, t_1 + \sigma \sin \beta \, t_2)} |\sigma| \, d\sigma d\beta$$

where $P_\beta^\xi(j\sigma) = \mathcal{F}\left\{p_\beta^\xi(s)\right\}$ is the Fourier transform of the RT.

Some details about the discrete implementation of the RT follows.

If the image $\xi(t_1, t_2)$ is frequency limited in a circle of diameter D_Ω, it can be reconstructed by the samples of its RT taken at a spatial sampling interval $\Delta s \leq 2\pi/D_\Omega$,

$$p_\beta^\xi[n] = p_\beta^\xi(s)|_{s=n\cdot\Delta s} \ \ n = 0, \pm 1, \pm 2, \cdots$$

Moreover, if the original image is approximately limited in the spatial domain, i.e., it vanishes out of a circle of diameter D_t, the sequence $p_\beta^\xi[n]$ has finite length $N = 1 + D_t/\Delta s$. In a similar way, the RT can be sampled with respect to the angular parameter β considering M different angles $m\Delta\beta$, $m = 0, \cdots, M - 1$, with sampling interval $\Delta\beta$, namely

$$p_{\beta_m}^\xi[n] = p_{\beta_m}^\xi(s)|_{s=n\cdot\Delta s, \ \beta_m = m\cdot\Delta\beta}.$$

The angular interval $\Delta\beta$ can be chosen so as to assure that the distance between points $p_\beta^\xi[n]$ and $p_{\beta+\Delta\beta}^\xi[n]$ lying on adjacent diameters remains less than or equal to the chosen spatial sampling interval Δs, that is,

$$\Delta\beta \cdot \frac{D_t}{2} \leq \Delta s.$$

The above condition is satisfied when $M \geq \frac{\pi}{2} \cdot N \simeq 1.57 \cdot N$.

As for the edge image, each region is here modeled as obtained by ideal sampling of an original image $x_1(t_1, t_2)$, approximately spatially bounded by a circle of diameter D_t, and bandwidth limited in a circle of diameter D_Ω. Under the hypothesis that $N - 1 \geq D_t \cdot D_\Omega/2\pi$ and $M \geq \frac{\pi}{2} \cdot N$, the M, N samples

$$p_{\beta_m}^{x_1}[n], m = 0, \cdots M - 1, n = 0, \cdots N - 1$$

of the RT $p_\beta^{x_1}(s)$ allow the reconstruction of the image $x_1(t_1, t_2)$, and hence of any pixel of the selected region.

Let us now consider the case of blurred observations. In the edge image, the blur tends to flatten and attenuate the edge peaks, and to smooth the edge contours in directions depending on the blur itself. The effect of blur on the RT of the edge image regions is twofold. The first effect is that, since the energy of each edge is spread in the spatial domain, the maximum value of the RT is lowered. The second effect is that, since the edge width is typically thickened, it contributes to different tomographic projections, enhancing two triangular regions in the RT.

Stemming from this compact representation of the blur effect, we will design an effective nonlinearity aimed at restoring the original edges as described in Section 2.4.5.2.

2.4.5.2 Local Radon Transform of the Edge Image: Nonlinearity Design

The design of the nonlinearity will be conducted after having characterized the blur effect at the output of a first-order CHF bank. By choosing the form-factor σ_0 of the zero-order CH filter $\psi^{(0)}$ small enough the blur transfer function is approximately constant in the passband, and thus the blur effect on the low-pass component results negligible.

As for the first order CH filter's domain, the blur causes the edges in the spatial domain to be spread along directions depending on the impulse responses of the blurring filters. After having partitioned the edge image into small regions in order to perform a local RT as detailed in Section 2.4.5.1, each region has to be classified either as a strong edge area or a weak edge and textured area. Hence, the nonlinearity has to be adapted to the degree of "edgeness" of each region in which the image has been partitioned. The decision rule between the two areas is binary. Specifically, an area intersected by a "strong edge" is characterized by a RT whose coefficients assume significant values only on a subset of directions β_m. Therefore, a region is classified as a strong edge area by comparing $\max_m \sum_n (p_{\beta_m}^{\xi_1}[n])^2$ with a fixed threshold. If the threshold is exceeded, the area is classified as a strong edge area, otherwise this implies that either no direction is significant, which corresponds to weak edges, or every direction is equally significant, which corresponds to textured areas.

Strong Edges

For significant image edges, characterized by relevant energy concentrated in one direction, the nonlinearity can exploit the spatial memory related to the edge structure. In this case, as above discussed, we use the Radon transform of the edge image. Let us consider a limited area of the edge image $\hat{x}_1[n_1, n_2]$ intersected by an edge, and its RT $p_{\beta_m}^{\hat{x}_1}[n]$, with m, n chosen as discussed in Section 2.4.5.1.

The nonlinearity we present aims at focusing the RT both with respect to m and n, and it is given by:

$$p_{\beta_m}^{\tilde{x}_1}[n] = p_{\beta_m}^{\hat{x}_1}[n] \cdot g^{\kappa_g}(\beta_m) \cdot f_{\beta_m}^{\kappa_f}(n) \tag{2.53}$$

with

$$g(\beta_m) = \frac{\max_n(p_{\beta_m}^{\hat{x}_1}[n]) - \min_{\beta_k,n}(p_{\beta_k}^{\hat{x}_1}[n])}{\max_{\beta_k,n}(p_{\beta_k}^{\hat{x}_1}[n]) - \min_{\beta_k,n}(p_{\beta_k}^{\hat{x}_1}[n])} \tag{2.54}$$

and

$$f_{\beta_m}(n) = \frac{p_{\beta_m}^{\hat{x}_1}[n] - \min_n(p_{\beta_m}^{\hat{x}_1}[n])}{\max_n(p_{\beta_m}^{\hat{x}_1}[n]) - \min_n(p_{\beta_m}^{\hat{x}_1}[n])} \tag{2.55}$$

where $\max_n(p_{\beta_m}^{\hat{x}_1}[n])$ and $\min_n(p_{\beta_m}^{\hat{x}_1}[n])$ represent the maximum and the minimum values, respectively, of the RT for the direction β_m under analysis, and $\max_{\beta_k,n}(p_{\beta_k}^{\hat{x}_1}[n])$ and $\min_{\beta_k,n}(p_{\beta_k}^{\hat{x}_1}[n])$, with $k = 0, ..., M-1$, the global maximum and the global minimum, respectively, in the Radon domain. Therefore, for each point belonging to the direction β_m and having index n, the nonlinearity (2.53) weights the RT by two gain functions. Specifically, (2.54) assumes its maximum value (equal to 1) for the direction β_{Max} where the global maximum occurs and it decreases for the other directions. In other words (2.54) assigns a relative weight equal to 1 to the direction β_{Max} whereas it attenuates the other directions. Moreover, for a given direction β_m, (2.55) determines the relative weight of the actual displacement n with respect to the others by assigning a weight equal to 1 to the displacement where the maximum occurs and by attenuating the other locations. The factors κ_g and κ_f in (2.54) and (2.55) are defined as $\kappa_g = \kappa_0 \sigma_{w^{(k)}}^2$ and $\kappa_f = \kappa_1 \sigma_{w^{(k)}}^2$ being $\sigma_{w^{(k)}}^2$ the deconvolution noise variance and κ_0 and κ_1 two constants empirically chosen and set for our experiments equal to 2.5 and 0.5, respectively. The deconvolution noise variance $\sigma_{w^{(k)}}^2$ depends on both the blur and on the observation noise, and it can be estimated as $E\{|w^{(k)}[n_1, n_2]|^2\} \approx E\{|\tilde{x}_\Gamma^{(k)}[n_1, n_2] - \hat{x}[n_1, n_2]|^2\}$ when the algorithm begins to converge. The deconvolution noise variance gradually decreases at each iteration, which guarantees a gradually decreasing action of the nonlinearity as the iterations proceed.

The edge enhancement in the Radon domain is then described by the combined action of (2.54) and (2.55), since the first estimates the edge direction and the second performs a thinning operation for that direction.

To depict the effect of the nonlinearity (2.53) in the edge domain, the case of a straight edge is illustrated in Figure 2.19.

Weak Edges and Textured Regions

If the image is locally low contrast or does not exhibit any directional structure able to drive the nonlinearity, we use a spatially zero memory nonlinearity

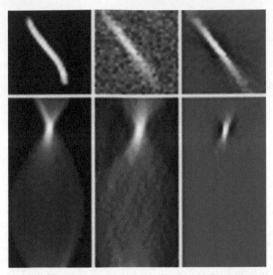

FIGURE 2.19: First row, from left to right: original edge, blurred edge, restored edge. Second row: corresponding discrete Radon Transforms.

acting pointwise on the edge image. Since the edge image is almost zero in every pixel corresponding to the interior of uniform regions, where small values are likely due to noise, the nonlinearity should attenuate low-magnitude values of $\hat{x}_1[n_1, n_2]$. On the other hand, high-magnitude values of $\hat{x}_1[n_1, n_2]$, possibly due to the presence of structures, should be enhanced. A pointwise nonlinearity performing the said operations is given in the following:

$$\tilde{x}_1[n_1, n_2] = (1 + 1/\alpha) \cdot \hat{x}_1[n_1, n_2] \cdot g\left(\left|\hat{x}_1[n_1, n_2]\right|\right)$$
$$g(\cdot) = 1 + \gamma \cdot \sqrt{1 + \alpha} \cdot \exp\left[-\frac{(\cdot)^2}{2} \cdot \frac{\alpha}{(1 + 1/\alpha)}\right]. \tag{2.56}$$

The magnitude of (2.56) is plotted in Figure 2.20 for different values of the parameter α.

The low gain zone near the origin is controlled by the parameter γ; the parameter α controls the enhancement effect on the edges. Both the parameters are set empirically. The nonlinearity (2.56) has been presented in [71], where the analogy of this nonlinearity with the MMSE estimator of spiky images in Gaussian observation noise is discussed.

To sum up, the adopted nonlinearity is locally tuned to the image characteristics. When the presence of an edge is detected, an edge thinning in the local RT of the edge image is performed. This operation, which encompasses a spatial memory in the edge enhancement, is performed directly in the RT domain since the image edges are compactly represented in this domain. When an edge structure is not detected, which may happen, for example, in

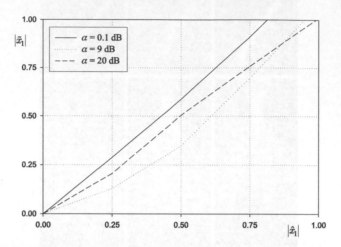

FIGURE 2.20:   Nonlinearity given by (2.56), employed for natural images deblurring, parameterized with respect to the parameter α for $\gamma = 0.5$.

textured or uniform regions, the adopted nonlinearity reduces to a pointwise edge enhancement. It is worth pointing out that, as extensively discussed in [68], the compact representation of an edge in the RT domain is related to the tuning between the size of the local RT transform with the band-pass of the edge extracting filter.

After the nonlinear estimate $\tilde{x}_1[n_1, n_2]$ has been computed, the estimate $\tilde{x}_\Gamma[n_1, n_2]$ is obtained by means of the inverse filter bank $\phi^{(0)}[n_1, n_2]$ and $\phi^{(1)}[n_1, n_2]$, i.e. (see Figure 2.18):

$$\tilde{x}_\Gamma[n_1, n_2] = \left(\phi^{(0)} * \hat{x}_0\right)[n_1, n_2] + \left(\phi^{(1)} * \tilde{x}_1\right)[n_1, n_2]. \qquad (2.57)$$

Let us remark that the nonlinear estimator modifies only the edge image magnitude, whereas the edge direction is restored in the linear estimation stage.

In Figure 2.21 and Figure 2.22 some of the images we have used for our experimentations are reported. The images are blurred using the blurring filters having the following impulse responses:

$$h_1[n_1, n_2] = \begin{bmatrix} 0\,0\,0\,1\,0\,0\,0 \\ 0\,0\,0\,1\,0\,0\,0 \\ 0\,0\,0\,1\,0\,0\,0 \\ 0\,0\,0\,1\,0\,0\,0 \\ 0\,0\,1\,0\,0\,0\,0 \end{bmatrix} \; ; \; h_2[n_1, n_2] = \begin{bmatrix} 0\,0\,0\,1\,0\,0\,0 \\ 0\,0\,0\,1\,0\,0\,0 \\ 0\,0\,0\,1\,0\,0\,0 \\ 0\,0\,1\,0\,0\,0\,0 \\ 0\,0\,1\,0\,0\,0\,0 \end{bmatrix} \; ;$$

FIGURE 2.21: "F16" image.

FIGURE 2.22: "Cameraman" image.

$$h_3[n_1, n_2] = \begin{bmatrix} 0 & 0 & 0 & 0 & 0 & 0 & 0 \\ 0 & 0 & 0 & 0 & 0 & 0 & 0 \\ 0.5 & 0.86 & 0.95 & 1 & 0.95 & 0.86 & 0.5 \\ 0 & 0 & 0 & 0 & 0 & 0 & 0 \\ 0 & 0 & 0 & 0 & 0 & 0 & 0 \end{bmatrix}.$$

In Figure 2.23 some details belonging to the "F16" image shown in Figure 2.21 are depicted. The corresponding blurred observations, affected by additive white Gaussian noise at SNR = 20 dB, obtained using the aforementioned blurring filters, are also shown along with the deblurred images. In Figure 2.24 the same images are reported for blurred images affected by additive

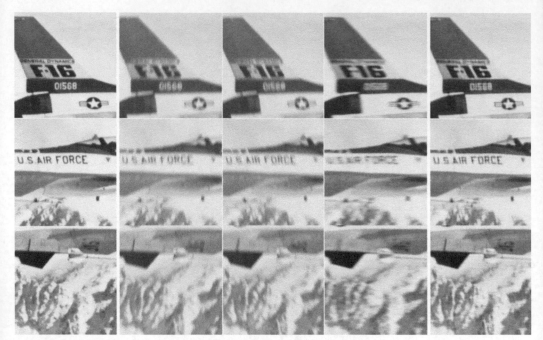

FIGURE 2.23: "F16" image. First column: details of the original image. Second, third, and fourth column: blurred observations of the original details. Fifth column: restored details. (SNR = 20 dB.)

white Gaussian noise at SNR = 40 dB. In Figure 2.25 the MSE, defined as

$$\mathrm{MSE} \overset{\mathrm{def}}{=} \frac{1}{N^2} \sum_{i,j=0}^{N-1} \Big(x[i,j] - \hat{x}[i,j] \Big)^2,$$

is plotted vs. the iteration number at different SNR values for the deblurred image.

Similar results are reported in Figure 2.26 for the "Cameraman" image shown in Figure 2.22 and the corresponding MSE is shown in Figure 2.25.

Note that the algorithm converges in few iterations toward a local minimum and then may even diverge. In fact, as well known [32], the global convergence of the Bussgang algorithm cannot be guaranteed. The choice of the desired restoration result can be made by visual inspection of the restored images obtained during the iterations, which is an approach, as also pointed out in [11], typical of blind restoration techniques.

It is worth noting that the deblurring algorithm gives images of improved visual quality, significantly reducing the distance, in the mean square sense, from the original unblurred image.

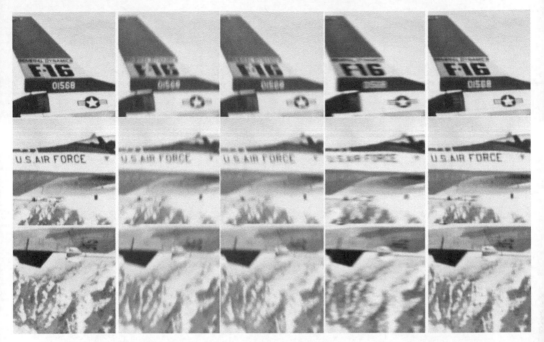

FIGURE 2.24: "F16" image. First column: details of the original image. Second, third, and fourth column: blurred observations of the original details. Fifth column: restored details. (SNR = 40 dB.)

2.5   Conclusions

In this chapter the Bussgang processes are briefly revised together with some basic related properties.

The single-channel Bussgang blind deconvolution method is described in a Bayesian framework. It is then applied to a texture synthesis-by-analysis method whose goal is the reproduction of textures perceptually indistinguishable from the given prototype. The method relies on the conjecture, supported by psychophysics experiments, that textures with the same statistical properties up to the second order give the same visual impression. With respect to single-stage filtered i.i.d. random fields approaches, this twin-stage technique seems to possess an increased capability of capturing textural spatial organization thanks to the replacement of i.i.d. sources with binary sources having a prescribed second-order distribution. For all the texture categories taken into account, significant quality results have been obtained.

The multichannel Bussgang blind deconvolution algorithm has been theorized, thus extending the single-channel case. The multichannel Wiener filter has been first derived, then the nonlinear estimator has been obtained for

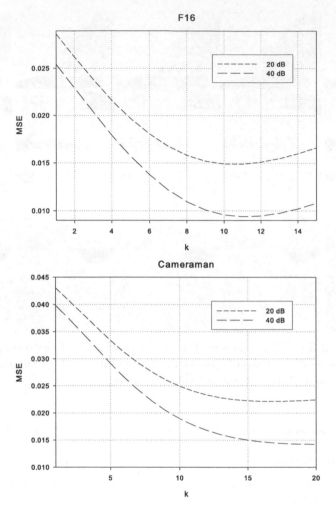

FIGURE 2.25: Natural images deconvolution: MSE vs. iteration number for "F16" and "Cameraman" images.

different image classes. In particular, both the cases of spatially uncorrelated images like binary text images and spiky images have been investigated. A Bayesian nonlinear estimator, acting in the spatial domain, has been derived for the case of uncorrelated images. For highly correlated natural images a different approach to the nonlinearity design, based on the representation of the image in a properly defined *edge* domain, has been described. Specifically, a suitable nonlinear processing has been designed in the Radon transform domain. Experimental results assessing the objective and visual qualities of the restored images have been provided.

FIGURE 2.26: Image "Cameraman." Left column, first, second, and third rows: observations. Fourth row: restored image (SNR = 20 dB). Right column, first, second, and third rows: observations. Fourth row: restored image

References

[1] *Digital Image Restoration*, A. Katsaggelos Ed., New York: Springer-Verlag, 1991.

[2] M. Banham and A. Katsaggelos, Digital Image Restoration, *IEEE Signal Processing Magazine*, pages 24–41, March 1997.

[3] R. Molina, On the hierarchical Bayesian approach to image restoration: application to astronomical images, *IEEE Transactions on Pattern Analysis and Machine Intelligence*, Vol. 16, pages 1122–1128, November 1994.

[4] R. Molina, J. Nunez, F. Cortijo, and J. Mateos, Image restoration in astronomy: a Bayesian review, *IEEE Signal Processing Magazine*, 18, pages 11-29, 2001.

[5] J. L. Starck and E. Pantin, Deconvolution in Astronomy: A Review, *Publications of the Astronomical Society of the Pacific*, 114, pages 1051-1069, October 2002.

[6] J. P. Muller, Ed., *Digital Image Processing in Remote Sensing*, Philadelphia: Taylor & Francis, 1988.

[7] A. Jalobeanu, L. Blanc-Fraud, and J. Zerubia, An Adaptive Gaussian Model for Satellite Image Deblurring, *IEEE Transactions on image processing*, Vol. 13, No. 4, pages 613–621, April 2004.

[8] O. Michailovich and D. Adam, A novel approach to the 2-D blind deconvolution problem in medical ultrasound, *IEEE Transactions on Medical Imaging*, 24, pages 86–104, January 2005.

[9] T. S. Ralston, D. L. Marks, F. Kamalabadi, and S. A. Boppart, Deconvolution methods for mitigation of transverse blurring in optical coherence tomography, *IEEE Transactions on Image Processing*, Vol. 14, No. 9, pages 1254–1264, September 2005.

[10] N. P. Galatsanos, V.Z. Mesarović, R. Molina, and A.K. Katsaggelos, Hierarchical Bayesian Image Restoration from Partially Known Blurs, *IEEE Transactions on Image Processing*, Vol.9, No.10, pages 1784–1797, October 2000.

[11] D. Kundur and D Hatzinakos, Blind Image Deconvolution, *IEEE Signal Processing Magazine*, pages 43–64, May 1996.

[12] D. Kundur and D. Hatzinakos, Blind image deconvolution revisited, *IEEE Signal Processing Magazine*, Vol. 13, No. 6, pages 61–63, Nov. 1996.

[13] R. L. Lagendijk, J. Biemond, and D. E. Boekee, Identification and restoration of noisy blurred images using the expectation-maximization algorithm, *IEEE Transactions on Acoustic, Speech, and Signal Processing*, Vol. 38, No. 7, pages 1180–1191, July 1990.

[14] S. J. Reeves and R. M. Mersereau, Blur identification by the method of generalized cross-validation, *IEEE Transactions on Image Processing*, Vol. 1, No. 3, pages 301–311, July 1992.

[15] V. Z. Mesarović, N. P. Galatsanos, and A. K. Katsaggelos, Regularized constrained total least squares image restoration, *IEEE Transactions on Image Processing*, Vol. 4, No. 8, pages 1096–1108, August 1995.

[16] T. F. Chan and C.-K. Wong, Total variation blind deconvolution, *IEEE Transactions on Image Processing*, Vol. 7, No. 3, pages 370–375, March 1998.

[17] L. Rudin, S. Osher, and E. Fatemi, Nonlinear total variation based noise removal algorithm, *Physica D*, Vol. 60, pages 259–268, 1992.

[18] D. P. K. Lun, T. C. Hsung, and T. W. Shen, Orthogonal discrete periodic Radon transform. Part II: applications, *Signal Processing*, Vol. 83, No. 5, May 2003.

[19] R. G. Lane and R. H. T. Bates, Relevance for blind deconvolution of recovering Fourier magnitude from phase, *Optical Communications*, Vol. 63, July 1987.

[20] J. Bardsley, S. Jefferies, J. Nagy, and R. Plemmons, A Computational Method for the Restoration of Images with an Unknown, Spatially-varying Blur, *Optics Express*, Vol. 14, No. 5, pages 1767-1782, March 2006.

[21] C. Vogel, T. Chan and R. Plemmons, Fast Algorithms for Phase Diversity-based Blind Deconvolution, In *Adaptive Optical System Technologies, Proceedings of SPIE,* Edited by D. Bonaccini and R. Tyson, Vol. 3353, pages 994–1005, 1998.

[22] A. C. Likas and N. P. Galatsannos, A variational approach for Bayesian blind image deconvolution, *IEEE Transactions on Signal Processing*, Vol. 52, No. 8, pages 2222–2233, August 2004.

[23] M. M. Bronstein, A. M. Bronstein, M. Zibulevsky, and Y. Y. Zeevi, Blind deconvolution of images using optimal sparse representations, *IEEE Transactions on Image Processing*, Vol. 14, No. 6, pages 726–736, June 2005.

[24] M. I. Gürelli and C. L. Nikias, EVAM: an eigenvector-based algorithm for multichannel blind deconvolution of input colored signals, *IEEE Transactions on Signal Processing*, Vol. 43, No. 1, pages 134–149, January 1995.

[25] G. Harikumar and Y. Bresler, Perfect blind restoration of images blurred by multiple filters: theory and efficient algorithm, *IEEE Transactions on Image Processing*, Vol. 8, pages 202–219, February 1999.

[26] G. B. Giannakis and R.W. Heath, Jr., Blind Identification of multi-channel FIR Blurs and Perfect Image Restoration, *IEEE Transactions on Image Processing*, Vol. 9, pages 1877–1896, November 2000.

[27] H. Pai and A.C. Bovik, On eigenstructure-based direct multichannel blind image restoration, *IEEE Transactions on Image Processing*, Vol. 10, No. 10, pages 1434–1446, October 2001.

[28] F. Sroubek and J. Flusser, Multichannel blind iterative image restoration, *IEEE Transactions on Image Processing*, Vol. 12, No. 9, pages 1094-1106, September 2003.

[29] F. Sroubek and J. Flusser, Multichannel blind deconvolution of spatially misaligned images, *IEEE Transactions on Image Processing*, Vol. 14, No. 7, pages 874–883, July 2005.

[30] V. Katkovnik, D. Paliy, K. Egiazarian, and J. Astola, Frequency domain blind deconvolution in multiframe imaging using anisotropic specially adaptive denoising. In *Proceedings of EUSIPCO 2006*, Florence, Italy, September 2006.

[31] J. Bussgang, Crosscorrelation functions of amplitude-distorted Gaussian signals. In *RLE Tech. Rep. 216*, 1952.

[32] R. Godfrey and F. Rocca, Zero memory nonlinear deconvolution. In *Geophysical Prospecting 29*, pages 189–228, 1981.

[33] G. Panci, C. Colonnese, P. Campisi, and G. Scarano, Blind equalization for correlated input symbols: a Bussgang approach, *IEEE Transactions on Signal Processing*, Vol. 53, No. 5, pages 1860-1869, May 2005.

[34] S. Bellini, Bussgang techniques for blind deconvolution and restoration. In *Blind Deconvolution*, S. Haykin ed., Prentice-Hall, 1994.

[35] Z. Ding and Y. Li, *Blind Equalization and Identification*, Marcel Dekker Inc., New York, 2001.

[36] P. Campisi and G. Scarano, A Multiresolution approach for texture synthesis using the Circular Harmonic Function, *IEEE Transactions on Signal Processing*, Vol. 11, pages 37–51, January 2002.

[37] D. H. Hubel and T. N. Wiesel, Brain mechanisms of vision, *Scientific Amererican*, Vol. 241, pages 150-162, September 1979.

[38] B. Julesz, Visual pattern discrimination, *IRE Transactions on Information Theory*, IT-8, pages 84–92, February 1962.

[39] J. Malik and P. Perona, Preattentive texture discrimination with early vision mechanisms, *Journal of the Optical Society of America A*, Vol. 7, pages 923–932, 1990.

[40] R. Chellappa and R.L. Kashyap, Texture synthesis using 2-D noncausal autoregressive models, *IEEE Transactions Acoustic Speech and Signal Processing*, Vol. 33, No. 1, pages 194-203, February 1985.

[41] J. K. Tugnait, Estimation of linear parametric models of non-Gaussian discrete random fields with applications to texture synthesis, *IEEE Transaction on Image Processing*, Vol. 3, No. 2, pages 109–127, March 1994.

[42] J. A. Cadzow, D. M. Wilkes, R. A. Peter II, and X. Li, Image texture synthesis-by-analysis using moving average models, *IEEE Transactions on Aerospace and Electronic System*, Vol. 29, No. 4, pages 1110–1121, October 1993.

[43] T. E. Hall and G. B. Giannakis, Bispectral analysis and model validation of texture images, *IEEE Transactions on Image Processing*, Vol. 4, No. 7, pages 996–1009, July 1995.

[44] L. M. Kaplan and C. C. Jay Kuo, Texture roughness analysis and synthesis via extended self-similar, (ESS) model, *IEEE Transactions on Pattern Analysis and Machine Intelligence*, Vol. 17, No. 11, pages 754–761, November 1995.

[45] L. M. Kaplan and C. C. Jay Kuo, An improved method for 2-D self-similar image synthesis, *IEEE Transactions on Image Processing*, Vol. 5, No.5, pages 754–761, May 1996.

[46] G. R. Cross and A. K. Jain, Markov random field texture models, *IEEE Transactions on Pattern Analysis and Machine Intelligence*, Vol. PAMI-5, pages 25–39, January 1983.

[47] I. M. Elfadel and R. W. Picard, Gibbs random fields, cooccurences, and texture modeling, *IEEE Transactions on Pattern Analysis and Machine Intelligence*, Vol. 16, No. 1, pages 24–37, January 1994.

[48] J. M. Francos, A. Z. Meiri, and B. Porat, A unified texture model based on a 2-D Wold-like decomposition, In *IEEE Transactions on Acoustic Speech and Signal Processing*, Vol. 41, No. 8, pages 2665–2678, August 1993.

[49] J. M. Francos, A. Nara, and J. W. Woods, Maximum likelihood parameter estimation of textures using a Wold-decomposition based model, *IEEE Transactions on Image Processing*, Vol. 4, No. 12, pages 1655–1666, December 1995.

[50] F. Liu and R. W. Picard, Periodicity, directionality, and randomness: Wold features for image modeling and retrieval, *IEEE Transactions on*

Pattern Analysis and Machine Intelligence, Vol. 18, No. 7, pages 722–733, July 1996.

[51] J. Portilla and E. P. Simoncelli, Texture modeling and synthesis using joint statistics of complex wavelet coefficients, Proc. of the IEEE Workshop on Statistical and Computational Theories of Vision, Fort Collins, CO, June 1999.

[52] G. Jacovitti, A. Neri, and G. Scarano, Texture synthesis-by-analysis with hard limited Gaussian process, *IEEE Transactions on Image Processing*, Vol. 7, No. 11, pages 1615–1621, November 1998.

[53] J. O. Van Vleck and D. Middleton, The spectrum of clipped noise, *Proceedings of IEEE*, Vol. 54, pages 2–19, January, 1966.

[54] B. Julesz and R. Bergen, Textons, the fundamental elements in preattentive vision and perception of textures, *Bell System Technical Journal*, Vol. 62, pages 1619–1645, July–August 1983.

[55] P. Campisi, A. Neri, and G. Scarano, Reduced complexity modeling and reproduction of colored textures, *IEEE Transactions on Image Processing*, Vol. 9, No. 3, pages 510–518, March 2000.

[56] P. Campisi, A. Neri, and D. Hatzinakos, A Perceptually Preattentive Lossless, Model based, Texture Compression Technique, *IEEE Transactions on Image Processing*, Vol. 9, No. 8, pages 1325–1336, August 2000.

[57] P. Campisi, A. Neri, G. Panci, and G. Scarano, Robust rotation-invariant texture classification using a model based approach, *IEEE Transactions on Image Processing*, Vol. 13, No. 6, pages 782–791, June 2004.

[58] P. Campisi, S. Colonnese, G. Panci, and G. Scarano, Reduced complexity rotation-invariant texture classification using a blind deconvolution approach, *IEEE Transactions on Pattern Analysis and Machine Intelligence*, Vol.28, No.1, pages 145–149, January 2006.

[59] T. Brodatz, *Textures: a photographic album for artists and designers*, Dover, 1966.

[60] G. Jacovitti, G. Panci, and G. Scarano, Bussgang-Zero Crossing equalization: an integrated HOS-SOS approach. In *IEEE Transactions on Signal Processing*, Vol. 49, pages 2798–2812, November 2001.

[61] G. Panci, P. Campisi, C. Colonnese, and G. Scarano, The Bussgang algorithm for multichannel blind image restoration. In *XI European Signal Processing Conference 2002, (EUSPICO 2002)*, Toulouse, France, 2002.

[62] A. Neri, G. Scarano, and G. Jacovitti, Bayesian iterative method for blind deconvolution, *Proc. Adaptive Signal Processing*, SPIE Vol. 1565, pages 196–208, San Diego (CA), July 1991.

[63] S. Colonnese, P. Campisi, G. Panci, and G. Scarano, Blind Image Deblurring Driven by Nonlinear Processing in the Edge Domain, *EURASIP Journal on Applied Signal Processing, Special Issue on Nonlinear Signal and Image Processing–Part II*, Vol. 2004, Issue 16, pages 2462–2475.

[64] S. R. Deans, *The Radon transform and some of its applications*, Krieger Publishing Company, 1993.

[65] G. Jacovitti and A. Neri, Multiresolution circular harmonic decomposition, *IEEE Transactions on Signal Processing*, Vol. 48, pages 3242–3247, November 2000.

[66] G. Jacovitti and A. Neri, Multiscale image features analysis with circular harmonic wavelets, *Proc. SPIE Wavelets Applications in Signal and Image Processing III*, Vol. 2569, pages 363–372, July 1995.

[67] C.A. Berenstein and A.V. Patrick, Exact deconvolution for multiple convolution operators—an overview, plus performance characterization for imaging sensors, *Proceedings of the IEEE*, Vol. 78, No. 4, pages 723–734, April 1990.

[68] J. L. Starck, E. J. Candes, and D. L. Donoho The curvelet transform for image denoising, *IEEE Transactions on Image Processing*, Vol. 11, No. 6, pages 678–684, June 2002.

[69] J.L. Starck, Image Processing by the Curvelet Transform, *Proc. of IWDC 2002*, Capri, Italy, September 2002.

[70] D. L. Donoho and M. R. Duncan, Digital curvelet transform: strategy, implementation and experiments, *Proc. of SPIE Wavelet Applications VII*, SPIE Vol. 4056, pages 12–29, 2000.

[71] G. Panci, P. Campisi, S. Colonnese, and G. Scarano, Multichannel blind image deconvolution using the Bussgang algorithm: spatial and multiresolution approaches, *IEEE Transactions on Image Processing*, Vol. 12, No. 11, pages 1324–1337, November 2003.

3

Blind Multiframe Image Deconvolution Using Anisotropic Spatially Adaptive Filtering for Denoising and Regularization

Vladimir Katkovnik, Karen Egiazarian, Jaakko Astola

Institute of Signal Processing, Tampere University of Technology, Tampere, Finland

e-mail: (katkov, karen, jta)@cs.tut.fi

Abstract

In this chapter an adaptive method for blind deblurring of noisy multiframe images is presented. The energy criterion consists of the fidelity and the inter-channel balancing terms. It is assumed that the blur operators have finite supports with support sizes approximately known. A recursive projection gradient algorithm is developed for minimization of the energy criterion. The incorporated anisotropic spatially adaptive regularization and filtering is based on local polynomial approximation (LPA) of the image and the blur operators. The adaptivity of this regularization exploits the paradigm of intersecting confidence intervals (ICI) applied for selection of adaptive varying scales (window sizes) of the LPA. This LPA–ICI algorithm is nonlinear and pixel wise spatially adaptive with respect to smoothness and irregularities in the image and blur operators. In order to deal with the image anisotropy more efficiently, a multiwindow directional version of the LPA–ICI algorithm is exploited using star-shaped adaptive neighborhoods for filtering. In this form the algorithm achieves high-quality image reconstruction with a quite accurate edge preservation. Most of the operations are performed in the frequency domain.

3.1   Introduction

3.1.1   Blind and Nonblind Inverse

Image processing from multiple frames (channels) aims for enhanced comprehensive quality. Classical fields of application are astronomy, remote sensing, medical imaging, where multisensor data of different spatial, temporal, and spectral resolutions are exploited as tools for image sharpening, improvement of registration accuracy, feature enhancement, and improved classification. Other examples can be seen in digital microscopy, where a particular specimen may be acquired at several different focus settings (none of which is known accurately), or in multispectral radar imaging through a diffracting or scattering medium, which has different transition characteristics at different frequencies.

The basic image restoration techniques involve two important concepts: channel model and inversion. The channel model includes all sorts of distortions responsible for image degradation; for example, atmospheric turbulence, the relative motion of object and camera, an out-of-focus camera, or variations in optical and electronic imaging components. Conventionally, the channel is modeled by linear blur operators and an additive noise. The blur operators state that the acquired images are indirect observations of the image of interest [1–3].

Compensation of the blur effects in image reconstruction requires an inversion of the blur operators. The ill-conditioning of these operators makes the inverse problem quite difficult even when the blur operators are perfectly known [4]. As the different channels can provide complementary information on the reconstructed image, the multiple-channel problem usually is less complex or at least less ill-conditioned than the single-channel one.

However, typically, the blur operators are completely or partially unknown. It is not practical also to assume availability of training images that could be exploited for estimation of the blur functions. When the blur is unknown, the image restoration is called a blind inverse problem and in the case of multiple frames it is a multiple-frame (or multiple-channel) blind inverse problem. It is a very difficult nonlinear problem since both the object image and the blurs are unknown.

In order to illustrate some principal difficulties of this problem note that for the shift-invariant blur operators the observations given in the frequency domain are products of the Fourier transforms of the image intensity and these operators. This is clear that reconstruction of the image and the blurs from these products cannot be achieved without special assumptions.

Theoretical breakthroughs in the multiple blind and nonblind deconvolution have been achieved in a number of works on perfect blur and image reconstruction [5–7]. It is shown in [6] that because of missing data at the image boundaries, perfect reconstruction in a nonblind problem is impossible (even in the noiseless case) unless there are at least three channels. When there are at least three channels, perfect reconstruction is not only possible in the absence of noise, but also that it can be done by finite impulse response (FIR) filtering.

Perfect blind reconstruction with FIR filters is studied in [5] and [7]. With blurs satisfying certain coprimeness requirements the existence and uniqueness of the solution is guaranteed under quite unrestrictive conditions, i.e., both the blur operators and the object image can be determined exactly in the absence of noise and stably estimated in its presence. These perfect reconstruction methods are algebraic and based on the eigenstructure and null-space techniques applied to matrices formed from the channel observations. Unfortunately, these methods are quite sensitive to noise and *a priori* information on the size of the blur operators.

In [8] a new null space-based approach has been formulated. The proposed restoration algorithm estimates directly the original image from the null space of a special matrix. This direct deconvolver estimation algorithm can also be formulated as a constrained optimization problem. By properly choosing the constraints, these optimization problems can be solved using matrix operations. In the noise-free case, it gives perfect reconstruction of the object image.

All the above-mentioned methods are algebraic and based on state-space observation modeling. A number of publications use different approaches, e.g., the design of a suitable criterion function with the algorithm derived from minimization of this function. In the line of these algorithms we wish to mention a nonlinear least square method proposed in [9]. This method is further developed in the paper [10] with the main emphasis on multichannel deblurring of spatially misaligned images.

A novel approximation of Wiener inverse filtering for multiframe blind deconvolution is developed in [11] utilizing the ideas of the Bussgang filtering algorithm.

3.1.2 Inverse Regularization

Regularization is one the most essential elements of nonblind as well as blind deconvolution. The classical techniques impose the quadratic functional restrictions on the image and blur operators and their derivatives. These form the class of the so-called Tichonov's regularizators [12]. This regularization helps to eliminate the effects of zeros in the transfer functions of the blur operators and to overcome the ill-conditioning of the estimation equations. This technique results in linear estimation equations but, unfortunately, often suffers from artifacts and poor quality of reconstruction.

Recently, novel and essentially nonlinear regularization techniques have been proposed to cope with the inverse problems. The inverse is obtained by minimizing energy functionals, including special nonquadratic penalty terms. These penalty terms are motivated by statistical or deterministic modeling of the image intensity function. In particular, the total variation (TV) and Mumford–Shah functionals are proposed as models of images with sharp changes in intensity [13].

A generalization of this approach has been done in [14] where a non-quadratic functional of the vector gradient is used as the regularizator. Mathematical results as well as simulation experiments have been reported in many publications. These techniques have been developed and studied for image denoising [15] and both for nonblind and blind deconvolution [16, 17]. It has been shown that with nonquadratic regularization quite special nonlinear edge-preserving filtering algorithms can be obtained [13, 14, 16, 18].

In [9, 10], optimization of nonquadratic energy functionals has been used for the multiframe blind deconvolution imaging. The energy criterion contains the standard quadratic fidelity term with residuals defined as differences between observations and their model predictions as well as a quadratic term evaluating the cross-channel balance of the estimates. Overall, the criterion is nonquadratic as TV or Mumford–Shah penalty functionals are included as additive regularizator terms. It has been shown that the proposed algorithm using this regularization performs quite well on noiseless and noisy images.

Minimizing an energy criterion composed from quadratic fidelity and non-quadratic penalty terms leads to the nonlinear Euler partial differential equation. The nonlinearity of this equation is defined by the penalty functionals included in the criterion [18]. Thus, the nonlinearity of this filtering is completely defined by the penalty functionals introduced for regularization.

Limitations of the variational approach follow from formulation of the filtering as an optimization problem with a global energy criterion. It follows that the filtering equations are identical for all parts of the image. More specifically, in the recursive filtering at each pixel the calculated image reconstruction is a nonlinear function of the image estimates in the neighboring pixels, and the form of this nonlinear function is the same for all parts of the image.

It has been shown in [19] that in order to be efficient, the variational approach should be used with varying regularization parameters that can be different for different parts of the image. This modification of the standard variational technique allows a dramatic improvement in performance.

The basic motivation of the classical Tichonov's regularization is to overcome ill-conditioning of the inverse problem and, in this way, attenuate the noisy components in the solution. In the variational approach based on nonquadratic penalty functionals the philosophy is very different, as overcoming of ill-conditioning is replaced by an essentially different filtering approach with the main emphasis on the functional modeling of signals.

In this generalized interpretation the term *regularization* loses its initial meaning and can be viewed as a special form of nonlinear filtering. We wish to note that any linear or nonlinear filtering also can be treated as a regularization if it is applied to inverse problems.

Recall that interpretation of the inversion, including regularization, as a special sort of filtering is not new at all. In particular, this subject is discussed in detail in the standard textbook on digital image processing by Gonzalez and Woods [20].

In this chapter we exploit the term *regularization* in this *filtering* interpretation, and as a generalized regularizator we use a nonlinear, spatially adaptive filter initially developed for image denoising and further applied to nonblind image deblurring.

This adaptive filter originated from recent results in statistical nonparametric regression estimation. It is based on minimax estimation and statistical hypothesis testing. The estimator works in a pointwise manner with a varying adaptive neighborhood. The filter is nonlinear and adaptively different for each pixel of the image.

Let us mention a few fundamental results in theoretical basis for these methods: *Lepski's approach, adaptive weights smoothing, local change point,* and *intersection of confidence interval rule.*

The *Lepski's approach* algorithm is used for the adaptive, varying size neighborhood selection. The algorithm searches for the largest local vicinity of the point of estimation where the observation model assumptions fit well to the data. The estimates are calculated for a few window sizes and compared. The adaptive window size is defined as the largest one of those for which the estimate does not differ significantly from the estimates corresponding to the smaller sizes. Theoretical results show that this sort of algorithm has quite exciting asymptotic properties in terms of the convergence rate, asymptotic accuracy, and adaptivity over various classes of functions [21–26].

Adaptive weights smoothing defines adaptive shape neighborhoods used for estimation by a special recursive procedure [27, 28].

The *local change point* algorithm has been developed for the general class of maximum likelihood estimates and gives an efficient tool to deal with non-Gaussian observations [29].

The *intersection of confidence interval* (ICI) rule belongs to the class of Lepski's algorithms and proves to be efficient for image denoising and nonblind single-channel deconvolution [30–32].

In this chapter we apply the ICI rule for blind multichannel deconvolution and show that it is able to produce excellent results. This filtering is applied to both the image and the PSFs which makes the difference between these estimated objects more essential.

The first results concerning this novel approach to the blind multichannel image deconvolution have been presented in [33].

3.2   Observation Model and Preliminaries

Consider a 2D single-input multiple-output (SIMO) linear spatially invariant imaging system. Such a system could model multiple cameras, multiple focuses of a single camera, or acquisition of images from a single camera through a changing medium.

The input to this system is an image $x[n_1, n_2]$, $n_1, n_2 \in U$, with a finite discrete grid $U = \{n_1, n_2 : n_1 = 1, 2, ..., N_1, n_2 = 1, 2, ..., N_2\}$ of the size $N_1 \times N_2$. This image is distorted by finite impulse response blurs modeled by point spread functions (PSF) h_i, $i = 1, ..., M$.

It is assumed that h_i are spatially invariant discrete filters also defined on the regular grid.

The 2D discrete convolutions of the input and the PSFs are degraded with an additive white Gaussian noise to produce observed output images:

$$y_i[n_1, n_2] = (x * h_i)[n_1, n_2] + \sigma_i w_i[n_1, n_2], \quad i = 1, ..., M. \quad (3.1)$$

We assume that the noise field in each channel is uncorrelated with the noise fields from other channels and that $w_i[n_1, n_2]$ have the standard Gaussian distribution $\mathcal{N}(0, 1)$.

The parameters σ_i are standard deviations of the noise in the channels.

The problem is to reconstruct both the image x and the PSFs h_i, $i = 1, ..., M$ from the observations $\{y_i[n_1, n_2], n_1, n_2 \in U, i = 1, ..., M\}$.

In this way we consider the deterministic setting of the problem with both the image and the PSFs deterministic but unknown.

As the blind deconvolution problem is ill-posed with respect to both the image and the blurring operators, a joint regularization technique is commonly used in order to regularize both x and h_i.

In particular, the estimates for x and h_i can be found by considering the optimization problem

$$\min_{x,\mathbf{h}} J(x, \mathbf{h}),$$

$$J(x, \mathbf{h}) = \sum_{i=1}^{M} \frac{1}{\sigma_i^2} \sum_{n_1,n_2} (y_i[n_1, n_2] - (x * h_i)[n_1, n_2])^2 + \tag{3.2}$$

$$\lambda_1 \sum_{i,j=1}^{M} d_{ij}^2 \sum_{n_1,n_2} ((y_i * h_j)[n_1, n_2] - (y_j * h_i)[n_1, n_2])^2 + \lambda_2 r(x) + \lambda_3 \sum_{i=1}^{M} r(h_i),$$

where $\mathbf{h} = (h_1, ..., h_M)^T$.

The first summand in J is the fidelity term used to measure the divergence between the model and the observations. The weights $1/\sigma_i^2$ for the channel's errors routinely follow from the maximum likelihood for the Gaussian noise in (3.1).

The second summand in J is used for channel equalization. Indeed, if the observations are noiseless, i.e. $y_i = (x * h_i)$, we can see that

$$(y_i * h_j) = ((x * h_i) * h_j) = ((x * h_j) * h_i) = (y_j * h_i) \tag{3.3}$$

and the cross-term $\sum_{i,j=1}^{M} d_{ij}^2 \sum_{n_1,n_2} ((y_i * h_j)[n_1, n_2] - (y_j * h_i)[n_1, n_2])^2 = 0$. For the true h_i the cross-term is equal to zero, while it is different from zero for approximate estimates and serves as a measure of the divergence between the true PSFs and their estimates.

The parameter λ_1 is used as the comparative value of the fidelity term (with the weight equal to 1) and the cross-channel equalization terms (with the weight equal to λ_1).

According to the maximum likelihood for the noisy data the weights d_{ij}^2 are equal to the inverse variances of the corresponding differences. Denote the variance operator as *var*. Then these variances are calculated as follows:

$$var\{(y_i * h_j)[n_1, n_2] - (y_j * h_i)[n_1, n_2]\} =$$
$$var\{\sigma_i(w_i * h_j)[n_1, n_2]\} + var\{\sigma_j(w_j * h_i)[n_1, n_2]\} =$$
$$\sigma_i^2 \sum_{n_1,n_2} h_j^2[n_1, n_2] + \sigma_j^2 \sum_{n_1,n_2} h_i^2[n_1, n_2].$$

Thus

$$d_{ij}^2 = \frac{1}{\sigma_i^2 \sum_{n_1,n_2} h_j^2[n_1, n_2] + \sigma_j^2 \sum_{n_1,n_2} h_i^2[n_1, n_2]}. \tag{3.4}$$

The last two terms in (3.2) are for Tichonov's regularization of both the signal x and the PSFs h_i. The typical choices for $r(x)$ and $r(h_i)$ are $r(x) = \sum_{n_1,n_2} x^2[n_1, n_2]$ or $r(x) = \sum_{n_1,n_2} ||\partial_{\mathbf{n}} x[n_1, n_2]||^2$, where $\partial_{\mathbf{n}} x$ is the gradient of x on $\mathbf{n} = (n_1, n_2)^T$. Other linear functionals of x also can be used in this sort of regularization.

The boundness of the regularization term is linked with the smoothness of the solution. The larger the regularization parameters λ_2, λ_3 the smoother the solutions x and h_i.

As it is discussed in Section 3.1.2, recently nonquadratic regularization has become very popular; in particular, the total variation norm [13] with $r(x) = \sum_{n_1,n_2} ||\partial_\mathbf{n} x[n_1, n_2]||$ and the Mumford–Shah energy $r(x) = \sum_{n_1,n_2} \phi(||\partial_\mathbf{n} x[n_1, n_2]||)$ having the form of a nonlinear function ϕ of the vector gradient [14]. These nonquadratic functionals are designed to penalize oscillations in images reflected in large derivatives. In the same time the total variation norm as well as the Mumford-Shah energy allow discontinuities in x, thus making them superior to Tichonov's quadratic regularization.

For multichannel blind deconvolution the criterion (3.2) with nonquadratic TV and Mumford–Shah penalties has been utilized in the papers [9] and [10].

Our approach is different from those used in [9] and [10] in two important aspects.

First of all we start from the quadratic version of the functional (3.2). This allows a frequency domain approach and as a result efficient and fast calculations for large images.

Second, the regularization and filtering are introduced as a nonlinear adaptive procedure external with respect to the basic quadratic minimization procedure. We use the LPA–ICI adaptive scale filtering [30]– [32] of the x and h_i estimates instead of the nonquadratic penalty terms explicitly defined in the minimization criterion.

3.3   Frequency Domain Equations

For a discrete signal $x[n_1, n_2]$ defined on the grid U its discrete Fourier transform (DFT) is calculated as

$$\mathcal{F}\{x\} = X[f_1, f_2] = \sum_{n_1=0}^{N_1-1} \sum_{n_2=0}^{N_2-1} x[n_1, n_2] W_{N_1}^{f_1 n_1} W_{N_2}^{f_2 n_2}$$

where W_N is the Nth root of unity,

$$W_N = e^{-j2\pi/N}, \, j = \sqrt{-1},$$

and the frequency $f = (f_1, f_2)$ is 2D discrete with values

$$F = \{f_1, f_2 : 0 \le f_1 \le N_1 - 1, \, 0 \le f_2 \le N_2 - 1\}. \tag{3.5}$$

The corresponding capital letters are used for DFT of all functions, $Y_i[f_1, f_2] = \mathcal{F}\{y_i\}$, $H_i[f_1, f_2] = \mathcal{F}\{h_i\}$.

Then, the Parseval theorem, in particular, for the image variable is of the form

$$\sum_{n_1, n_2} x^2[n_1, n_2] = \frac{1}{N_1 N_2} \sum_{f \in F} |X(f_1, f_2)|^2.$$

With $r(x) = \sum_{n_1, n_2} x^2[n_1, n_2]$ and $r(h_i) = \sum_{n_1, n_2} h_i^2[n_1, n_2]$ we can write the criterion (3.2) in the form

$$J(X, \mathbf{H}) = \sum_{i=1}^{M} \frac{1}{\sigma_i^2} \sum_{f \in F} |Y_i - X H_i|^2 + \lambda_1 \sum_{i,j=1}^{M} \sum_{f \in F} d_{ij}^2 |Y_i H_j - Y_j H_i|^2 + (3.6)$$

$$\lambda_2 \sum_{f \in F} |X|^2 + \lambda_3 \sum_{i=1}^{M} \sum_{f \in F} |H_i|^2,$$

where $\mathbf{H} = (H_1, ..., H_M)^T$,

$$d_{ij}^2 = \frac{N_1 N_2}{\sigma_i^2 \sum_{f \in F} |H_j|^2 + \sigma_j^2 \sum_{f \in F} |H_i|^2}, \quad d_{ij}^2 = d_{ji}^2. \tag{3.7}$$

For the sake of simplicity we do not show in these formulas the frequency argument of the variables. Note that in (3.6) and (3.7) $|X|^2 = X \cdot X^*$, where the star $(^*)$ stands for the complex conjugate.

In the following analysis and calculation of the derivatives of J the weights d_{ij}^2 are fixed, assuming that they are updated in the corresponding recursive procedures. Besides, we omitted in (3.6) and (3.7) the irrelevant common factor $1/(N_1 N_2)$.

For the complex variables X and H_i the gradient of the criterion J is defined by the derivatives $\partial_{X^*} J$, $\partial_{H_1^*} J$, ..., $\partial_{H_M^*} J$ and the necessary unconstrained minimality conditions for J can be written as follows:

$$\partial_{X^*} J = 0,$$
$$\partial_{H_i^*} J = 0, \quad i = 1, ..., M.$$

Simple manipulations give

$$\partial_{X^*} J(X, \mathbf{H}) = -\sum_{i=1}^{M} \frac{1}{\sigma_i^2}(Y_i - H_i X) H_i^* + \lambda_2 X = 0 \tag{3.8}$$

and

$$\partial_{H_i^*} J(X, \mathbf{H}) = \frac{1}{\sigma_i^2}(Y_i - H_i X)(-X^*) + \tag{3.9}$$

$$2\lambda_1 \sum_{j=1, j \neq i}^{M} d_{ji}^2 (Y_j H_i - Y_i H_j) Y_j^* + \lambda_3 H_i = 0.$$

Solving (3.8) with respect to X and (3.9) with respect to H_i we obtain a set of the equations for X and $\mathbf{H}$:

$$X = \frac{\sum_{i=1}^{M} Y_i H_i^* / \sigma_i^2}{\sum_{i=1}^{M} |H_i|^2 / \sigma_i^2 + \lambda_2}, \tag{3.10}$$

$$H_i = \frac{Y_i X^* / \sigma_i^2 + 2\lambda_1 Y_i \sum_{j=1, j \neq i}^{M} d_{ji}^2 H_j Y_j^*}{|X|^2 / \sigma_i^2 + 2\lambda_1 \sum_{j=1, j \neq i}^{M} d_{ji}^2 |Y_j|^2 + \lambda_3}, \quad i = 1, ... M. \tag{3.11}$$

For the analysis of the properties of J with respect to X and $\mathbf{H}$ we use the Hessian matrix (matrix of second derivatives) obtained by differentiation of (3.8) and (3.9).

This matrix can be represented in the following structured form,

$$G = \begin{pmatrix} G_{XX^*} & G_{\mathbf{H}X^*} \\ (G_{\mathbf{H}X^*}^*)^T & G_{\mathbf{H}^*\mathbf{H}^T} \end{pmatrix}, \tag{3.12}$$

where

$$G_{XX^*} = \partial_X \partial_{X^*} J = \sum_{i=1}^{M} \frac{1}{\sigma_i^2} |H_i|^2 + \lambda_2, \tag{3.13}$$

$$G_{\mathbf{H}^*\mathbf{H}^T} = \partial_{\mathbf{H}^*} \partial_{\mathbf{H}^T} J = \tag{3.14}$$
$$\begin{pmatrix} G_{H_1^* H_1} & -2\lambda_1 d_{12}^2 Y_2^* Y_1 & \cdots & -2\lambda_1 d_{1M}^2 Y_M^* Y_1 \\ -2\lambda_1 d_{21}^2 Y_1^* Y_2 & G_{H_2^* H_2} & & -2\lambda_1 d_{2M}^2 Y_M^* Y_2 \\ \vdots & & \ddots & \vdots \\ -2\lambda_1 d_{M1}^2 Y_1^* Y_M & -2\lambda_1 d_{M2}^2 Y_2^* Y_M & \cdots & G_{H_M^* H_M} \end{pmatrix},$$

with the diagonal terms

$$G_{H_i^* H_i} = \partial_{H_i} \partial_{H_i^*} J = \frac{1}{\sigma_i^2} |X|^2 + 2\lambda_1 \sum_{j=1, j \neq i}^{M} d_{ji}^2 |Y_j|^2 + \lambda_3, \tag{3.15}$$

and

$$G_{\mathbf{H}X^*} = \partial_{X^*} \partial_{\mathbf{H}^T} J = \left(X H_1^* / \sigma_1^2, \, X H_2^* / \sigma_2^2, ..., \, X H_M^* / \sigma_M^2 \right).$$

3.4   Projection Gradient Optimization

We define the estimates of the image and the PSFs as a solution of the following constrained optimization problem,

$$(\hat{y}, \mathbf{h}) = \arg \min_{x \in Q_x, h_j \in Q_{h_j}} J(x, \mathbf{h}), \tag{3.16}$$

where the admissible sets Q_x for x and $Q_{h_j} = Q_{h_j}^0 \cap Q_{h_j}^1 \cap Q_{h_j}^+$ for h_j are defined as follows:

$$Q_x = \{x : 0 \le x \le 1\}, \tag{3.17}$$

$$Q_{h_j}^0 = \{h_j = 0 : n_1 > \Delta_{1,j}, \, n_2 > \Delta_{2,j}\}, \tag{3.18}$$

$$Q_{h_j}^1 = \{h_j : \sum_{n_1, n_2} h_j[n_1, n_2] = 1\}, \tag{3.19}$$

$$Q_{h_j}^+ = \{h_j : h_j[n_1, n_2] \ge 0\}. \tag{3.20}$$

The condition (3.17) means that the image intensity is nonnegative and belongs to the segment $[0, 1]$.

The conditions (3.18) $Q_{h_j}^0$ for with integers $\Delta_{1,j}$, $\Delta_{2,j}$ define the size of the finite support of the *PSF* estimate as the rectangle $\Delta_{1,j} \times \Delta_{2,j}$. These conditions do not assume that the supports of the true PSFs are known.

The condition (3.19) $Q_{h_j}^1$ assumes normalization of the *PSF* estimates and the last condition (3.20) means that the h_j is nonnegative.

The constraints (3.17), (3.19), and (3.20) can all be enforced as the one is in (3.16) which means, in particular, that the finite support PSFs are nonnegative and normalized. However, these constraints can be dropped if they are not applicable.

The functional $J(x, \mathbf{h})$ is quadratic on x for a fixed $\mathbf{h}$ and quadratic on $\mathbf{h}$ for a fixed x. The sets Q_x, $Q_{h_j}^0$, $Q_{h_j}^1$, and $Q_{h_j}^+$ are convex. It follows that (3.16) is a partial constrained convex optimization problem, where *partial* means optimization separate on the variables x or $\mathbf{h}$.

In general, $J(x, \mathbf{h})$ considered as a function of both variables x and $\mathbf{h}$ jointly can be nonconvex. Note that this convexity of J is valid provided that the weights d_{ij}^2 are fixed.

The partial convexity of $J(x, \mathbf{h})$ means that the gradient algorithms can be used at least for finding a local constrained extremum for the problem (3.16).

We exploit for the constrained optimization (3.16) the concept of the projection gradient and design a recursive optimization algorithm built with the main calculations in the frequency domain. This frequency domain technique makes all calculations simple and transparent.

The recursive projection gradient algorithm can be written in the following basic form,

$$X^{(k)} = P_{Q_x}\{X^{(k-1)} - \alpha_k \partial_{X^*} J(X^{(k-1)}, \mathbf{H}^{(k-1)})\}, \tag{3.21}$$

$$H_j^{(k)} = P_{Q_{h_j}}\{H_j^{(k-1)} - \beta_k \partial_{H_j^*} J(X^{(k)}, \mathbf{H}^{(k-1)})\}, \tag{3.22}$$

$$k = 1, ...$$

where the gradient components are given in (3.8) and (3.9) and some initialization $(X^{(0)}, \mathbf{H}^{(0)})$ is assumed.

For a convex set Q the projection $P_Q\{v\}$ is as an element $u \in Q$ closest to v:

$$P_Q\{v\} = \arg \min_{u \in Q}(||v - u||^2),$$

where $|| \cdot ||^2$ means the Euclidean norm.

For the sets (3.17) through (3.20) this results in simple operations

$$P_{Q_x}\{x\} = \max\{0, \min(1, x)\}, \tag{3.23}$$

$$P_{Q_{h_j}^0}\{h_j\} = \begin{cases} h_j \text{ if } 1 \le n_1 \le \Delta_{1,j} \text{ and } 1 \le n_2 \le \Delta_{2,j}, \\ 0, \text{ otherwise} \end{cases} \tag{3.24}$$

$$P_{Q_{h_j}^1}\{h_j\} = h_j / \sum_{n_1, n_2} h_j, \tag{3.25}$$

$$P_{Q_{h_j}^+}\{h_j\} = \begin{cases} h_j \text{ if } h_j \ge 0 \\ 0, \text{ otherwise.} \end{cases} \tag{3.26}$$

The algorithm in (3.21) and (3.22) assumes two-stage calculations. First, $\tilde{X}^{(k)}$ and $\tilde{H}_j^{(k)}$ are found without projection

$$\tilde{X}^{(k)} = X^{(k-1)} - \alpha_k \partial_{X^*} J(X^{(k-1)}, H^{(k-1)}), \tag{3.27}$$

$$\tilde{H}_j^{(k)} = H_j^{(k-1)} - \beta_k \partial_{H_j^*} J(X^{(k)}, H^{(k-1)}). \tag{3.28}$$

Second, these values are projected on the sets Q_x, Q_{h_j}:

$$X^{(k)} = P_{Q_x}\{\tilde{X}^{(k)}\},$$
$$H_j^{(k)} = P_{Q_{h_j}}\{\tilde{H}_j^{(k)}\}.$$

This projection requires the inverse Fourier transform $x^{(k)} = \mathcal{F}^{-1}\{X^{(k)}\}$ and $h_j^{(k)} = \mathcal{F}^{-1}\{H_j^{(k)}\}$ with the projections calculated according to (3.23) through (3.26).

In the frequency domain the criterion J (3.6) is convex with respect to X provided that H_j are fixed and convex with respect to H_j, $j = 1, ..., M$, provided X is fixed. Then, under some natural assumptions, the projection gradient algorithm converges with respect to the variables $X^{(k)}$ for $H_j^{(k)} = H_j$ and with respect to the variables $H_j^{(k)}$ for $X^{(k)} = X$.

The step-size parameters $\alpha_k, \beta_k > 0$ in (3.27) and (3.28) influence the convergence and the convergence rate of the algorithm [34].

The frequency domain procedures (3.21) and (3.22) make steps in the direction to the minimum of J with respect to the corresponding variables. It is important that in the frequency domain these steps are performed for all frequencies $f \in F$ in parallel.

The criterion J as a function of X and H_j has a different scale behavior for different frequencies. In particular, it is true for the ill-conditioned inverse problems. In order to enable stable iterations the step sizes α_k and β_k should be very small, and then the convergence rate becomes very slow even in the partial optimizations given by the iterations on X and H_j only.

There are standard recipes how to improve this poor convergence [34]. In frequency domain the behavior of J on the variables X and H_j is defined by the Hessian in (3.12) through (3.15).

The convergence rate of the recursive projection gradient algorithm can be significantly improved using the diagonal of the matrix G for normalization (rescaling) of the gradient components in the projection gradient procedure in (3.21) and (3.22):

$$X^{(k)} = P_{Q_x}\{X^{(k-1)} - \alpha_k \frac{1}{G_{X^*X}} \partial_{X^*} J_X(X^{(k-1)}, \mathbf{H}^{(k-1)})\}, \quad (3.29)$$

$$H_j^{(k)} = P_{Q_{h_j}}\{H_j^{(k-1)} - \beta_k \frac{1}{G_{H_j^* H_j}} \partial_{H_j^*} J_{H_j}(X^{(k)}, \mathbf{H}^{(k-1)})\}, \quad (3.30)$$

where G_{X^*X} and $G_{H_j^* H_j}$ are given in (3.13) and (3.15).

Inserting in (3.29) and (3.30) the expressions for the gradient in (3.8) and (3.9) and for the Hessian G in (3.12) through (3.15) we obtain the projection gradient algorithm in the following final form:

$$X^{(k)} = P_{Q_x}\{(1 - \alpha_k)X^{(k-1)} + \alpha_k \frac{\sum_{j=1}^{M} Y_j H_j^{*(k-1)}/\sigma_j^2}{\sum_{j=1}^{M} |H_j^{(k-1)}|^2/\sigma_j^2 + \lambda_2}\}, \quad (3.31)$$

$$H_j^{(k)} = P_{Q_{h_j}}\{(1 - \beta_k)H_j^{(k-1)} + \quad (3.32)$$

$$\beta_k \frac{Y_j X^{*(k)}/\sigma_j^2 + 2\lambda_1 Y_j \sum_{i=1, i \neq j}^{M} d_{ij}^2(k-1) H_i^{(k-1)} Y_i^*}{|X^{(k)}|^2/\sigma_j^2 + 2\lambda_1 \sum_{i=1, i \neq j}^{M} d_{ij}^2(k-1)|Y_i|^2 + \lambda_3}\},$$

where $d_{ij}^2(k-1)$ is calculated according to the formula in (3.7):

$$d_{ij}^2(k-1) = \frac{N_1 N_2}{\sigma_i^2 \sum_{f \in F} |H_j^{(k-1)}|^2 + \sigma_j^2 \sum_{f \in F} |H_i^{(k-1)}|^2}.$$

For $\alpha_k = 1$ and $\beta_k = 1$ and without projections the algorithms (3.31) and

(3.32) become the alternating recursive minimization procedure:

$$X^{(k)} = \frac{\sum_{i=1}^{M} Y_i H_i^{*(k-1)}/\sigma_i^2}{\sum_{i=1}^{M} |H_i^{(k-1)}|^2/\sigma_i^2 + \lambda_2}, \tag{3.33}$$

$$H_i^{(k)} = \frac{Y_j X^{*(k)}/\sigma_j^2 + 2\lambda_1 Y_j \sum_{i=1,i\neq j}^{M} d_{ij}^2(k-1)H_i^{(k-1)}Y_i^*}{|X^{(k)}|^2/\sigma_j^2 + 2\lambda_1 \sum_{i=1,i\neq j}^{M} d_{ij}^2(k-1)|Y_i|^2 + \lambda_3}, \quad i = 1,...M. \tag{3.34}$$

It is interesting to note that this last procedure is a recursive solution of the equations (3.10) and (3.11).

Comparing the recursive projection algorithm in (3.31) and (3.32) and the alternating recursive algorithm in (3.33) and (3.34) we can see that the recursive projection algorithm deals more carefully with the iterations, updating them only partially as the step sizes a_k, β_k are always smaller than 1. In general, it allows us to improve the convergence of the algorithm.

3.5   Anisotropic LPA–ICI Spatially Adaptive Filtering

The adaptive algorithm described in this chapter is based on the pointwise nonparametric regression estimation presented in a number of publications [30, 35, 36]. A kernel filter g_θ is derived from the local polynomial approximation (LPA) of the image intensity in a rotating directional nonsymmetric window of varying size. The nonlinearity of the method is incorporated by the ICI adaptive choice of the window size, allowing us to get near optimal quality of image and edge recovery.

The directionality of the filter is mainly defined by the nonsymmetric windows used in the LPA.

This technique provides a number of valuable benefits:

- Unlike many other methods which start from the continuous domain and are then discretized, this technique works directly in the discrete space domain and is applicable for multidimensional data;

- The designed kernels are truly multivariate, nonseparable, and anisotropic with arbitrary orientations, widths, and lengths;

- The desirable smoothness of the kernels along and across the main direction is enabled by the corresponding vanishing moment conditions;

- The kernel support can be flexibly shaped to any desirable geometry in order to capture geometrical structure and pictorial information. In this way a special design can be done, particularly for complex objects and applications.

3.5.1 Motivation

Points, lines, edges, and textures are present in all images. To deal with these features, oriented and directional filters are used in many vision and image processing tasks, such as edge detection, texture and motion analysis, and so on.

The key question is how to design a kernel for a specified direction. A good initial idea arises from a definition of the *right-hand directional derivative* for the direction defined by the angle θ,

$$\partial_{+\theta} x[n_1, n_2] = \lim_{\rho \to 0^+} \left(x(n_1 + \rho \cos \theta, \ n_2 + \rho \sin \theta) - x(n_1, n_2) \right) / \rho.$$

Whenever x is a differentiable function, elementary calculations give the well-known result

$$\partial_{+\theta} x[n_1, n_2] = \partial_\theta x[n_1, n_2] = \cos \theta \cdot \partial_{n_1} x[n_1, n_2] + \sin \theta \cdot \partial_{n_2} x[n_1, n_2]. \quad (3.35)$$

Thus, in order to find the derivative for any direction θ it suffices to estimate the two derivatives on n_1 and n_2 only. This concept has been exploited and generalized by the so-called steerable filters [37].

Although continuous models of the discrete image intensity are widely used in image processing, estimates such as (3.35) are too rough in order to be useful for those applications where sharpness and detail are of first priority. For discrete images lacking global differentiability or continuity the only reliable way to obtain accurate directional anisotropic information is to calculate variations of x in the desired direction θ and, say, to estimate the directional derivative by the finite difference counterpart of $\partial_{+\theta} x[n_1, n_2]$.

In more general terms this means that the estimation or image analysis should be based on directional kernels, templates, or atoms which are quite narrow and concentrated in desirable directions. Since points, lines, edges, and textures can exist at all possible positions, orientations, and scales, one would like to use families of filters that can be tuned to all orientations, scales, and positions.

Recent development shows an impressive success of methods for directional image and multivariate signal processing. In particular, narrow multidirectional items are the building blocks of the new ridgelet and curvelet transforms [38].

The main intention of our approach is to obtain in a data-driven way the largest local vicinity of the estimation point in which the underlying model fits the discrete data. We assume that this vicinity is a star-shaped set, which can be approximated by some sectorial segmentation with, say, K nonoverlapping sectors, and use special directional kernels with supports embedded in these sectors. The kernels are equipped with univariate scale parameters defining the size of the supports in the sectors.

The ICI rule is exploited K times, once for each sector, in order to find the optimal pointwise adaptive scales for each sector's estimates, which are then combined into the final estimate. In this way, we reduce the 2D scale selection problem to a multiple univariate one.

Below we briefly outline the basic ideas of the LPA-ICI algorithm with reference to some of the publications on this subject.

3.5.2 Sectorial Neighborhoods

Let U^* be an optimal neighborhood of a pixel where the image intensity can be well fitted by the polynomial model (see Figure 3.1a). Introduce a sectorial (angular) partition of the unit disc centered at the estimation point (Figure 3.1b), where the angle θ_i names the corresponding sector, and use the optimal lengths a^* of these sectors in order to approximate U^* (Figure 3.1c). If the set U^* is a star-shaped body as it is in Figure 3.1a, then, with a sufficient number of angular sectors, a good approximation of U^* can be achieved.

3.5.3 Adaptive Window Size

Assume that g_{a,θ_i} is a compactly supported kernel such that $\operatorname{supp} g_{a_i} = S_{\theta_i}^a$ for all values of the scale parameter a defining the length of the sector. Let us use generic notations x and $\hat{x}$ for the signal to be filtered and its estimate.

Then, the introduced anisotropic directional estimator has the following form

$$\hat{x}[n_1, n_2] = \sum_{i=1}^{K} \lambda_i[n_1, n_2] \hat{x}_{a,\theta_i}[n_1, n_2], \qquad (3.36)$$
$$\hat{x}_{a,\theta_i}[n_1, n_2] = (g_{a,\theta_i} * x)[n_1, n_2],$$

where $\lambda_i[n_1, n_2] \geq 0$, $\sum_{i=1}^{K} \lambda_i[n_1, n_2] = 1$ are weights and g_{a,θ_i} is the kernel defined by the order of the LPA and the corresponding window function.

Let $A = \{a_1, ..., a_J\}$ be a set of sector sizes and let the estimates $\hat{x}_{a,\theta_i}[n_1, n_2]$ be calculated for all $a \in A$. The ICI rule is applied in order to find the adaptive scale $a_{j,i}^+[n_1, n_2]$ minimizing the mean squared error for the estimate $\hat{x}_{a,\theta_i}[n_1, n_2]$, $a \in A$. These adaptive scales are found for all $x \in U$ and for all θ_i.

Thus, we obtain the directional adaptive estimates $\hat{x}_{a_i^+[n_1, n_2], \theta_i}[n_1, n_2]$ with the optimized sizes of the sectors used for estimation. For each n_1, n_2 and each j we have K (according to the number of sectors) directional adaptive estimates.

The formula (3.36) determines how these estimates are aggregated in the final formula:

$$\hat{x}[n_1, n_2] = \sum_{i=1}^{K} \lambda_i[n_1, n_2] \cdot \hat{x}_{a_i^+[n_1, n_2], \theta_i}[n_1, n_2], \qquad (3.37)$$
$$\hat{x}_{a_i^+[n_1, n_2], \theta_i}[n_1, n_2] = (g_{a_i^+[n_1, n_2], \theta_i} * x)[n_1, n_2].$$

The weights $\lambda_i[n_1, n_2]$ are selected according to the formula

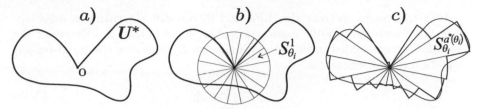

FIGURE 3.1: A neighborhood of the estimation point: a) the optimal estimation set U^*, b) the unit ball segmentation, c) the sectorial approximation of U^*.

$$\lambda_i[n_1, n_2] = \sigma_i^{-2}[n_1, n_2] / \sum_j \sigma_j^{-2}[n_1, n_2], \qquad (3.38)$$

where $\sigma_i^2[n_1, n_2]$ are the variances of the estimates $\hat{x}_{a_i^+[n_1,n_2],\theta_i}[n_1, n_2]$.

Here we use a linear aggregation of the estimates with the inverse variances of the estimates as the weights. Note that the weights $\lambda_i[n_1, n_2]$ in (3.38) are data-driven adaptive and depend on the adaptive $a_i^+[n_1, n_2]$ [30, 35, 36, 39].

3.5.4 LPA-ICI Filtering

The LPA, ICI, and aggregation define the LPA–ICI filtering procedure as follows:

- *The* LPA *estimation.* The inputs are the observations x and the observation variance σ_x^2. The outputs are the estimates $\hat{x}_{a,\theta_i}[n_1, n_2]$ calculated for all scales $a \in A$ and for all directions θ_i. Besides, the variances $\sigma_{a,\theta_i}^2[n_1, n_2]$ of these estimates are also calculated.

- *The* ICI *rule.* The inputs are the estimates $\hat{x}_{a,\theta_i}[n_1, n_2]$ and the variances $\sigma_{a,\theta_i}^2[n_1, n_2]$. The outputs are the adaptive sector size estimates $\hat{x}_{a_i^+[n_1,n_2],\theta_i}[n_1, n_2]$ and the variances $\sigma_i^2[n_1, n_2] = \sigma_{a_i^+[n_1,n_2],\theta_i}^2[n_1, n_2]$ of these estimates.

The ICI algorithm uses the confidence intervals

$$\mathcal{D}_i(a) = \{\hat{x}_{a,\theta_i}[n_1, n_2] - \Gamma\sigma_{a,\theta_i}[n_1, n_2], \ \hat{x}_{a,\theta_i}[n_1, n_2] + \Gamma\sigma_{a,\theta_i}[n_1, n_2]\},$$
$$(3.39)$$

where Γ is a parameter of the ICI rule.

- *The aggregation of the directional estimates.* The inputs are the adaptive directional estimates $\hat{x}_{a_i^+[n_1,n_2],\theta_i}[n_1, n_2]$ and the variances $\sigma_i^2[n_1, n_2]$ of these estimates. The outputs are the final estimates $\hat{x}[n_1, n_2]$.

It is convenient to treat this LPA–ICI directional algorithm as an adaptive filter with two inputs x and σ_x^2 and a single output $\hat{x}$. Denoting the calculations embedded in this algorithm as an $\mathcal{LI}$ operator the input-output equation can be written as

$$\hat{x} = \mathcal{LI}\left\{x, \sigma_x^2\right\}. \tag{3.40}$$

3.6   Blind Deconvolution Algorithm

3.6.1   Main Procedure

Now we are in position to describe the developed blind multichannel deconvolution algorithm.

(1) *Initialization*:

We use the Gaussian density for $h_j^{(0)}$ and the mean of the observed images $x^{(0)} = \sum_{j=1}^{M} y_j[n_1, n_2]/M$ as the initial estimates. The shifts of the Gaussian $h_j^{(0)}$ can be estimated in order to compensate possible misalignment of the observed images. This problem is addressed later in this section.

Iterate given $x^{(k-1)}$ and $h_j^{(k-1)}$ for $k = 1, \ldots.$

(2) *Image estimation.*

According to (3.31) the image estimate before projection is calculated as

$$X^{(k)} = (1 - \alpha_k)X^{(k-1)} + \alpha_k \frac{\sum_j Y_j H_j^{*(k-1)}/\sigma_j^2}{\sum_j |H_j^{(k-1)}|^2/\sigma_j^2 + \lambda_2}. \tag{3.41}$$

(3) *Image filtering.*

Filter $X^{(k)}$ by the LPA–ICI algorithm:

(3a) calculate $x^{(k)} = \mathcal{F}^{-1}\{X^{(k)}\}$;

(3b) calculate the estimate of the standard deviation $\sigma_{x^{(k)}}$ of the noise in $x^{(k)}$ using the median estimate of the signal's differences

$$\hat{\sigma} = \{\operatorname*{median}_s(\left|x_s^{(k)} - x_{s-1}^{(k)}\right| : s = 2, .., n_1 n_2)\}/(\sqrt{2} \cdot 0.6745), \tag{3.42}$$

where $x_s^{(k)}$ is a columnwise representation of the image. The normalizing constant $\sqrt{2} \cdot 0.6745$ in (3.42) makes this estimate unbiased, $E\{\hat{\sigma}\} = \sigma$, for Gaussian additive noise provided that the signal differences have zero mean value;

(3c) filter $x^{(k)}$ according to

$$x^{(k)} \triangleq \mathcal{LI}\left\{x^{(k)}, \sigma_{x^{(k)}}\right\}. \tag{3.43}$$

(4) *Image projection.*
(4a) Project $x^{(k)}$ onto the segment $[0, 1]$, $x^{(k)} \triangleq P_{Q_x}\{x^{(k)}\}$ according to (3.23):

$$x^{(k)} \triangleq \max\{0, \min(1, x^{(k)})\}; \tag{3.44}$$

(4b) calculate the *DFT* of $x^{(k)}$: $X^{(k)} = \mathcal{F}\{x^{(k)}\}$.
(5) *PSF estimation.*
  Iterate given $H_j^{(k_1-1)}$ for $k_1 = 1, ...T$ with $H_j^{(0)} = H_j^{(k-1)}$.
Calculate $H_j^{(k_1)}$ according to (3.32) before filtering and projection

$$H_j^{(k_1)} = (1 - \beta_k)H_j^{(k_1-1)} + \tag{3.45}$$
$$\beta_k \frac{Y_j X^{*(k)}/\sigma_j^2 + 2\lambda_1 Y_j \sum_{i=1, i\neq j}^M d_{ij}^2(k-1)H_i^{(k_1-1)}Y_i^*}{|X^{(k)}|^2/\sigma_j^2 + 2\lambda_1 \sum_{i=1, i\neq j}^M d_{ij}^2(k-1)|Y_i|^2 + \lambda_3},$$
$$j = 1, ..., T,$$

with the outputs $H_j^{(k)} = H_j^{(T)}$.
(6) *PSF filtering.*
  Filter $H_j^{(k)}$ by the LPA–ICI algorithm:
(6a) calculate $h_j^{(k)} = \mathcal{F}^{-1}\{H_j^{(k)}\}$;
(6b) calculate the standard deviation of the noise in $h_j^{(k)}$ as in (3.42);
(6c) filter $h_j^{(k)}$ according to the LPA-ICI algorithm:

$$h_j^{(k)} \triangleq \mathcal{LI}\left\{h_j^{(k)}, \sigma_{h_j^{(k)}}\right\}, j = 1, ..., M. \tag{3.46}$$

(7) *PSF projection.*
(7a) Set $h_j^{(k)}$ to zero outside of the support (3.24):

$$h_j^{(k)} = \begin{cases} h_j^{(k)} & \text{if } 1 \leq n_1 \leq \Delta_{1,j} \text{ and } 1 \leq n_2 \leq \Delta_{2,j}, \\ 0 & \text{otherwise}; \end{cases} \tag{3.47}$$

(7b) normalize the positive estimates $h_j^{(k)} > 0$ (3.25):

$$h_j^{(k)} \triangleq h_j^{(k)} / \sum_{n_1, n_2} h_j^{(k)}[n_1, n_2]; \tag{3.48}$$

(7c) calculate $H_j^{(k)} = \mathcal{F}\{h_j^{(k)}\}$.
Repeat steps (2) through (7) up to the convergence of the algorithm.
  The final iterations define the estimates of the image $\hat{x}$ and of the *PSFs* $\hat{h}_j$, $j = 1, ..., M$.
  Note that the LPA-ICI filtering regularization is embedded in the recursive algorithm introduced originally in the form (3.31) and (3.32). This LPA-ICI

filtering is produced in the spatial domain and requires backward and forward DFT of the frequency domain estimates $H_j^{(k)}$, $X^{(k)}$.

In this algorithm the formula (3.42) for the standard deviation is quite approximate as it is derived for the uncorrelated additive noise in the signal, while the estimates $h_j^{(k)}$ and $x^{(k)}$ have correlated noises.

Note that the T iterations on $H_j^{(k_1)}$ are given in (3.45) as a cycle embedded in the main recursive procedure. This is done in order that the convergence rates of the algorithm agree with respect to the variables X and H_j. The number T of cycles on k_1 is one of the design parameters of the algorithm.

3.6.2   Image Alignment

We use a part of the loss function (3.6) as the criterion for estimation of shifts between images in different channels:

$$J_{shift} = \sum_{j,i=1}^{M} \sum_{f \in F} d_{ij}^2 |Y_i H_j - Y_j H_i|^2, \qquad (3.49)$$

$$d_{ij}^2 = \frac{1}{|H_j|^2 \sigma_i^2 + |H_i|^2 \sigma_j^2}.$$

Here the functions $H_j = \mathcal{F}\{h_j\}$ serve as special test PSFs which can be different from the ones exploited in the above modeling. We assume that these h_j are symmetric with respect to the origin of the coordinates. The parameter $\mathbf{b}_j = (b_{1,j}, b_{2,j})$ defines a shift of h_j with respect to the common origin. The shifted test PSFs are defined in the form $h_j(n_1 - b_{1,j}, n_2 - b_{2,j})$. Assume that $\mathbf{b_1} = \mathbf{0}$, then the relative shifts of other test PSFs are estimated by minimization of J_{shift} on $\mathbf{b}_j$, $j = 2, ..., M$:

$$(\hat{\mathbf{b}}_j, j = 2, ...,) = \arg \min_{\mathbf{b}_j, j=2,...,M} J_{shift}. \qquad (3.50)$$

Our experiments with the Gaussian test PSFs show good results, in particular for $M = 3$. The accuracy of estimation is within a few pixels. This sort of error in the shift is easily compensated by the main blind deconvolution algorithm.

3.7   Identifiability and Convergence

3.7.1   Perfect Reconstruction

Let us start from the perfect reconstruction problem when the data are noiseless. The unique accurate solution (perfect reconstruction of the image and PSFs) exists provided that we are looking for the finite support *PSF* estimates of the support size equal to the size of the true PSFs [5] through [7]. If the estimates are oversized, then, in general there are an infinite number of solutions satisfying the homogeneous equations

$$Y_j - XH_j = 0, \tag{3.51}$$

$$Y_i H_j - Y_j H_i = 0, \ i, \ j = 1, ..., M, \tag{3.52}$$

giving zero value to the criterion J provided that $\lambda_2 = \lambda_3 = 0$ in (3.6).

Using the regularization parameters λ_2, $\lambda_3 > 0$ can enable the existence of a unique regularized solution of the problem (3.16) even when the estimates of PSFs are oversized. With λ_2, $\lambda_3 \to 0$ this regularized solution converges to one of the solutions of (3.51) and (3.52) with minimum norms of x and h_j. It may happen that this minimum norm solution differs from the true x and h_j functions. Thus, regularization, being a universal tool for inverse problems, sometimes can be quite misleading.

The practical success of all of the algorithms developed for blind deconvolution is heavily dependent on initialization and on incorporation of additional prior information on the image and PSFs. For the blur operators this information mainly concerns the size and the shift of PSFs [9, 10, 40].

Theoretical analysis of the convergence for the blind deconvolution is a difficult problem which can be done only for some special cases. One of the results in this area can be found in [41] where a single-channel system is studied with the quadratic regularization for estimation.

Here we wish to mention also a recent paper [42] where a single-channel blind deconvolution is studied. *A priori* assumptions on the image and PSF are used in order to guarantee the strong convexity of the regularized criterion and the unique solution of the blind deconvolution problem. An interesting noniterative algorithm is developed for this setting of the problem where the solution, of course, depends on the *a priori* assumptions.

3.7.2 Hessian and Identifiability

Despite the complexity of the detailed theoretical analysis, useful information can be derived in quite a simple way. Here we will consider the partial convexity of J with respect to the variables X and H_j which are analyzed based on the Hessian of J.

3.7.2.1 Estimation of X

The corresponding Hessian (scalar G_{X*X}) is given by the formula (3.13). The criterion J is convex with respect to X provided $G_{X*X} > 0$. This means that

$$G_{X*X} = \sum_j \frac{1}{\sigma_j^2} |H_j|^2 > 0, \ f \in F. \tag{3.53}$$

If the PSFs are fixed, the criterion J is strongly convex with respect to X and the procedure (3.31) converges to the minimum of J with respect to X corresponding to the given H_j.

The inequality $\sum_j \frac{1}{\sigma_j^2}|H_j|^2 > 0$, $f \in F$ is a standard channel disparity condition known also as the coprimeness requirements. It says that there are no zeros common for all PSFs simultaneously. If $\sum_j \frac{1}{\sigma_j^2}|H_j|^2 = 0$ for one of the frequencies, this frequency of the image is unobserved in Y_j, $j = 1, ..., M$, and the image contents cannot be accurately restored for this frequency even in the noiseless case.

The inequality (3.53) can be treated as a condition of the sufficient excitation of all modes (or frequencies) of the image X.

3.7.2.2   Estimation of H_j

The corresponding Hessian matrix $G_{\mathbf{H}^*\mathbf{H}^T}$ is given in (3.14) and (3.15). This matrix is Hermitian and all its eigenvalues are real. The criterion J is convex with respect to H_j, $j = 1, ...M$, provided that this matrix is positive definite, i.e., $G_{\mathbf{H}^*\mathbf{H}^T} > 0$ for all $f \in F$.

In order to prove that this matrix is positive definite provided some conditions, we consider the quadratic form corresponding to $G_{\mathbf{H}^*\mathbf{H}^T}$ and show that this quadratic form is positive definite.

This quadratic form is as follows:

$$Q_{\mathbf{H}^*\mathbf{H}^T} = \mathbf{b}^H G_{\mathbf{H}^*\mathbf{H}^T} \mathbf{b} = \tag{3.54}$$

$$\sum_{l=1}^{M} \frac{|b_l|^2}{\sigma_l^2}|X|^2 + 2\lambda_1 \left(\sum_{l=1}^{M} |b_l|^2 \sum_{k=1, k \neq l}^{M} d_{kl}^2 |Y_k|^2 - \sum_{k,l=1, k \neq l}^{M} d_{kl}^2 \cdot Y_k^* b_k^* \cdot Y_l b_l \right),$$

where $\mathbf{b}$ is a complex valued vector, $\mathbf{b} \in R^M$.
Further, as $d_{kl}^2 = d_{lk}^2$

$$d_{kl}^2|b_l|^2|Y_k|^2 + d_{lk}^2|b_k|^2|Y_l|^2 - (d_{kl}^2 Y_k^* b_k^* Y_l b_l + d_{lk}^2 Y_l^* b_l^* Y_k b_k) = \tag{3.55}$$
$$d_{kl}^2(b_l Y_k^* - b_k Y_l^*)(b_l Y_k^* - b_k Y_l^*)^* = d_{kl}^2|b_l Y_k^* - b_k Y_l^*|^2 \geq 0.$$

It follows from (3.54) that

$$Q_{\mathbf{H}^*\mathbf{H}^T} = \sum_{l=1}^{M} \frac{|b_l|^2}{\sigma_l^2}|X|^2 + 2\lambda_1 \sum_{k,l=1, k \neq l}^{M} d_{kl}^2|b_l Y_k^* - b_k Y_l^*|^2.$$

This quadratic form is positive definite provided that it is positive and takes a zero value only for $\mathbf{b} = 0$. It is obvious that $Q_{\mathbf{H}^*\mathbf{H}^T} > 0$ for any $||\mathbf{b}|| \neq 0$ provided that

$$|X|^2 > 0, \ f \in F. \tag{3.56}$$

This is a condition of the sufficient excitation of H_j by the input signal X. It enables the convexity of J as well as the identifiability of all PSFs for all $f \in F$. If $|X|^2 = 0$ for some f the information about H_j on this frequency does not exist in the observations Y_j and cannot be reconstructed accurately.

3.7.3 Conditioning and Convergence Rate

The partial convergence rates of the recursive procedures (3.21) and (3.22) with respect to X and H_j (analyzed separately) are defined by the corresponding Hessians, or more accurately by their eigenvalues, which are real because G_{X^*X} is real and the Hessian $G_{H^*H^T}$ is the Hermitian matrix $G_{H^*H^T} = (G^*_{H^*H^T})^T$.

The convergence and the convergence rate of the joint optimization on X and H_j are defined by the full Hessian matrix G (3.12). If this matrix is definite positive at the point where the criterion J achieves the minimum value, this solution is stable and the convergence rate of the algorithm is defined by the eigenvalues of G. While the analytical study of the matrix G is quite complex the eigenvalues of G can be easily checked numerically. We found in our experiments that if the conditions (3.53) and (3.56) are fulfilled and the regularization parameters λ_2, $\lambda_3 > 0$, the eigenvalues of the Hessian G are also positive.

Then the matrix G is positive definite and the convergence of the projection gradient algorithm can be guaranteed. However, the convergence and the convergence rate depend on the conditioning number of the matrix G.

This conditioning number of the matrix $G(f)$ for the frequency f is calculated as the ratio of the maximum and minimum eigenvalues,

$$\rho(f) = \frac{\max_k \lambda_k(f)}{\min_k \lambda_k(f)}, \tag{3.57}$$

where $\lambda_k(f)$, $k = 1, ..., M$ are eigenvalues of $G(f)$.

Let us assume for simplicity that the step-size parameters α, β in (3.21) and (3.22) are equal to each other, i.e., $\alpha = \beta$. Then, this recursive algorithm in (3.21) and (3.22) is stable provided that [34]

$$\beta(f) < \frac{2}{\max_k \lambda_k(f)}. \tag{3.58}$$

The optimal convergence rate of this procedure for the frequency f is defined as

$$q(f) = \frac{\max_k \lambda_k(f) - \min_k \lambda_k(f)}{\max_k \lambda_k(f) + \min_k \lambda_k(f)} = \frac{\rho(f) - 1}{\rho(f) + 1} \tag{3.59}$$

with the optimal selection of the step-size parameter

$$\beta(f) = \frac{2}{\max_k \lambda_k(f) + \min_k \lambda_k(f)}. \tag{3.60}$$

FIGURE 3.2:   The ratio of the conditioning numbers of the Hessian matrix after and before normalization. The center of the image corresponds to $f = 0$. The white and black correspond to larger and smaller values of this ratio. The range of the ratio values is $[1.5 \times 10^{-8}, 1]$.

If the matrix $G(f)$ is ill-conditioned for the frequency f, then $\max_k \lambda_k(f) \gg \min_k \lambda_k(f)$ and the optimal convergence rate becomes very slow, with $q(f) \approx 1$ achieved provided that the step-size $\beta(f)$ is very small.

The algorithm in (3.21) and (3.22) has the same value of the step-size parameter for all frequencies. Thus the conditions (3.58) through (3.60) should hold for all frequencies. It can be enabled when in the above formulas $\max_k \lambda_k(f)$ and $\min_k \lambda_k(f)$ are replaced by $\max_{k,f} \lambda_k(f)$ and $\min_{k,f} \lambda_k(f)$, respectively.

Thus the frequency-dependent conditioning number $\rho(f)$ is changed for the invariant one:

$$\rho = \frac{\max_{k,f} \lambda_k(f)}{\min_{k,f} \lambda_k(f)}.$$

The convergence condition (3.58) becomes

$$\beta < \frac{2}{\max_{k,f} \lambda_k(f)}$$

with the optimal convergence rate defined as

$$q = \frac{\max_{k,f} \lambda_k(f) - \min_{k,f} \lambda_k(f)}{\max_{k,f} \lambda_k(f) + \min_{k,f} \lambda_k(f)} = \frac{\rho - 1}{\rho + 1}$$

and achieved for

$$\beta = \frac{2}{\max_{k,f} \lambda_k(f) + \min_{k,f} \lambda_k(f)}.$$

Obviously, the situation with the procedure simultaneous for all frequencies, results in much more restrictive conditions on β and in a much slower, even optimal convergence rate.

The normalization of the gradient by the diagonal of the Hessian G as it is done in (3.29) and (3.30) allows us to improve the convergence rate.

It can be shown that for this procedure the convergence and convergence rate are defined by the matrix

$$\tilde{G} = (diag(G))^{-1/2} \cdot G \cdot (diag(G))^{-1/2},$$

where $diag(G)$ means the diagonal matrix which is composed from the diagonal items of the matrix G.

The optimal convergence rate for the matrix $\tilde{G}$ is defined as

$$q_{\tilde{G}} = \frac{\max_{k,f} \lambda_{k,\tilde{G}}(f) - \min_{k,f} \lambda_{k,\tilde{G}}(f)}{\max_{k,f} \lambda_{k,\tilde{G}}(f) + \min_{k,f} \lambda_{k,\tilde{G}}(f)}, \tag{3.61}$$

where $\lambda_{k,\tilde{G}}$ are eigenvalues of the matrix $\tilde{G}$.

Usually the conditioning of the normalized matrix $\tilde{G}$ is much better than the conditioning of the original matrix G.

Figure 3.2 illustrates this fact. It shows the ratio $q_{\tilde{G}}(f)/q(f)$ of the conditioning numbers of the matrices $\tilde{G}$ and G for different frequencies. The center of the image corresponds to $f = 0$. The black and white show smaller and larger values of this ratio respectively. The range of the values in this ratio is $[1.5 \times 10^{-8}, 1]$. In Figure 3.2 we can see black for nearly all pixels. It says that the conditioning of the matrix $\tilde{G}$ is significantly improved. For some isolated frequencies we have the ratio values close to 1 (white pixels in Figure 3.2). It says that for these particular frequencies the normalization of the gradient is not efficient.

The conditioning improvement for almost all frequencies is sufficient for a significant improvement of the algorithm performance.

Imaging in Figure 3.2 is done for the 256×256 "Cameraman" test image and a three-channel imaging system used in simulations in Section 3.8.

Note that the above analysis based on the eigenvalues of the Hessian matrix is equally valid for the unconstrained as well as for the constrained optimization implemented in the considered projection gradient algorithm [34].

A theoretical analysis of convergence becomes impossible if we wish to take into consideration that the weights d_{kl}^2 depend on the estimates and the highly nonlinear adaptive filtering and regularization are used. In this case only simulation is able to provide explicit evaluation of the algorithm.

3.8   Simulations

We consider three-channel observations with the following PSFs:

1. Box-car 9×9 uniform (h_1);

2. Box-car 7×7 uniform rotated by $45°$ (h_2);

3. "Inverse-polynomial" $h_3(x_1, x_2) = (1 + x_1^2 + x_2^2)^{-1}$, $x_1, x_2 = -7, \ldots, 7$.

The level of the noise in the observation sets y_j, $j = 1, 2, 3$, is characterized in dB by $BSNR$ (blurred SNR):

$$BSNR_i = \qquad\qquad\qquad\qquad\qquad\qquad\qquad\qquad (3.62)$$

$$10 \log_{10} \frac{1}{\sigma_i^2 N_1 N_2} \sum_{n_1, n_2} ((h_i * x)[n_1, n_2] - \frac{1}{N_1 N_2} \sum_{n_1, n_2} (h_i * x)[n_1, n_2])^2.$$

For simulation we fix the value of $BSNR_j$ and calculate the standard deviations of the noise in the corresponding channels using the formula (3.62). For our experiments we use the MATLAB® test images (8 bit gray scale): "Boats" (512×512), "Lena" (512×512), "Cameraman" (256×256), "Test-pat1" (256×256), and "Text" (256×256).

3.8.1   Criteria and Algorithm Parameters

3.8.1.1   Criteria

(1) Root mean squared error ($RMSE$):

$$RMSE = \sqrt{\frac{1}{N_1 N_2} \sum_{n_1, n_2} (x[n_1, n_2] - \hat{x}[n_1, n_2])^2}$$

(2) Signal-to-noise ratio (SNR) in dB:

$$SNR = 10 \log_{10}(\sum_{n_1, n_2} |x[n_1, n_2]|^2 / \sum_{n_1, n_2} |x[n_1, n_2] - \hat{x}[n_1, n_2]|^2)$$

(3) Improvement in SNR ($ISNR$) in dB with respect to the noise

$$ISNR = 20 \log_{10}(\hat{\sigma}/RMSE),$$

where $\hat{\sigma}$ is an estimate of the observation standard deviation
(4) Peak signal-to-noise ratio ($PSNR$) in dB:

$$PSNR = 20 log_{10}(max_{n_1, n_2}|x[n_1, n_2]|/RMSE)$$

(5) Mean absolute error (MAE):

$$MAE = \frac{1}{n} \sum_x |y(x) - \hat{y}(x)|$$

(6) Maximum absolute error (MAX):

$$MAX = \max_x |y(x) - \hat{y}(x)|$$

There is no one-to-one link between the visual image quality and the above criteria. Different criteria show quite different optimal values for the design parameters. Thus, a visual inspection, which of course is quite subjective, continues to be the most important final performance criterion.

3.8.1.2 Parameters of the Basic Recursive Procedure

The parameters $\lambda_1, \lambda_2, \lambda_3$ define the weights of the criterion components in (3.6). The regularization parameters λ_2 and λ_3 actually do not influence the results and we take them as small as 10^{-7}. It demonstrates that the regularization terms in (3.6) do not work and necessary regularization effects are produced by the LPA-ICI filtering.

The parameter λ_1 balancing the fidelity and the channel equalization terms is considered as a design parameter of the algorithm. It is found that $\lambda_1 = 1.2$ gives good results in a variety of scenarios.

The size of the PSF supports in (3.18) is $\Delta_{1,i} = \Delta_{2,i} = 21$, which is larger than the maximum support of the PSFs defining the observation blurs.

The step sizes $\beta_k = 0.6$ in the recursions for PSFs and $\alpha_k = 0.9$ for the recursion on the image signal X.

The total number of the algorithm iterations is about $10 \div 20$. The number of embedded iterations on H_i is $T = 7$.

3.8.1.3 Parameters of the LPA–ICI Filtering

For filtering the image estimates we use narrow directional linewise kernels of the order $m = 0$ with a uniform LPA window function.

The length of the kernel supports is equal to a and the set of the lengths used is given by

$$A_x = \{1, 2, 3, 5, 7, 11\}. \tag{3.63}$$

The ICI rule selects the best kernel size, comparing the estimates of the supports a_j from $A_x = \{a_1, ..., a_5\}$ where a_j is defined in (3.63).

The estimates are calculated as the sample means of the observations covered by the kernel supports. The ICI rule is used for image reconstruction with the threshold $\Gamma = 0.9$. The estimates and the adaptive scales $a_i^+[n_1, n_2]$ are found for eight directions $\theta_i = (i - 1)\pi/4$, $i = 1, ..., 8$. These ICI adaptive directional estimates are fused in the final one using the multiwindow method (3.37).

FIGURE 3.3:   Three-channel noisy observations of "Boats."

For filtering the *PSF* estimates we use wider quadrant support kernels also of the order $m = 0$ and with a uniform LPA window function. The supports are squares of the size $a_j \times a_j$ with the values of these a_j given by the set $A_h = \{1, 2, 3, 5, 7, 11\}$. With the uniform *LPA* window function these estimates are mean samples of the observations included in the kernels.

The ICI rule for the PSFs is used with the threshold $\Gamma = 1.5$. The estimates and the adaptive scales $a_i^+[n_1, \, n_2]$ are found for four quadrants. These ICI adaptive directional estimates are fused in the final one using the multiwindow method (3.37).

The parameters Γ used in the ICI rule for the image and PSFs estimates can be treated as regularization parameters of the algorithm. Larger values of Γ enable stronger smoothing of the estimate and result in the biasedness of the estimate. At the same time it gives better removal of random (or high-frequency) components of the image.

3.8.2   Illustrative Results

Image estimation: "Boats"   A test image is the 512×512 "Boats" image corrupted by an additive zero-mean Gaussian noise. The noisy images with $BSNR = 40 \; dB$ can be seen in Figure 3.3.

The central panel of Figure 3.4 shows the true image and results obtained after the 10 iterations of the algorithm. Eight surrounding panels show the ICI adaptive scales $a_i^+[n_1, \, n_2]$ for the corresponding eight directions θ_i. Thus, we can see the adaptive scales for directional estimates looking at the horizontal and vertical directions, i.e. to *East, North, West*, and *South*, as well as four diagonal directions *North-East, North-West, South-West, South-East*.

White and black correspond to large- and small-scale (window-size) values, respectively. The adaptive scales delineate the image intensity very well. This delineation is obviously directional as the obtained scales depend on the corresponding directions. The noise effects are seen in the adaptive scales $a_i^+[n_1,$

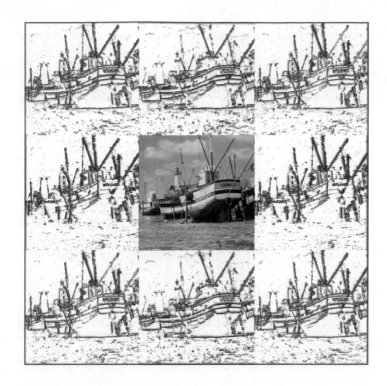

FIGURE 3.4: ICI adaptive directional scales $\hat{a}_j^+(x)$, $\theta_j = (j-1)\pi/4$, $j = 1, ..., 8$, for the "Boats" test image. The true image is shown in the central panel.

$n_2]$ as black isolated points where ICI erroneously takes smaller values of the scale.

The eight narrow kernels exploited here allow us to build the estimates highly sensitive with respect to image details and improve the quality of filtering and regularization.

Figure 3.5 demonstrates the obtained estimates. The central panel shows the final fused estimate calculated from the sectorial ones according to the multiwindow estimate formula (3.37). The surrounding panels show the sectorial directional adaptive scale estimates $\hat{x}_{a_i^+[n_1,n_2],\theta_i}[n_1, n_2]$, $i = 1, ..., 8$, corresponding to the adaptive scales given in Figure 3.4 for the relevant directions.

The directional nature of the adaptive estimates $\hat{x}_{a_i^+[n_1,n_2],\theta_i}[n_1, n_2]$ can be noticed since the corresponding directions are seen as a linewise background of this imaging. The multiwindow fusing allows us to delete and smooth these

FIGURE 3.5: LPA–ICI adaptive scale directional estimates $\hat{x}_{a_j^+(x)}(x)$, $\theta_j = (j-1)\pi/4$, $j = 1,...,8$. The central panel shows the multiwindow fused estimate $\hat{x}$ with $ISNR = 7.52\ dB$, $RMSE = 6.37$, $SNR = 26.71\ dB$, and $PSNR = 32.04\ dB$.

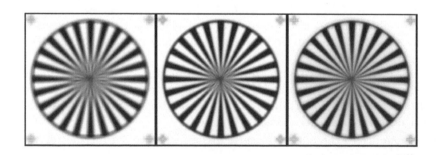

FIGURE 3.6: Observations of the noisy blurred "Testpat1" images.

FIGURE 3.7: ICI adaptive directional scales $a_i^+(x)$, $\theta_i = (i-1)\pi/4$, $i = 1, ..., 8$, for the "Testpat1" image. The true image is shown in the central panel.

directional line effects and obtain a good-quality final estimate.

The numerical results in Table 3.1 are given for the considered eight-directional estimates. They show the criteria values for all eight directional and the final estimates.

The criterion values for the final estimate compared with the directional sectorial ones show a definite improvement in the fused estimate. In particular, we have for $ISNR$ the value of about $6.80\ dB$ for the sectorial estimates while for the fused estimate $ISNR \simeq 7.50\ dB$. The fusing works very well for all criteria in Table 3.1.

Adaptive weight fusing is an efficient tool for accuracy improvement. Visually, the improvement effects are also clear from the comparison of the directional and final estimates.

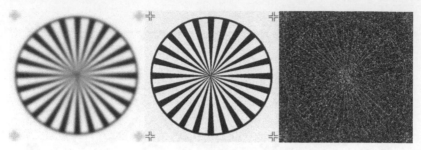

FIGURE 3.8: Initial guess, calculated as the mean of the three observed images, the estimate, and estimation errors.

TABLE 3.1: "Boats" image: criteria values for the eight directional and final multiwindow estimates.

	θ_1	θ_2	θ_3	θ_4	θ_5	θ_6	θ_7	θ_8	Final Est.
ISNR,dB	6.92	6.95	6.70	6.96	6.95	6.93	6.80	6.88	7.52
SNR,dB	26.11	26.14	25.89	26.16	26.15	26.12	25.99	26.07	26.71
PSNR,dB	31.45	31.47	31.22	31.49	31.48	31.45	31.33	31.41	32.04
RMSE	6.82	6.80	7.00	6.78	6.79	6.81	6.91	6.85	6.37
MAE	4.85	4.84	5.02	4.84	4.83	4.87	4.99	4.90	4.36
MAX	80.04	80.04	80.04	80.04	80.04	80.04	74.39	80.04	78.79

Image estimation: "Testpat1"   Observations of the 256×256 image "Testpat1" image corrupted by additive zero-mean Gaussian noise ($BSNR = 40\ dB$) are shown in Figure 3.6.

The central panel of Figure 3.7 shows the true image. The eight surrounding panels show the ICI adaptive scales $a_i[n_1,\ n_2]$ for the corresponding eight directions θ_i. The directional nature of the estimates is clearly demonstrated in these images. We may observe a great deal of difference in the adaptive scales for the eight directional estimates looking to the horizontal, diagonal, and vertical directions.

The initial guess as well as the estimate and the estimation errors are shown in Figure 3.8. Note that the reconstruction is nearly perfect, in particular in the difficult central part of the image.

The three true PSFs and their estimates are given in Figure 3.9. The estimates definitely correspond well to the true PSFs.

The difference between the directional and the final fused estimates is much more significant for the binary "Testpat1" image than for the much more complex texture of the "Boats" image.

The numerical results for "Testpat1" are shown in Table 3.2.

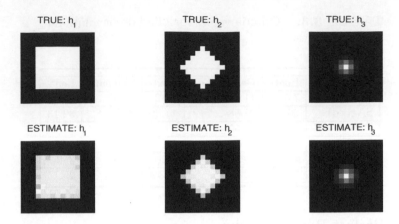

FIGURE 3.9: The estimates and true PSFs of the three-channel imaging system.

TABLE 3.2: "Testpat1" image: criteria values for the eight directional and final multiwindow estimates.

	θ_1	θ_2	θ_3	θ_4	θ_5	θ_6	θ_7	θ_8	Final Est.
ISNR, dB	11.97	11.96	12.38	11.84	12.30	11.73	12.32	11.97	14.59
SNR, dB	23.31	23.31	23.72	23.18	23.65	23.08	23.66	23.31	25.94
PSNR, dB	25.31	25.30	25.72	25.17	25.64	25.07	25.66	25.31	27.93
RMSE	13.83	13.84	13.19	14.04	13.31	14.21	13.28	13.83	10.22
MAE	9.01	9.21	7.63	9.11	7.64	9.29	7.61	9.01	5.19
MAX	111.1	99.03	112.4	108.2	127.5	133.9	113.2	111.1	90.24

PSF estimation The used frequency domain estimation of the PSFs assume a finite support of the estimate included in the projection operation as nulling the PSF outside of the rectangular support of the given size. We studied the effect of this operation and found that the algorithm yields a good accuracy even when the size of the estimate support is significantly larger than the support of the true PSFs. A more accurate support size in the estimates results in faster convergence of the algorithm.

The parameter λ_1 influences the accuracy of estimation by balancing it between the image and the PSFs.

TABLE 3.3:   Criteria values for blind deconvolution of the test-images.

	Boats	Cameraman	Text	Testpat1	Lena
ISNR, dB	7.52	11.94	8.12	14.59	10.64
SNR, dB	26.71	27.88	11.40	25.94	32.13
PSNR, dB	32.04	33.55	22.32	27.93	37.80
RMSE	6.37	5.3	19.52	10.22	3.28
MAE	4.36	3.42	8.20	5.19	2.45
MAX	78.79	43.90	203.4	90.24	25.15

3.8.3   Perfect Reconstruction

In order to test perfect reconstruction we ran the algorithm with a very small level of noise ($BSNR = 60\ dB$). It gives perfect results for all our test images.

Without the LPA–ICI filter the performance of the recursive algorithm degrades quickly. This confirms that the adaptive LPA–ICI algorithm works not only as a noise filter but also as the regularizator of the inverse problem.

Figures 3.10 and 3.11 illustrate this role of the filter. The curves in these figures show $SNRs$ for estimates of the image x_{est} and PSFs h_i as functions of the number of iterations. Solid curves are for the algorithm with the LPA-ICI filtering and regularization and dash-dot curves are for the algorithm without this filtering and regularization.

Figures 3.10 and 3.11 are given for $BSNR = 40\ dB$ and $BSNR = 60\ dB$, respectively. We can see in Figure 3.10 that the curves corresponding to the estimates with the filtering are rapidly growing functions. Without this filtering for first iterations the $SNRs$ keep values close to the initial ones and later start to decrease.

For the lower level of noise, $BSNR = 60\ dB$, in Figure 3.11 the difference between the curves with filtering and without filtering becomes more essential, with faster growing SNR of the estimates with the LPA–ICI filtering.

Without the LPA–ICI filter there is no convergence of the algorithm and there is no good accuracy of image reconstruction.

3.8.4   Numerical Results

The blurring enabled by the used PSFs is quite strong as can be seen in particular from Figure 3.3. Nevertheless the performance of the algorithms is very good as may be concluded from the criterion values in Table 3.3. These results are obtained after 15 iterations of the recursive algorithm.

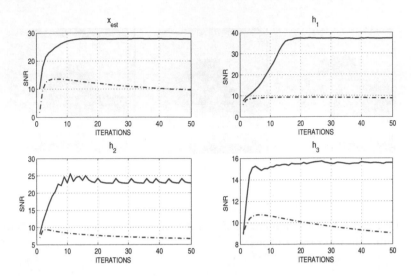

FIGURE 3.10: SNR in dB for estimates of the image x_{est} and PSFs h_i as functions of the number of iterations. Solid curves are for the algorithm with the LPA-ICI filtering and regularization and dash-dot curves for the algorithm without this filtering and regularization. $BSNR = 40\ dB$.

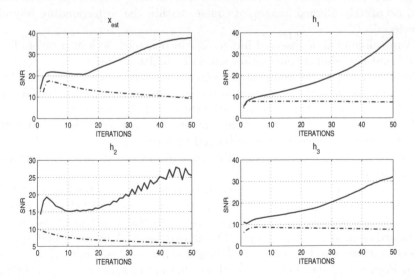

FIGURE 3.11: SNR in dB for estimates of the image x_{est} and PSFs h_i as functions of the number of iterations. Solid curves are for the algorithm with the LPA-ICI filtering and regularization and dash-dot curves for the algorithm without this filtering and regularization. $BSNR = 60\ dB$.

FIGURE 3.12:   A fragment (128×128) of "Cameraman" image, blurred, noisy, and shifted observations.

3.8.5   Image Alignment

If the observed images are shifted and the sizes of the shifts are known, then the corresponding shifts of the PSFs allow us to compensate these shifts exactly if they are integers given in numbers of pixels.

In this case there is no difference between the problem with the shifted and accurately aligned images, of course, within the corresponding boundary effects.

A misalignment of observed images y_i can become a serious problem for the blind and nonblind deconvolution when the shifts are unknown.

Similarly to [10] for image alignment we use a two step procedure. At the first step, the possible integer shifts are estimated by the algorithm in (3.49) and (3.50). Usually this gives an estimate within few pixel errors.

In the second step our blind deconvolution procedure compensates these small-size shifts which are not restricted to integers.

In this section we demonstrate the work of our algorithms with estimation of the shifts as the first step.

The initial data are three-channel noisy observations of the fragment (128×128) of "Cameraman" image shifted in the second and third channel by $sh_2 = [-5, -5]$ and $sh_3 = [-10, -6]$ respectively. In this experiment we used the standard circular shifts of images. The algorithm in (3.49) and (3.50) gives the integer estimates of these shifts as follows: $sh_2^0 = [-6, -5]$ and $sh_3^0 = [-9, -5]$.

In Figure 3.12 the shifted noisy three-channel images y_1, y_2, y_3 ($BSNR = 40$ dB) are shown.

The estimates of the PSFs are shown in Figure 3.13. We may note that they are shifted with respect to the true PSFs as in estimation these shifts compensate the shifts of the observed images. The initial guess and the estimate of "Cameraman" can be seen in Figure 3.14. The quality of this reconstruction

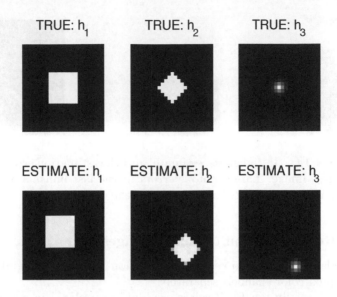

FIGURE 3.13: The true PSFs and their shifted estimates.

is very good. However, the error image in Figure 3.14 shows that this estimate is shifted with respect to the true unshifted "Cameraman" image.

In estimation with unknown shifts the algorithm estimates only relative shifts of the observed images and the true position of the image cannot be found. Thus, the estimates of the image and PSFs can appear with some uncontrolled shifts, and these shifts compensate each other in prediction of the observations in the corresponding channels.

The ICI adaptive scales obtained in the last iteration of this experiment are shown in Figure 3.15. The accurate directional delineation of the image boundaries is quite obvious.

Figure 3.16 demonstrates the obtained results. The central panel shows the final fused estimate calculated from the sectorial ones according to the multiwindow estimate formula (3.37). The surrounding panels show the sectorial directional adaptive scale estimates $\hat{x}_{a_i^+[n_1,n_2],\theta_i}[n_1,\ n_2]$, $i = 1, ..., 8$, corresponding to the adaptive scales given in Figure 3.15 for the relevant directions.

The directional nature of the adaptive estimates $\hat{x}_{a_i^+[n_1,n_2],\theta_i}[n_1,\ n_2]$ can be noticed in Figure 3.16, since the corresponding directions are seen as a linewise background of this imaging. The multiwindow fusing allows us to delete and smooth these directional line effects and obtain a good-quality final estimate.

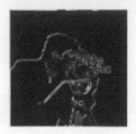

FIGURE 3.14: Initial guess, estimate, and the estimation errors for "Cameraman" test image. Large contour errors in the last image appear because of the shift of the estimate with respect to the original image.

3.8.6   Reconstruction of Color Images

A standard approach to color deconvolution is to apply the algorithm to the corrupted color channels separately.

However, usually the RGB colors are highly correlated in natural images. The opponent color space transformation is used in order to decorrelate color signals [43]:

$$\begin{pmatrix} x_{I_1} \\ x_{I_2} \\ x_{I_3} \end{pmatrix} = \begin{pmatrix} 1/3 & 1/3 & 1/3 \\ 1/2 & 0 & -1/2 \\ 1/4 & -1/2 & 1/4 \end{pmatrix} \begin{pmatrix} x_R \\ x_G \\ x_B \end{pmatrix}. \tag{3.64}$$

Here the indexes I_1, I_2, and I_3 correspond to the achromatic and two opponent color components respectively and x_R, x_G, and x_B are the intensities of the RGB colors in the initial data. These transformations are performed for all M channels of the multichannel imaging system. In the transformed color space the achromatic signal x_{I_1} has a higher SNR while the opponent signals x_{I_2}, x_{I_3} have lower $SNRs$ with emphasized image details.

It is assumed that the blurring operator h_j is the same for all color data R (red), G (green), and B (blue) in the j-th channel. We use the developed algorithm with the parameter values discussed above for gray scale image reconstruction. The results are illustrated in Figure 3.17. The channel data and the initial guess calculated as the mean of the channel observations look quite similar in Figure 3.17(a) through (d). In comparison with the true image in Figure 3.17(f) the reconstruction in the opponent color space (Figure 3.17(e)) is nearly perfect. We do not show the reconstruction in RGB as visually it is very similar to the reconstruction in the opponent color space.

A numerical comparison of the RGB and opponent color space reconstructions is done in Table 3.4. In this table the first and the second figure after the slash correspond to the opponent and RGB color space reconstructions, respectively. All these numerical results are in favor of the opponent color space. Only the maximum error for the blue color is better for the RGB reconstruction.

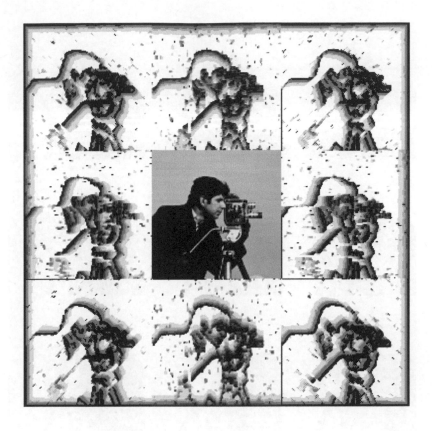

FIGURE 3.15: ICI adaptive directional scales $a_i^+(x)$, $\theta_i = (i-1)\pi/4$, $i = 1, ..., 8$ for the "Cameraman" test image. The true image is shown in the central panel.

FIGURE 3.16: LPA-ICI adaptive scale directional estimates $\hat{x}_{a_j^+(x)}(x)$, $\theta_j = (j-1)\pi/4$, $j = 1, ..., 8$. The central panel shows the multiwindow fused estimate.

3.9   Conclusions

We have developed an iterative algorithm for multichannel blind deconvolution that minimizes a quadratic functional of residuals. Nonlinear regularization emphasizing specific features (local smoothness and anisotropy) of the image and the PSFs is implemented by nonlinear scale adaptive LPA-ICI filtering. The use of the recursive projection gradient algorithm allows us to

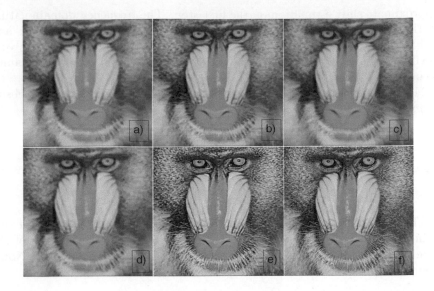

FIGURE 3.17: Reconstruction of the "Baboon" color image in the opponent color space: noisy blurred observations channels 1(a), 2(b), 3(c), initial guess (d), reconstruction (e), true image (f). (**See color insert following page 140.**)

TABLE 3.4: Criteria values of color reconstruction for blind deconvolution of the "Baboon" image.

	R	G	B
ISNR, dB	**4.69**/3.87	**4.93**/4.06	**4.74**/4.14
SNR, dB	**20.52**/19.71	**18.59**/17.72	**17.66**/17.06
PSNR, dB	**25.15**/24.34	**23.96**/23.09	**23.43**/22.83
RMSE	**14.09**/15.47	**16.16**/17.87	**17.18**/18.42
MAE	**10.51**/11.34	**11.99**/13.22	**12.84**/13.77
MAX	**81.38**/94.57	**95.87**/96.78	122.0/**85.30**

implement the main calculations in the frequency domain. It is shown by simulations that the algorithm is stable and gives good estimates of the image as well as a good localization and shape estimation of PSFs.

For noiseless observations the algorithm demonstrates perfect reconstruction of the image and PSFs. However, for all cases a good initial guess is of importance. We show that an inaccurate registration of channel images within the shifts of the images can be compensated by a two-step procedure with the approximate shift estimation on the first step. The frequency domain nature of the algorithm has at least two important benefits. First, it is fast and can work with large images. Second, the frequency domain approach is very flexible and allows us to easily design algorithms for different multichannel scenarios.

MATLAB® codes of the developed algorithm are available in the website *http://www.cs.tut.fi/~lasip/*.

Acknowledgments

This work was supported by the Academy of Finland, project No. 213462 (Finnish Centre of Excellence program 2006–2011). In part, the work of Dr. Vladimir Katkovnik is supported by a *Visiting Fellow* grant from the Nokia Foundation.

References

[1] M. R. Banham and A. K. Katsaggelos, "Digital image restoration," *IEEE Signal Processing Magazine*, vol. 14, pp. 24–41, 1997.

[2] N. Miura and N. Baba, "Segmentation based multiframe blind deconvolution of solar images," *Journal of the Optical Society of America A*, vol. 12, no. 9, pp. 1858–1866, 1995.

[3] T. Schulz, "Multiframe blind deconvolution of astronomical images," *Journal of the Optical Society of America A*, vol. 10, no. 5, pp. 1064–1073, 1993.

[4] A. K. Katsaggelos, Ed., *Digital Image Restoration*. Berlin, Germany: Springer-Verlag, 1991.

[5] G. B. Giannakis and R. W. Heath, Jr., "Blind identification of multi-channel FIR blurs and perfect image restoration," *IEEE Transactions on Image Processing*, vol. 9, pp. 1877–1896, 2000.

[6] G. Harikumar and Y. Bresler, "Perfect blind restoration of images blurred by multiple filters: Theory and efficient algorithm," *IEEE Transactions on Image Processing*, vol. 8, pp. 202–219, Feb. 1999.

[7] G. Harikumar and Y. Bresler, "Exact image deconvolution from multiple FIR blurs," *IEEE Transactions on Image Processing*, vol. 8, no. 6, pp. 846–862, 1999.

[8] H. T. Pai and A. Bovik, "Exact multichannel blind image restoration," On eigenstructure-based direct multichannel blind image restoration," *IEEE Transactions on Image Processing*, vol. 10, no. 10, pp. 1434–1446, 2001.

[9] F. Sroubek and J. Flusser, "Multichannel blind iterative image restoration," *IEEE Transactions on Image Processing*, vol. 12, no. 9, pp. 1094–1106, 2003.

[10] F. Sroubek and J. Flusser, "Multichannel blind deconvolution of spatially misaligned images," *IEEE Transactions on Image Processing*, vol. 14, no. 7, pp. 874-883, 2005.

[11] G. Panci, P. Campisi, S. Colonnese, and G. Scarano, "Multichannel blind image deconvolution using the bussgang algorithm: Spatial and multiresolution approaches," *IEEE Transactions on Image Processing*, vol. 12, no. 11, pp. 1324–1337, 2003.

[12] A. N. Tikhonov and V.Y. Arsenin, *Solution of Ill-posed Problems*. New York: Wiley, 1977.

[13] L. Rudin, S. Osher, and E. Fatemi, "Nonlinear total variation based noise removal algorithms," *Physica D*, vol. 60, pp. 259–268, 1992.

[14] D. Mumford and J. Shah, "Optimal approximation by piecewise smooth functions and associated variational problems," *Communications on Pure and Applied Mathematics*, vol. 42, pp. 577–685, 1989.

[15] C. Vogel and M. Oman, "Fast, robust total variation-based reconstruction of noisy blurred images," *IEEE Transactions on Image Processing*, vol. 7, no. 6, pp. 813–824, 1998.

[16] T. Chan and C. Wong, "Total variation blind deconvolution," *IEEE Transactions on Image Processing*, vol. 7, pp. 370–375, 1998.

[17] Y. L. You and M. Kaveh, "Blind image restoration by anisotropic regularization," *IEEE Transactions on Image Processing*, vol. 8, no. 3, pp. 396–407, 1999.

[18] T. Chan, S. Osher, and J. Shen, "The digital TV filter and nonlinear denoising," *IEEE Transactions on Image Processing*, vol. 10, no. 10, pp. 231–241, 2001.

[19] W. Vanzella, F. A. Pellegrino, and V. Torre, "Self-adaptive regularization," *IEEE Transactions on Pattern Analysis and Machine Intelligence*, vol. 26, no. 6, pp. 804–809, 2004.

[20] R. C. Gonzalez and R. E. Woods, *Digital Image Processing,* Prentice Hall, Upper Saddle River, New Jersey, 2002.

[21] A. Goldenshluger and A. Nemirovski, "On spatial adaptive estimation of nonparametric regression," *Mathematical Methods of Statistics*, vol. 6, pp. 135-170, 1997.

[22] O. Lepskii, "On one problem of adaptive estimation in Gaussian white noise," *SIAM Theory of Probability and Its Applications*, vol. 35, no. 3, pp. 454-466, 1990.

[23] O. Lepskii, "Asymptotically minimax adaptive estimation I: Upper bounds. Optimally adaptive estimates," *SIAM Theory of Probability and Its Applications*, vol. 36, pp. 682-697, 1991.

[24] O. Lepski and V. Spokoiny, "Optimal pointwise adaptive methods in nonparametric estimation," *The Annals of Statistics*, vol. 25, no. 6, pp. 2512–2546, 1997.

[25] O. Lepski, E. Mammen, and V. Spokoiny, "Ideal spatial adaptation to in-homogeneous smoothness: An approach based on kernel estimates with variable bandwidth selection," *The Annals of Statistics*, vol. 25, no. 3, 929–947, 1997.

[26] A. Nemirovski, Topics in non-parametric statistics, *Lecture notes in mathematics*, 1738, pp. 85-277, New York: Springer, 2000.

[27] J. Polzehl and V. Spokoiny, "Adaptive weights smoothing with applications to image restoration," *Journal of the Royal Statistical Society: Series B*, vol. 62, pp. 335–354, 2000.

[28] J. Polzehl and V. Spokoiny, "Image denoising: pointwise adaptive approach," *The Annals of Statistics*, vol. 31, no. 1, pp. 30-57, 2003.

[29] D. Mercurio and V. Spokoiny, "*Estimation of time-dependent volatility via local change point analysis*," WIAS, Preprint 904, 2004.

[30] V. Katkovnik, K. Egiazarian, and J. Astola, "Adaptive window size image de-noising based on intersection of confidence intervals (ICI) rule," *Journal of Mathematical Imaging and Vision*, vol. 16, pp. 223-235, 2002.

[31] V. Katkovnik, K. Egiazarian, and J. Astola, "A spatially adaptive non-parametric regression image deblurring," *IEEE Transactions on Image Processing*, vol. 14, no. 10, pp. 1469–1478, 2005.

[32] V. Katkovnik, A. Foi, K. Egiazarian, and J. Astola, "Directional varying scale approximations for anisotropic signal processing," *Proceedings of EUSIPCO 2004*, pp. 101-104, 2004.

[33] V. Katkovnik, D. Paliy, K. Egiazarian, and J. Astola, "Frequency domain blind deconvolution in multiframe imaging using anisotropic specially adaptive denoising," *Proceedings of EUSIPCO 2006*, 2006.

[34] B. T. Polyak. *Introduction to optimization.* Optimization Software Inc, 1987.

[35] V. Katkovnik, "A new method for varying adaptive bandwidth selection," *IEEE Transactions on Signal Processing,* vol. 47, no. 9, pp. 2567-2571, 1999.

[36] V. Katkovnik, K. Egiazarian, and J. Astola. *Local Approximation Techniques in Signal and Image Processing.* SPIE PRESS, Bellingham, Washington, USA, 2006.

[37] W. T. Freeman and E. H. Adelson, "The design and use of steerable filters," *IEEE Trans. Pattern Analysis and Machine Intelligence,* vol. 13, no. 9, pp. 891-906, 1991.

[38] J. Starck, E.J. Candes, and D.L. Donoho, "The curvelet transform for image denoising," *IEEE Transactions on Image Processing,* vol. 11, N. 6, pp. 670-684, 2002.

[39] A. Foi, Katkovnik, V., Egiazarian, K. & Astola, J., A novel anisotropic local polynomial estimator based on directional multiscale optimizations, *Proc. of the 6th IMA Conf. Math. in Signal Processing,* Great Britain: Cirencester, 2004.

[40] D. Kundur and D. Hatzinakos, "Blind image deconvolution," *IEEE Signal Processing Magazine,* vol. 13, pp. 43–64, May 1996.

[41] T. Chan and C. Wong, "Convergence of the alternating minimization algorithm for blind deconvolution," *Linear Algebra and its Applications,* vol. 316, no. 1–3, pp. 259–285, 2000.

[42] J. Justen and R. Ramlau, "A non-iterative regularization approach to blind deconvolution," *Inverse Problems,* vol. 22, pp. 771–800, 2006.

[43] K. N. Plataniotis, A. N. Venetsanopulos, *Color Image Processing and Applications,* Berlin and Heidelberg: Springer-Verlag, 2000.

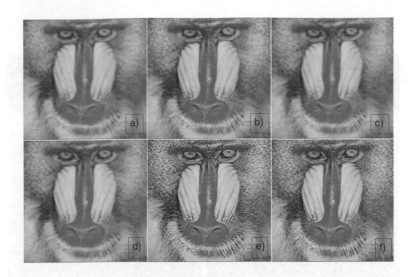

FIGURE 3.17. See caption, page 135.

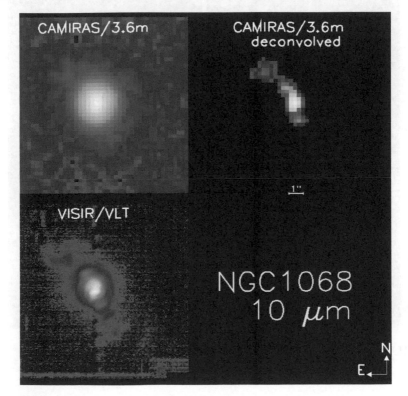

FIGURE 7.1. See caption, page 279.

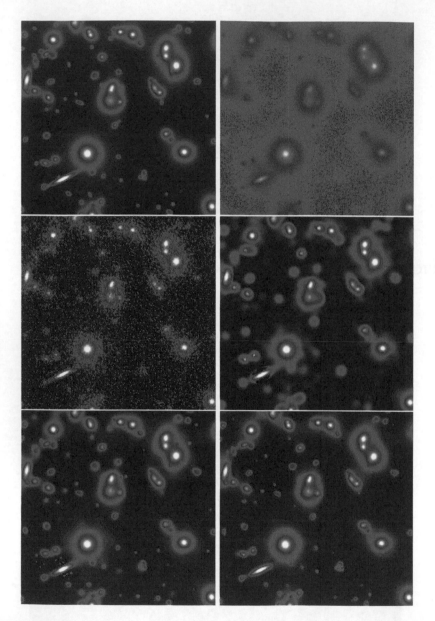

FIGURE 7.2. See caption, page 294.

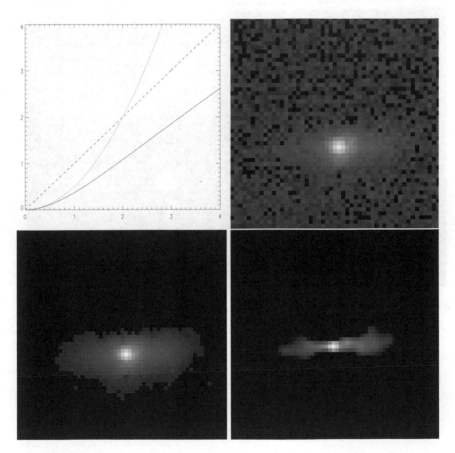

FIGURE 7.3. See caption, page 297.

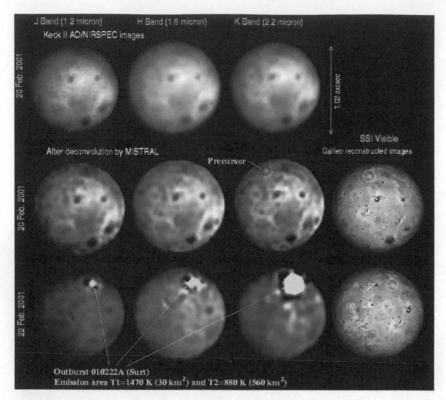

FIGURE 7.4. See caption, page 305.

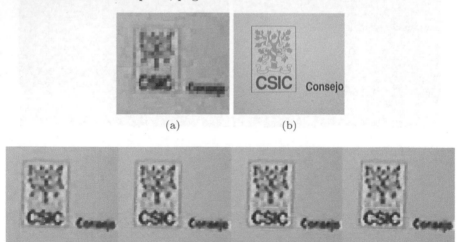

FIGURE 8.12. See caption, page 343.

4

Bayesian Methods Based on Variational Approximations for Blind Image Deconvolution

Aristidis Likas and Nikolas P. Galatsanos
Department of Computer Science
University of Ioannina GR 45110, Ioannina Greece
e-mail: (arly, galatsanos)@cs.uoi.gr

Abstract

In this chapter the blind image deconvolution (BID) problem is addressed using the Bayesian framework. In order to solve the Bayesian model that appears in BID we present a methodology based on variational approximations, which can be viewed as a generalization of the expectation maximization (EM) algorithm. We present three algorithms to solve the Bayesian BID model that can be implemented in the discrete Fourier domain which makes them very efficient even for very large images. We demonstrate with numerical experiments that these algorithms yield promising improvements as compared to previous BID algorithms. Furthermore, the proposed methodology is quite general with potential application to other Bayesian models for this and other imaging problems.

4.1 Introduction

The blind image deconvolution (BID) problem is a difficult and challenging problem because the observed image does not uniquely define the convolved signals. Nevertheless there are many applications where the observed images have been blurred either by an unknown or a partially known point spread function (PSF). Such examples can be found in astronomy and remote sensing where the atmospheric turbulence cannot be exactly measured, in medical imaging where the PSF of different instruments has to be measured and thus is subject to errors, in photography where the PSF of the lens used to obtain the image is unknown or approximately known, and so on.

A plethora of methods has been proposed to address this problem (see [1] for an old survey on this problem). Since in BID the observed data are not sufficient to specify the convolved functions, most recent methods attempt to incorporate in the BID algorithm some prior knowledge about these functions. Since it is very hard to track the properties of the PSF and the image simultaneously, several BID methodologies attempt to impose constraints on the image and the PSF in an alternating fashion. In other words, such approaches cycle between the image and the PSF estimation steps. In the image estimation step the image is estimated assuming that the PSF is fixed to its last estimate from the PSF estimation step. In the PSF estimation step the PSF is estimated assuming the image to be fixed to its last estimate from the image estimation step. This decouples the nonlinear observation model in BID into two linear observation models that are easy to solve. A number of algorithms of this nature have been proposed that use a deterministic framework to introduce prior knowledge in the form. For example, convex sets, "classical" regularization, regularization with anisotropic diffusion functionals, and fuzzy soft constraints have been proposed (see [5 – 7] and [15], respectively).

A probabilistic framework using maximum likelihood (ML) estimation was applied to the BID problem in [2, 3] and [4] using the expectation maximization (EM) algorithm [11]. ML estimation requires the maximization of the data likelihood $p(g; \theta)$ with respect to the unknown parameters θ where g represents the observations. This likelihood on many occasions cannot be directly found and its computation is greatly facilitated by the introduction of "hidden" variables s that are subsequently integrated out, or marginalized. These variables act as links that connect the observations to the unknown parameters via Bayes' law. The choice of hidden variables is problem dependent. However, with the proper choice of hidden variables $p(g|s, \theta)$ should be easy to write and then the likelihood is obtained by the following marginalization:

$$p(g; \theta) = \int p(g, s; \theta) \, ds = \int p(g|s, \theta) \, p(s) \, ds.$$

Despite the introduction of the hidden variables, in many cases the above integral is either impossible to compute in closed form or yields a marginal likelihood that is complicated so that its direct maximization is difficult. In such cases, iterative algorithms can be used to maximize either the marginal likelihood or a lower bound of it. Algorithms that do the former are the Expectation Maximization (EM) algorithms, and algorithms that do the latter are the so-called variational EM algorithms (see, for example, [11], [16], and [19]). Both of these algorithms are guaranteed to converge—to a local maximum of the marginal likelihood in the case of the EM, and to a local maximum of a strict lower bound on the marginal likelihood in the case of the variational EM.

In [2, 3] and [4] the unknown PSF is considered as part of the deterministic unknown parameters θ. Thus these works are identical in spirit and only differ in the prior model used for the image. As a result ML estimates based on the EM were used to estimate the PSF. However, this formulation does not allow the incorporation of prior knowledge about the PSF, since it is considered deterministic,. This is a major drawback since prior knowledge is necessary to reduce the degrees of freedom of the available observations in BID. As a result, in order to make these algorithms in [2, 3] and [4] work in practice, a number of deterministic constraints were imposed on the PSF such as the size of the PSF support and PSF symmetry. Although these constraints make intuitive sense, strictly speaking, they cannot be justified theoretically by the ML-EM framework.

In [8, 9] and [10] the Bayesian formulation is used for a special case of the BID problem where the PSF was assumed partially known. In this case the PSF was assumed to be given by the sum of a known deterministic component and an unknown stochastic component. In [9] and [10] the dependence of the observations on the unknown image was bypassed by integrating out the unknown image. More specifically, a Laplace approximation of the Bayesian integral that appears in this formulation was used. In spite of this, as reported in [9] the accuracy of the obtained estimates of the statistics of the errors in the PSF and the image could vary significantly depending on the initialization. Thus, with the Bayesian approach in [10], prior knowledge about either the statistics of the error in the PSF or the image was introduced in the form of hyperpriors.

The Bayesian framework is a very powerful and flexible methodology for estimation and detection of problems because it provides a structured way to include prior knowledge concerning the quantities to be estimated. Furthermore, both the Bayesian methodology and its application to practical problems have recently experienced explosive growth (see, for example, [12, 13] and [14]). In spite of this, the application of this methodology to the BID problem remains elusive mainly due to the nonlinearity of the observation model. This makes intractable the computation of the joint probability density function (PDF) of the image and the PDF given the observations. One way to bypass this problem is to employ in a Bayesian framework the technique of alternating between estimating the image and the PSF while keeping the other constant as previously described. The main advantage of such strategy is that it linearizes the observation model and then it is easy to apply the Bayesian framework. However, clearly this is a suboptimal strategy. Another approach to bypass this problem could be to use Markov Chain Monte Carlo (MCMC) techniques to generate samples from this elusive conditional PDF and then estimate the required parameters from the statistics of those samples. However, MCMC techniques are notoriously computational intensive, and furthermore there is no universally accepted criterion or methodology to decide when to terminate [13].

In what follows we propose to use a new methodology termed "variational" to address the Bayesian BID problem that bypasses the previously described shortcomings. The proposed approach is a generalization of both the ML framework in [2 – 4] and [24] and the partially known PSF model in [8, 9] and [10]. The variational methodology that we use was first introduced in the machine learning community to solve Bayesian inference problems with complex probabilistic models (see, for example, [17, 19, 20, 22] and [23]). In the machine learning community the term *graphical models* has been coined in such cases, since a graph can be used to represent the dependencies among the random variables of the models, and the computations required for Bayesian inference can be greatly facilitated based on the structure of this graph. It has also been shown that the variational approach can be viewed as a generalization of the EM algorithm [16]. In [21] a similar methodology to the variational, which is termed ensemble learning, is used by Miskin and MacKay to address BID in a Bayesian framework. However, the approach in [21] uses a different model for both the image and the PSF. This model assumes that the image pixels are independent identically-distributed and thus does not capture the between-pixel correlations of natural images. Furthermore, our model allows simplified calculations in the frequency domain. This greatly facilitates the implementation for high-resolution images. In [25] an alternative variational framework for blind image deconvolution has been presented in which a different prior model for the PSF is used. Furthermore, in [25] hyperpriors are used. The material in this chapter is based on work presented in [26].

The rest of this chapter is organized as follows: in Section 4.2 we provide the background on variational methods; in Section 4.3 we present the Bayesian model that we propose for the BID problem and the resulting variational functional; in Section 4.4 two iterative algorithms are presented that can be used to solve for this model and we provide numerical experiments indicating the superiority of the proposed algorithms as compared to previous BID approaches; finally, in Section 4.5 we provide our conclusions and suggestions for future work.

4.2 Background on Variational Methods

The variational framework constitutes a generalization of the well-known Expectation Maximization (EM) algorithm for likelihood maximization in Bayesian estimation problems with "hidden variables." The EM algorithm has been proved a valuable tool for many problems, since it provides an elegant approach to bypass difficult optimization and integrations required in Bayesian estimation problems. In order to efficiently apply the EM algorithm two requirements should be fulfilled [11]: i) In the E-step we should be able to compute the conditional PDF of the "hidden variables" given the observation data. ii) In the M-step, it is highly preferable to have analytical formulas for the update equations of the parameters. Nevertheless, in many problems it is not possible to meet the above requirements, and several variants of the basic EM algorithm have emerged. For example, a variant of the EM algorithm called the "generalized EM" (GEM) proposes a partial M-step in which the likelihood always improves. In many cases partial implementation of the E-step is also used. An algorithm along such lines was investigated in [16].

The most difficult situation for applying the EM algorithm emerges when it is not possible to specify the conditional PDF of the hidden variables given the observed data that are required in the E-step. In such cases the implementation of the EM algorithm is not possible. This significantly restricts the range of problems where EM can be applied. To overcome this serious shortcoming of the EM algorithm, the variational methodology was developed [17]. It can be shown that EM naturally arises as a special case of the variational methodology.

Assume an estimation problem where x and s are the observed and hidden variables, respectively, and θ is the model parameter to be estimated. All PDFs are parameterized by the parameters θ, i.e. $p(x;\theta)$, $p(s,x;\theta)$, and $p(s|x;\theta)$, and we omit θ for brevity in what follows. For an *arbitrary* PDF $q_s(s)$ of the hidden variables s it is easy to show that:

$$\log p(x) + E_q\left(\log q(s)\right) = E_q\left(\log p(x,s)\right) + E_q\left(\log q(s)\right) - E_q\left(\log p(s|x)\right)$$

where E_q denotes the expectation with respect to $q(s)$. The above equation can be written as

$$L\left(\theta\right) + E_q\left(\log q(s)\right) = E_q\left(\log p(x,s)\right) + KL\left(q(s)||p(s|x)\right),$$

where $L(\theta) = \log p(x;\theta)$ the likelihood of the unknown parameters and $KL(q(s)||p(s|x))$ is the Kullback–Leibler distance between $q(s)$ and $p(s|x)$.

Rearranging the previous equation we obtain:

$$\mathbf{F}(q,\theta) = L(\theta) - KL(q(s)||p(s|x)) = E_q(\log p(x,s)) + H(q) \qquad (4.1)$$

where $H(q)$ is the entropy of $q(s)$. From Equation (4.1) it is clear that $\mathbf{F}(q,\theta)$ provides a *lower bound* for the likelihood of θ parameterized by the family of PDFs $q(s)$, since $KL\left(q(s)||p(s|x)\right) \geq 0$. When $q^*(s) = p(s|x;\theta)$ the lower bound becomes exact: $\mathbf{F}(q^*,\theta) = L(\theta)$.

However, the previous framework allows, based on Equation (1), to find a local maximum of $L(\theta)$ *using an arbitrary* PDF $q(s)$. This is a very useful generalization because it bypasses one of the main restrictions of EM: that of exactly knowing $p(s|x)$. The variational method works to maximize the lower bound of $\mathbf{F}(q,\theta)$ with respect to both θ and q. This is justified by a theorem in [16] stating that, if $\mathbf{F}(q,\theta)$ has a local maximum at $q^*(s)$ and θ^*, then $L(\theta)$ has a local maximum at θ^*. Furthermore, if $\mathbf{F}(q,\theta)$ has a global maximum at $q^*(s)$ and θ^*, then $L(\theta)$ has a global maximum at θ^*. Consequently the variational EM approach can be described as follows:

$$\text{E-step: } q^{(t+1)} = \arg\max_q \mathbf{F}(q,\ \theta^{(t)})$$

$$\text{M-step: } \theta^{(t+1)} = \arg\max_\theta \mathbf{F}(q^{(t+1)},\theta).$$

This iterative approach increases at each iteration t the value of the bound $\mathbf{F}(q,\theta)$ until a local maximum is attained. Using this framework the EM algorithm in [11] can be viewed as a special case of the variational EM approach when $q^*(s) = p(s|x;\theta)$ [16].

4.3   Variational Blind Deconvolution

4.3.1   Variational Functional $\mathbf{F}(q,\theta)$

In what follows we apply the variational approach to the Bayesian formulation of the *blind deconvolution* problem. The observations are given by:

$$g = h * f + w = H \cdot f + w = F \cdot h + w \tag{4.2}$$

and we assume the $N \times 1$ vector g to be the observed variables, the $N \times 1$ vectors f and h are the hidden variables, w is Gaussian noise, and H and F the $N \times N$ convolution matrices. We assume Gaussian PDFs for the priors of f and h. In other words, we assume $p(f) = N(\mu_f, \Sigma_f)$, $p(h) = N(\mu_h, \Sigma_h)$ and $p(w) = N(0, \Sigma_w)$. Thus, the parameters are $\theta = [\mu_f, \Sigma_f, \mu_h, \Sigma_h, \Sigma_w]^T$. The dependencies of the parameters and the random variables for the BID problem can be represented by the graph in Figure 4.1.

The key difficulty with the above Bayesian model for the blind deconvolution problem is that the posterior PDF $p(f,h|g;\theta)$ of the hidden variables f and h given the observations g *is intractable*. In other words it is not possible to come up with it using Bayes since it is impossible to marginalize the joint $p(f,h,g;\theta)$ with respect to the hidden variables. This fact makes impossible the direct application of the EM algorithm. However, with the *variational*

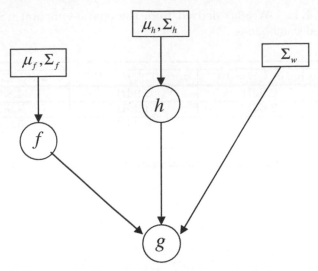

FIGURE 4.1: The graphical model describing the data generation process for the blind deconvolution problem considered in this chapter.

approximation described in the previous section it is possible to bypass this difficulty. More specifically, we *select a factorized form for* $q(s)$ that employs Gaussian components,

$$q(s) = q(h, f) = q(h)q(f) = N(m_{f^q}, C_{f^q})N(m_{h^q}, C_{h^q}), \qquad (4.3)$$

where $\theta_q = [m_{f^q}, C_{f^q}, m_{h^q}, C_{h^q}]^T$ are the parameters of $q(s)$.

This choice for $q(s)$ can be justified since the variational formulation based on it is *tractable* and has been used extensively with success in variational formulations [16, 17], and [19]. Furthermore, it provides a variational bound $\mathbf{F}(q, \theta)$ (Equation 1) that can be specified analytically in the Discrete Fourier Transform (DFT) domain if circulant covariance matrices are used. From the right-hand side of Equation (4.1) we have

$$\mathbf{F}(q, \theta) = E_q(\log(p(x, s))) + H(q) \qquad (4.4)$$

where $p(x, s) = p(g, f, h) = p(g|f, h) \cdot p(f) \cdot p(h)$ with $p(g|h, f) = N(h * f, \Sigma_w)$.

The variational approach requires the computation of the expectation (Gaussian integral) in Equation (4.4) with respect to $q(s)$. In order to facilitate computations for large images, we will assume circulant convolutions in Equation (4.2) and that matrices Σ_f, Σ_h, Σ_w, C_{f^q}, and C_{h^q} are circulant. This facilitates implementation in the DFT domain. Computing the expectation $E_q(\log(p(g, f, h)))$ as well as the entropy of $q(s)$ we can write the result in the DFT domain as (the derivation is described in Appendix A):

TABLE 4.1: Weights derived from the spatio-temporal CSF for the different video subbands.

	$\sigma = 10^{-2}, \beta = 10^{-4}$	$\sigma = 10^{-3}, \beta = 10^{-4}$	$\sigma = 10^{-4}, \beta = 10^{-4}$
VAR1	2.6dB	3.9dB	4.8dB
VAR2	2.6dB	3.9dB	4.9dB
VAR3	2.5dB	2.9dB	3.0dB
PKN	2.1dB	3.0dB	no convergence
ITW	2.5dB	2.8dB	1.64dB

$$F(q,\theta) =$$

$$C - \frac{1}{2}\sum_{k=0}^{N-1} \left(\log \Lambda_w(k) + \log \Lambda_f(k) + \log \Lambda_h(k) + \log S_{f^q}(k) + \log S_{h^q}(k)\right)$$

$$-\frac{1}{2}\sum_{k=0}^{N-1} \frac{\overbrace{\frac{1}{N}\left(|G(k)|^2 - 2Re\left\{M_{f^q}(k)M_{h^q}(k)G^*(k)\right\}\right)}^{A_1(k)}}{\Lambda_w(k)}$$

$$-\frac{1}{2}\sum_{k=0}^{N-1} \frac{\overbrace{N\left(S_{f^q}(k) + \frac{1}{N}|M_{f^q}(k)|^2\right)\left(S_{h^q}(k) + \frac{1}{N}|M_{h^q}(k)|^2\right)}^{A_2(k)}}{\Lambda_w(k)}$$

$$-\frac{1}{2}\sum_{k=0}^{N-1} \frac{\overbrace{\left(S_{f^q}(k) + \frac{1}{N}|M_{f^q}(k)|^2\right) + \frac{1}{N}\left(|M_f(k)|^2 - 2Re\left\{M_f^*(k)M_{f^q}(k)\right\}\right)}^{B(k)}}{\Lambda_f(k)}$$

$$-\frac{1}{2}\sum_{k=0}^{N-1} \frac{\overbrace{\left(S_{h^q}(k) + \frac{1}{N}|M_{h^q}(k)|^2\right) + \frac{1}{N}\left(|M_h(k)|^2 - 2Re\left\{M_h^*(k)M_{h^q}(k)\right\}\right)}^{C(k)}}{\Lambda_h(k)}$$

$$(4.5)$$

where $S_{f^q}(k)$, $S_{h^q}(k)$, $\Lambda_f(k)$, $\Lambda_h(k)$, and $\Lambda_w(k)$ are the eigenvalues of the $N \times N$ circulant covariance matrices C_{f^q}, C_{h^q}, Σ_f, Σ_h, and Σ_w, respectively. Also, $G(k)$, $M_{f^q}(k)$, and $M_{h^q}(k)$ are the DFT coefficients of the vectors g, m_{f^q}, and m_{h^q}, respectively.

(a)   (b)

(c)   (d)

FIGURE 4.2: Trevor images referring to Table 4.1 with $\sigma = 10^{-3}$, $\beta = 10^{-4}$. (a) Degraded image (b) Iterative Wiener (ITW), improvement in the signal-to-noise ratio $(ISNR) = 2.8\,dB$. (c) Partially Known (PKN), $ISNR = 3.0\,dB$. (d) VAR2, $ISNR = 3.9\,dB$.

4.3.2 Maximization of the Variational Bound $\mathbf{F}(q, \theta)$

In analogy to the conventional EM framework, the maximization of the variational bound $\mathbf{F}(q, \theta)$ can be implemented in two steps as described in the end of Section 4.2. In the **E-step**, the parameters $\theta_q = [m_{f^q}, C_{f^q}, m_{h^q}, C_{h^q}]^t$ of $q(s)$ are updated. Three approaches are proposed for this update. The first approach (called VAR1) is based on the direct maximization of $\mathbf{F}(q, \theta)$ with respect to the parameters θ_q. It can be easily shown that such maximization can be performed analytically by setting the gradient of $\mathbf{F}(q, \theta)$ with respect to each parameter equal to zero, thus the updates for $m_{f^q}^{(t+1)}$, $C_{f^q}^{(t+1)}$, $m_{h^q}^{(t+1)}$ and $C_{h^q}^{(t+1)}$ can be obtained. The details of this approach and the resulting equations are given in Appendix B.

In the second approach (called VAR2) we assume that $q(f) = p(f|g; h)$ and $q(h) = p(h|g; f)$. When h or f are assumed known, the observation model in Equation (4.2) is linear. Thus, for Gaussian priors on h, f and Gaussian noise w, the conditionals of h and f given the observations are Gaussians $p(f|h,g) = N(m_{f/g}, C_{f/g})$, $p(h|f,g) = N(m_{h/g}, C_{h/g})$ with known means and covariances which are given by (see [3] and [4])

$$
\begin{aligned}
m_{f/g} &= \mu_f + \Sigma_f \cdot H^t \cdot (H \cdot \Sigma_f \cdot H^t + \Sigma_n)^{-1} \cdot (g - H \cdot \mu_f) \\
C_{f/g} &= \Sigma_f - \Sigma_f \cdot H^t \cdot (H \cdot \Sigma_f \cdot H^t + \Sigma_n)^{-1} \cdot H \cdot \Sigma_f
\end{aligned} \tag{4.6}
$$

$$
\begin{aligned}
m_{h/g} &= \mu_h + \Sigma_h \cdot F^t \cdot (F \cdot \Sigma_h \cdot F^t + \Sigma_n)^{-1} \cdot (g - F \cdot \mu_h) \\
C_{h/g} &= \Sigma_h - \Sigma_h \cdot F^t \cdot (F \cdot \Sigma_h \cdot F^t + \Sigma_n)^{-1} \cdot F \cdot \Sigma_h
\end{aligned} \tag{4.7}
$$

Therefore we set $m_{f^q}^{(t+1)} = m_{f/g}$, $C_{f^q}^{(t+1)} = C_{f/g}$, $m_{h^q}^{(t+1)} = m_{h/g}$, and $C_{h^q}^{(t+1)} = C_{h/g}$.

Since in the above equations we do not know the values of h and f, we use their current estimates $m_{h^q}^{(t)}$ and $m_{f^q}^{(t)}$. It must also be noted that all computations take place in the DFT domain. A disadvantage of this approach is that the update equations of the parameters θ_q do not theoretically guarantee the increase of the variational bound $\mathbf{F}(q, \theta)$. Nevertheless, the numerical experiments have shown that this is not a problem in practice, since in all experiments the update equations resulted in an increase of $\mathbf{F}(q, \theta)$. See, for example, Figure 4.5 (b).

In the **M-step**, the parameters θ_q are considered fixed and Equation (4.5) is maximized with respect to the parameters θ leading to the following update equations:

$$
\mu_f^{(t+1)} = \mu_{f^q}^{(t+1)}, \quad \mu_h^{(t+1)} = \mu_{h^q}^{(t+1)}, \quad \Sigma_f^{(t+1)} = C_{f^q}^{(t+1)} \text{ and } \Sigma_h^{(t+1)} = C_{h^q}^{(t+1)} \tag{4.8}
$$

for both approaches VAR1 and VAR2. The covariance of the noise is updated for the VAR1 and VAR2 approaches according to

$$
\begin{aligned}
\Lambda_w^{(t+1)}(k) &= \tfrac{1}{N} \left(|G(k)|^2 - 2Re \left\{ M_{f^q}^{(t+1)}(k) M_{h^q}^{(t+1)}(k) G^*(k) \right\} \right) \\
&+ N \left(S_{f^q}^{(t+1)}(k) + \tfrac{1}{N} \left| M_{f^q}^{(t+1)}(k) \right|^2 \right) \left(S_{h^q}^{(t+1)}(k) + \tfrac{1}{N} \left| M_{h^q}^{(t+1)}(k) \right|^2 \right)
\end{aligned} \tag{4.9}
$$

for $k = 0, 1 \ldots N - 1$, where $\Lambda_w^{(t+1)}(k)$, $S_{f^q}^{(t+1)}(k)$, $S_{h^q}^{(t+1)}(k)$, $M_{f^q}^{(t+1)}(k)$, $M_{h^q}^{(t+1)}(k)$ and $G(k)$ are defined as in Equation (4.5). The detailed derivations of the formulas for the parameter updates of our models are given in Appendix B.

In the third approach (called VAR3) the optimization of $\mathbf{F}(q, \theta)$, at each iteration, is done in an alternating fashion in *two stages* assuming either f or h to be constant. More specifically, at the first stage, f is assumed a random variable and the parameters associated with f are updated while h is considered a known constant. In the second stage the reverse happens. More specifically, at the E-step of the first stage, since h is assumed deterministic, we have that $q(s) = q(f)$ and from Equation (4.1) the new variational bound can be written as

$$\mathbf{F}_f(q(f), \theta) = E_{q(f)}(\log p(g, f)) + H(q(f)) \tag{4.10}$$

where $\theta = [\mu_f, \Sigma_f, \Sigma_w]^T$. $\mathbf{F}_f(q, \theta)$ can be easily obtained from $\mathbf{F}(q, \theta)$ in Equation (4.5) by replacing $M_{h^q}(k)$ with $H(k)$, setting $S_{h^q}(k) = 0$, and dropping all the terms that contain $\Lambda_h(k)$. From Equation (4.1) it is clear that in this case setting $q(s) = q(f) = p(f|g; h)$ (given by Equation (4.6)) leads to maximization of $\mathbf{F}_f(q, \theta)$ with respect to $q(f)$. In the M-step of the first stage, in order to maximize $\mathbf{F}_f(q, \theta)$ with respect to θ it suffices to maximize $E_{p(f/g;h)}(\log p(g, f))$ since the entropy term is not a function of θ. Thus, the first stage reduces to the "classical" EM for the linear model $g = Hf + w$, also known as the "iterative Wiener filter"; see, for example, [3]. In the second stage of the VAR3 method, the role of f and h is interchanged and the computations are similar. In other words, the variational bound $\mathbf{F}_h(q, \theta)$ (where $\theta = [\mu_h, \Sigma_h, \Sigma_w]^T$) is obtained from $\mathbf{F}(q, \theta)$ in Equation (4.5) by replacing $M_{f^q}(k)$ with $F(k)$, $S_{f^q}(k) = 0$, and dropping all the terms that contain $\Lambda_f(k)$. The parameters of $p(h|g; f)$ in this case are updated by Equation (4.7).

As for the VAR3 approach the M-step updates specified in Equation (4.8) still hold for both stages. However, the update of $\Lambda_w^{(t+1)}(k)$ in the stage where h is considered deterministic and known is obtained from Equation (4.9) by following the same rules as the ones used to obtain $\mathbf{F}_f(q, \theta)$ from $\mathbf{F}(q, \theta)$. This yields the update

$$\Lambda_w^{(t+1)}(k) = \frac{1}{N} \left(|G(k)|^2 - 2Re\left\{ M_{f^q}^{(t+1)}(k) H(k) G^*(k) \right\} \right) \\ + |H(k)|^2 \left(S_{f^q}^{(t+1)}(k) + \frac{1}{N} \left| M_{f^q}^{(t+1)}(k) \right|^2 \right). \tag{4.11}$$

Similarly the update $\Lambda_w(k)$ in the stage where f is considered deterministic and known is

$$\Lambda_w^{(t+1)}(k) = \frac{1}{N} \left(|G(k)|^2 - 2Re\left\{ F(k) M_{h^q}^{(t+1)}(k) G^*(k) \right\} \right) \\ + |F(k)|^2 \left(S_{h^q}^{(t+1)}(k) + \frac{1}{N} \left| M_{h^q}^{(t+1)}(k) \right|^2 \right). \tag{4.12}$$

It is worth noting that the VAR3 approach, since it uses linear models, can be also derived without the variational principle by applying the "classical" EM (iterative Wiener filter) twice, once for f using as the data generation model $g = Hf + w$ with H known, and once for h using as the data generation model $g = Fh + w$ with F known. From a Bayesian inference point of view, clearly VAR3 *is suboptimal* since it alternates between the assumptions that f is random and h deterministic and vice-versa, as we have explained in the introduction.

4.4 Numerical Experiments

In our experiments we used a simultaneously autoregressive (SAR) model [18] for the image; in other words we assumed $p(f) \propto (\alpha)^{\frac{N-1}{2}} \exp\left(-\frac{1}{2}\alpha \|Qf\|^2\right)$ where Q is the circulant matrix such that Qf represents the convolution of the image f with the discrete Laplacian operator. This model basically assumes that the output $Qf = \varepsilon$ of the Laplacian operator is an independent, identically distributed Gaussian signal. For h we assume $p(h) = N(m_h, \beta^2 I)$, and for the noise $p(n) = N(0, \sigma^2 I)$. Therefore the parameters to be estimated are α, m_h, β, and σ^2.

The following five approaches have been implemented and compared:

- the variational method VAR1;

- the variational method VAR2 (with $q(f) = p(f|h, g)$ and $q(h) = p(h|f, g)$);

- the variational approach VAR3 in which h and f are estimated in an alternating fashion. Since the VAR3 approach, in contrast with the VAR1 and VAR2 methods, does not use a "full Bayesian" model, it serves as the comparison benchmark for the value of such model;

- the Bayesian approach for partially known blurs (PKN) as described in [9];

- the iterative Wiener filter (ITW) as described in [3] where only the parameters α and σ^2 are estimated. The ITW, since it does not attempt to estimate the PSF, is expected to give always inferior results.

As a metric of performance for both the estimated image and the PSF the improvement in the signal-to-noise ratio ($ISNR$) was used. This metric is defined for the image as $ISNR_f = \log 10 \frac{\|f-g\|^2}{\|f-\hat{f}\|^2}$, where $\hat{f}$ is the restored image and for the PSF as $ISNR_h = \log 10 \frac{\|h-h_{in}\|^2}{\|h-\hat{h}\|^2}$ where h_{in} and $\hat{h}$ are the initial guess and the estimate of the PSF, respectively. Two series of experiments were performed: first with PSFs that were partially known, in other words corrupted with random error, and second with PSFs that were completely unknown.

4.4.1 Partially Known Case

Since in many practical cases the PSF is not completely unknown, in this series of experiments we consider that the PSF is partially known [8–10], i.e., it is the sum of a deterministic component and a random component: $h = h_0 + \Delta h$. The Bayesian model that we use in this chapter includes the partially known PSF case as a special case. Thus, in this experiment we compared the proposed variational approaches with previous Bayesian formulations designed for this problem. The deterministic component h_0 was selected to have a Gaussian shape with support 31×31 pixels given by the formula $h_0(k, m) = \exp\left(\frac{k^2}{\sigma_X^2} + \frac{m^2}{\sigma_Y^2}\right)$ with $k, m = -15 \dots 15$ that is also normalized to one such that $\sum_{k=-15}^{15} \sum_{m=-15}^{15} h_0(k, m) = 1$. The width and the shape of the Gaussian are defined by the variances which were set $\sigma_X^2 = \sigma_Y^2 = 20$. For the random component Δh we used white Gaussian noise with $p(\Delta h) = N(0, \beta^2 I)$. In these experiments, since $m_h = h_0$ is known, the parameters to be estimated are α, β, and σ^2.

The following three cases were examined where in each case a degraded image was created by considering the following values for the noise and the PSF: i) $\sigma = 10^{-2}$, $\beta = 10^{-4}$; ii) $\sigma = 10^{-3}$, $\beta = 10^{-4}$; iii) $\sigma = 10^{-4}$, $\beta = 10^{-4}$. In all experiments and for all tested methods the initial values of the parameters were $\widehat{\alpha} = 500, \widehat{\beta^2} = 10^{-7}, \widehat{\sigma^2} = 10^{-5}$. The obtained $ISNR_f$ values of the restored images are summarized in Table 4.1. Table 4.1 clearly indicates the superior restoration performance of the proposed variational methods (VAR1 and VAR2) as compared with both the partially known (PKN) method and the VAR3 approach. As expected, the improvement becomes more significant when the standard deviation of the PSF noise β becomes comparable to the standard deviation of the additive noise σ. Also, as the noise in the PSF becomes larger the benefits of compensating for the PSF increase as compared to using the ITW. It must be noted that, as also reported in [9], the PKN method is very sensitive to initialization of β and σ and it did not converge in the third experiment. It is also interesting to mention that the first two variational schemes provide similar reconstruction results in all tested cases. In Figure 4.2 we provide the images for the case $\sigma = 10^{-3}$, $\beta = 10^{-4}$.

4.4.2 Unknown Case

In this series of experiments we assumed that the PSF is unknown; however, an initial estimate is available. In this experiment an additional image was used to test the proposed algorithm. The initial estimate is the PSF that was used for restoration with the iterative Wiener (ITW) and as initialization of the PSF mean for the three variational (VAR1, VAR2, VAR3) methods. More specifically, the degraded image was generated by blurring with a Gaussian shaped PSF h_{true} as before and additive Gaussian noise with variance $\sigma_g^2 = 10^{-6}$. The initial PSF estimate h_{init}^1 was also assumed Gaussian shaped but with different variance than those used to generate the images. Furthermore, the support of the true PSF is unknown. For this experiment the unknown parameters to be estimated are α, m_h, β, and σ^2. The PKN method was not tested for this set of experiments since it is expected to yield suboptimal results because it is based on a different PSF model. Two cases were examined and the results are presented in Table 4.2 along with the obtained *ISNR* values after 500 iterations of the algorithm. The PSF initializations h_{init}^1 and h_{init}^2 for these two experiments were chosen such that $\left\| h_{true} - h_{init}^1 \right\| = \left\| h_{true} - h_{init}^2 \right\|$, where h_{true} is the true PSF which we are trying to infer.

In Figures 4.3 and 4.4 we provide the images for cases 1 and 2 of Table 4.2. In Figures 4.5 (a–c) we show the values of the function $\mathbf{F}(q, \theta)$ and the values of $ISNR_f$ and $ISNR_h$ for the VAR1 approach as a function of the iteration number. In Figures 4.5 (d–f) we show a cross section in the middle of the 2-D estimated, initial and true PSFs for the VAR1, VAR2, and VAR3 approaches. Everything in Figure 4.4 refers to case 1 of Table 4.2. In Figures 4.6 (a–c) we show the value of the function $\mathbf{F}(q, \theta)$ as a function of the iteration number for the VAR1, VAR2 and VAR3 approaches, respectively, for case 2 of Table 4.2. In Figures 4.6 (d–f) we show a cross section in the middle of the 2-D estimated, initial, and true PSFs for the VAR1, VAR2, and VAR3 approaches, respectively, for the case 2 of Table 4.2. In Figure 4.7 we show the images which resulted from the experiments tabulated in Table 4.3 case 1, where the "Lena" image has been used.

From this set of numerical experiments it is clear that the VAR1 approach is superior to both the VAR2 and VAR3 approaches in terms of both $ISNR_f$ and $ISNR_h$. This is expected since both VAR2 and VAR3 are suboptimal in a certain sense: VAR2 since in the E-step it does not explicitly optimize $\mathbf{F}(q, \theta)$ with respect to $q(s)$, and VAR3 since it does not use the "full Bayesian" as previously explained. Nevertheless, in all our experiments all methods increased monotonically the variational bound $\mathbf{F}(q, \theta)$. This is somewhat surprising since the VAR2 method does not optimize $\mathbf{F}(q, \theta)$ in the E-step and the VAR3 method optimizes $\mathbf{F}_f(q, \theta)$ and $\mathbf{F}_h(q, \theta)$ in an alternating fashion.

4.5 Conclusions and Future Work

In this chapter the blind image deconvolution (BID) problem was addressed using a Bayesian model with priors for both the image and the point spread function. Such a model was deemed necessary to reduce the degrees of freedom between the estimated signals and the observed data. However, for such a model, even with the simple Gaussian priors that were used in this chapter, it is impossible to write explicitly the probabilistic law that relates the convolving functions given the observations required for Bayesian inference. To bypass this difficulty a variational approach was used and we derived three algorithms which solved the proposed Bayesian model. We demonstrated with numerical experiments that the proposed variational BID algorithms provide superior performance in all tested scenarios compared with previous methods. The main shortcoming of the variational methodology is the fact that there is no analytical way to evaluate the tightness of the variational bound. Recently methods based on Monte Carlo sampling and integration have been proposed to address this issue [23]. However, the main drawback of such methods is on the one hand computational complexity and on the other hand convergence assessment of the Markov Chain. Thus, clearly this is an area where more research is required in order to implement efficient strategies to evaluate the tightness of this bound. Furthermore, further research on methods to optimize this bound is also necessary. In spite of this, the proposed methodology is quite general and it can be used with other Bayesian models for this and other imaging problems.

APPENDIX A: Computation of the Variational Bound $\mathbf{F}(q,\theta)$

From Equation (4.1) we have that

$$\mathbf{F}(q,\theta) = E_q\left(\log p(x,s)\right) + H(q) \tag{A.1}$$

where $q(s) = q(f)q(h) = N(m_{f^q}, C_{f^q})N(m_{h^q}, C_{h^q})$, $p(x,s) = p(g,f,h) = p(g|f,h) \cdot p(f) \cdot p(h)$ (with $p(g|h,f) = N(h*f, \Sigma_w)$) and $H(q) = E_q(\log q(s))$ the entropy of $q(s)$.

The implementation of the variational EM requires the computation of the Gaussian integrals appearing in Equation (A.1). The integrand of the first part of Equation (A.1) is given by

$$\log p(g,f,h) = \log p(g|f,h) + \log p(f) + \log p(h)$$

$$= K_1 - \frac{1}{2}\left\{ \log|\Sigma_w| + \underbrace{(g - h*f)^t \Sigma_w^{-1} (g - h*f)}_{b1} + \log|\Sigma_f| \right.$$

$$\left. + \underbrace{(f - \mu_f)^t \Sigma_f^{-1} (f - \mu_f)}_{b2} + \log|\Sigma_h| + \underbrace{(h - \mu_h)^t \Sigma_h^{-1}(h - \mu_h)}_{b3} \right\} \tag{A.2}$$

where K_1 is a constant. The terms that are not constant in this integration with respect to the hidden variables are called $E_q(b_i)$ with $i = 1, 2, 3$. These terms can be computed as

$$
E_q\left(b_1\right) = E_q\left(\underbrace{g^t \Sigma_w^{-1} g}_{I_1} - \underbrace{(h * f)^t \Sigma_w^{-1} g - g^t \Sigma_w^{-1}(h * f)}_{I_2} \overset{+}{} \underbrace{(h * f)^t \Sigma_w^{-1}(h * f)}_{I_3}\right).
$$
(A.3)

These are the terms that must be integrated with respect to $q(h)$ and $q(f)$. The last one using the interchangeability of the convolution and its matrix vector representation is given by

$$
\begin{aligned}
E_q\left(I_3\right) &= \iint (h * f)^t \Sigma_w^{-1}(h * f) \cdot q(h) \cdot q(f) \cdot df \cdot dh \\
&= \int \left(\int f^t \cdot H^t \cdot \Sigma_w^{-1} \cdot H \cdot f \cdot q(f) \cdot df \right) \cdot q(h) \cdot dh \\
&= \int \left(\int trace\left(H^t \cdot \Sigma_w^{-1} \cdot H \cdot f \cdot f^t\right) \cdot q(f) \cdot df \right) \cdot q(h) \cdot dh \\
&= \int trace\left(H^t \cdot \Sigma_w^{-1} \cdot H \cdot \left(C_{f^q} + m_{f^q} m_{f^q}^t\right)\right) \cdot q(h) \cdot dh.
\end{aligned}
$$
(A.4)

To compute this integral we resort to the fact that these matrices are circulant and have common eigenvectors given by the discrete Fourier transform (DFT). Furthermore, for a circulant matrix C it holds that $WCW^{-1} = \Lambda$, where Λ is the diagonal matrix containing the eigenvalues and W the DFT matrix. This decomposition can be also written as $\frac{1}{N} WCW^* = \Lambda$, where W^* denotes the conjugate since $W^{-1} = \frac{1}{N} W^*$; see, for example, [3]. Using these properties of circulant matrices we can write

$$
\begin{aligned}
E_q\left(I_3\right) &= \int \sum_{k=0}^{N-1} \left(\frac{|H(k)|^2}{\Lambda_w(k)} \left(S_{f^q}(k) + \frac{1}{N} |M_{f^q}(k)|^2 \right) \right) \cdot q(h) \cdot dh \\
&= \sum_{k=0}^{N-1} \frac{\left(S_{f^q}(k) + \frac{1}{N} |M_{f^q}(k)|^2 \right) \left(S_{h^q}(k) + \frac{1}{N} |M_{h^q}(k)|^2 \right)}{\Lambda_w(k)}.
\end{aligned}
$$
(A.5)

In the above equation $S_{f^q}(k)$, $S_{h^q}(k)$, and $\Lambda_w(k)$ are the eigenvalues of the covariance matrices C_{f^q}, C_{h^q}, and Σ_w. The $M_{f^q}(k)$ and $M_{h^q}(k)$ are the DFTs of the vectors m_{f^q} and m_{h^q}, respectively. The remaining terms $E_q\left(I_1 + I_2\right)$ of (A.3) can be computed similarly:

$$
E_q\left(I_2 + I_1\right) = \frac{1}{N} \sum_{k=0}^{N-1} \left\{ \frac{|G(k)|^2 + M_{f^q}(k) M_{h^q}(k) G^*(k) + M_{f^q}^*(k) M_{h^q}^*(k) G(k)}{\Lambda_w(k)} \right\}.
$$
(A.6)

As a result, for the term $E_q(b_1)$ we can write

$$
E_q(b_1) = \sum_{k=0}^{N-1} \left\{ \frac{\frac{1}{N}\left(|G(k)|^2 + M_{f^q}(k)M_{h^q}(k)G^*(k) + M_{f^q}^*(k)M_{h^q}^*(k)G(k)\right)}{\Lambda_w(k)} \right.
$$
$$
\left. + \frac{N\left(S_{f^q}(k) + \frac{1}{N}|M_{f^q}(k)|^2\right)\left(S_{h^q}(k) + \frac{1}{N}|M_{h^q}(k)|^2\right)}{\Lambda_w(k)} \right\}. \quad (A.7)
$$

The other terms $E_q(b_2)$ and $E_q(b_3)$ are similarly computed as

$$
E_q(b_2) = \sum_{k=0}^{N-1} \left\{ \frac{\left(S_{f^q}(k) + \frac{1}{N}|M_{f^q}(k)|^2\right)}{\Lambda_f(k)} \right.
$$
$$
\left. + \frac{\frac{1}{N}\left(M_f^*(k)M_{f^q}(k) + M_f(k)M_{f^q}^*(k) + |M_f(k)|^2\right)}{\Lambda_f(k)} \right\} \quad (A.8)
$$

and

$$
E_q(b_3) = \sum_{k=0}^{N-1} \left\{ \frac{S_{h^q}(k) + \frac{1}{N}|M_{h^q}(k)|^2}{\Lambda_h(k)} \right.
$$
$$
\left. + \frac{\frac{1}{N}M_h^*(k)M_{h^q}(k) + M_h(k)M_{h^q}^*(k) + |M_h(k)|^2}{\Lambda_h(k)} \right\}. \quad (A.9)
$$

The computation of $H(q)$ is easy because of the Gaussian choice for $q(f)$ and $q(h)$. In essence we have to compute the sum of the entropies $E_q(J_i)$ with $i = 1, 2$ of two Gaussian PDFs which is given by

$$
H(q) = -E_q \left\{ \underbrace{\log q(f)}_{J_1} + \underbrace{\log q(h)}_{J_2} \right\} =
$$
$$
E_q(J_1) + E_q(J_2) = -C + N + \frac{1}{2}\left(\log|C_{f^q}| + \log|C_{h^q}|\right). \quad (A.10)
$$

Replacing (A.7) through (A.10) into (A.2) results in Equation (4.5) for $\mathbf{F}(q, \theta)$.

APPENDIX B: Maximization of $\mathbf{F}(q, \theta)$

We wish to maximize $\mathbf{F}(q, \theta)$ with respect to parameters θ_q and θ where θ_q are the parameters that define $q(\cdot)$. Since we are not bound by the EM

TABLE 4.2:   Final ISNRs of estimated images and PSF with the Trevor image.

	case 1			case 2		
	Generating PSF Initialization			Generating PSF Initialization		
	$\sigma_x^2 = 20, \sigma_y^2 = 20$	$\sigma_x^2 = 12, \sigma_y^2 = 12$		$\sigma_x^2 = 20, \sigma_y^2 = 20$	$\sigma_x^2 = 20, \sigma_y^2 = 20$
	$ISNR_f$	$ISNR_h$		$ISNR_f$	$ISNR_h$
VAR1	3.18dB	7.45dB		1.63dB	2.92dB
VAR2	1.8dB	-6.54dB		1.59dB	2.36dB
VAR3	2.24dB	-0.59dB		1.53dB	2.52dB
ITW	2.25dB	NA		-15.7dB	NA

TABLE 4.3:   Final ISNRs of estimated images and PSF with the Lena image.

	case 1			case 2		
	Generating PSF Initialization			Generating PSF Initialization		
	$\sigma_x^2 = 20, \sigma_y^2 = 20$	$\sigma_x^2 = 12, \sigma_y^2 = 12$		$\sigma_x^2 = 20, \sigma_y^2 = 20$	$\sigma_x^2 = 20, \sigma_y^2 = 20$
	$ISNR_f$	$ISNR_h$		$ISNR_f$	$ISNR_h$
VAR1	3.94dB	7.23dB		3.37dB	4.81dB
VAR2	2.37dB	-4.87dB		3.14dB	2.87dB
VAR3	2.68dB	-1.28dB		2.43dB	2.69dB
ITW	2.73dB	NA		-18.01dB	NA

framework that contains E and M steps, we can do this optimization in any way we wish. However, in analogy to the EM framework we have adopted the following two steps that we call E and M steps:

E-step (update of θ_q):

$$\theta_q^{t+1} = \arg \max_{\theta^q} \left\{ F \left(\theta_q, \theta^t \right) \right\}$$

M-step (update of θ):

$$\theta^{t+1} = \arg \max_{\theta} \left\{ F \left(\theta_q^{t+1}, \theta \right) \right\}$$

In the M-step, in order to find the parameters θ that maximize F, we need to find the derivatives $\frac{\partial F(q,\theta)}{\partial \theta}$ and set them to zero. From Equation (4.5) we have

$$\frac{\partial F (q, \theta)}{\partial \Lambda_w(k)} = 0 \Rightarrow \frac{1}{\Lambda_w(k)} - \frac{A_1(k) + A_2(k)}{(\Lambda_w(k))^2} = 0 \Rightarrow \Lambda_w(k) = A_1(k) + A_2(k)$$
$$\text{for } k = 0, 1 \ldots N - 1.$$

Similarly, we get $\Lambda_f(k) = B(k)$ and $\Lambda_h(k) = C(k)$ for $k = 0, 1 \ldots N - 1$. $\frac{\partial F(q,\theta)}{\partial M_f(k)} = 0 \Rightarrow M_f(k) = M_{f^q}(k)$ and $\frac{\partial F(q,\theta)}{\partial M_h(l)} = 0 \Rightarrow M_h(l) = M_{h^q}(l)$ for $k = 0, 1 \ldots N - 1$.

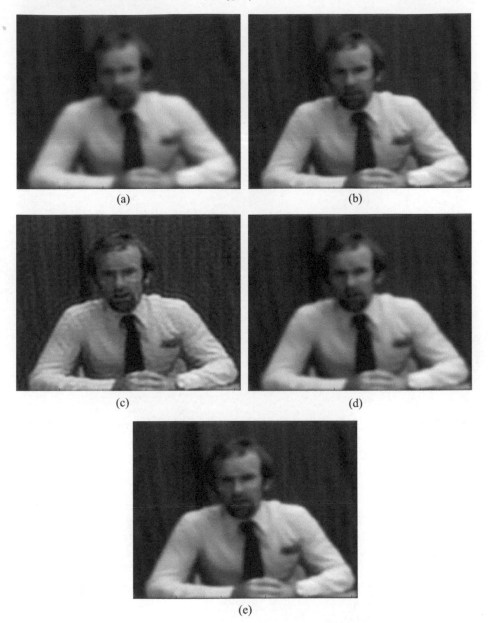

FIGURE 4.3: Trevor images referring to Table 4.2 case 1: (a) degraded; (b) ITW, $ISNR_f = 2.25 \ dB$; (c) VAR1, $ISNR_f = 3.18 \ dB$; (d) VAR2, $ISNR_f = 1.8 \ dB$; (e) VAR3, $ISNR_f = 2.24 \ dB$.

Thus we can compute the unknown parameters $\theta^{(t+1)}$ as

$$M_f^{(t+1)}(k) = M_{f^q}^{(t+1)}(k) \tag{B.1}$$

FIGURE 4.4: Trevor images referring to Table 4.2 case 2: (a) ITW, $ISNR_f = -15.7\ dB$; (b) VAR1, $ISNR_f = 1.63\ dB$; (c) VAR2, $ISNR_f = 1.59\ dB$; (d) VAR3, $ISNR_f = 1.53\ dB$.

$$M_h^{(t+1)}(k) = M_{h^q}^{(t+1)}(k) \tag{B.2}$$

$$\begin{aligned}
\Lambda_f^{(t+1)}(k) &= \left(S_{f^q}^{(t+1)}(k) + \frac{1}{N}\left| M_{f^q}^{(t+1)}(k)\right|^2 \right) \\
&\quad + \frac{1}{N}\left(\left| M_f^{(t+1)}(k)\right|^2 - 2Re\left\{ \left(M_f^{(t+1)}(k)\right)^* M_{f^q}^{(t+1)}(k) \right\} \right) \\
&= S_{f^q}^{(t+1)}(k)
\end{aligned} \tag{B.3}$$

For similar reasons

$$\Lambda_h^{(t+1)}(k) = S_{h^q}^{(t+1)}(k) \tag{B.4}$$

$$\begin{aligned}
\Lambda_w^{(t+1)}(k) &= \frac{1}{N}\left(|G(k)|^2 - 2Re\left\{ M_{f^q}^{(t+1)}(k) M_{h^q}^{(t+1)}(k) G^*(k) \right\} \right) \\
&\quad + \left(S_{f^q}^{(t+1)}(k) + \frac{1}{N}\left| M_{f^q}^{(t+1)}(k)\right|^2 \right)\left(S_{h^q}^{(t+1)}(k) + \frac{1}{N}\left| M_{h^q}^{(t+1)}(k)\right|^2 \right)
\end{aligned} \tag{B.5}$$

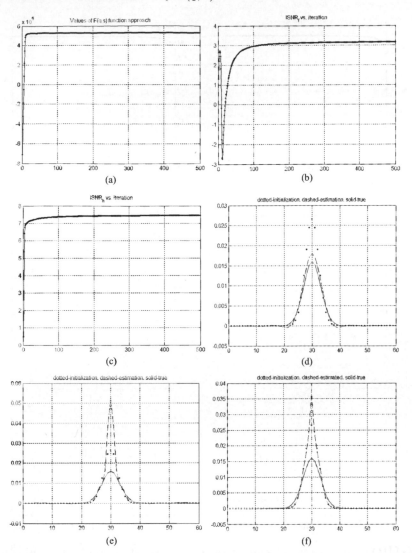

FIGURE 4.5: Table 4.2 case 1: (a) $\mathbf{F}(q, \theta)$ vs. iteration VAR1; (b) $ISNR_f$ vs. iteration VAR1; (c) $ISNR_h$ vs. iteration VAR1; (d) 1-D cross section of the 2-D PSFs VAR1; (e) 1-D cross section of the 2-D PSFs VAR2; (f) 1-D cross section of the 2-D PSFs VAR3.

for $k = 0, 1 \ldots N - 1$.

In our experiments we have used an SAR prior [12] for the image model, thus $p(f) \propto (\alpha)^{\frac{N-1}{2}} \exp\left(-\frac{1}{2} \|Qf\|^2\right)$, $p(h) = N(m_h, \beta I)$, and $p(n) = N(0, \sigma^2 I)$, where Q is the circulant matrix that represents the convolution with the Laplacian operator. Therefore, the unknown parameter vector θ to be estimated

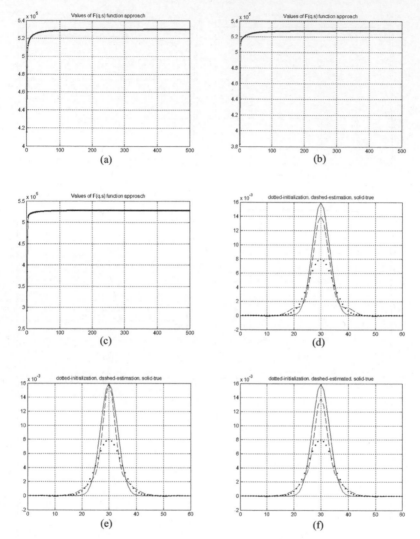

FIGURE 4.6: Table 4.2 case 2: (a) $\mathbf{F}(q, \theta)$ vs. iteration VAR1; (b) $\mathbf{F}(q, \theta)$ vs. iteration VAR2; (c) $\mathbf{F}(q, \theta)$ vs. iteration for VAR3; (d) 1-D cross section 2-D PSFs VAR1; (e) 1-D cross section 2-D PSFs VAR2; (f) 1-D cross section 2-D PSFs VAR3.

contains the parameters α, β, and σ^2. Because of the circulant properties it holds that: $\Lambda_f(k) = \frac{1}{\alpha|Q(k)|^2}$, $\Lambda_h(k) = \beta$, and $\Lambda_w(k) = \sigma^2$. Based on these assumptions, the general Equations (B.1) through (B.5) for the updates at the M-step take the specific form:

(a)

(b)

(c)

(d)

(e)

FIGURE 4.7: Lena images referring to Table 4.3 case 1: (a) degraded; (b) ITW, $ISNR_f = 2.73\ dB$; (c) VAR1, $ISNR_f = 3.94\ dB$; (d) VAR2, $ISNR_f = 2.37\ dB$; (e) VAR3, $ISNR_f = 2.68\ dB$.

M-step

$$\alpha^{(t+1)} = \left[\frac{1}{N-1}\sum_{k=0}^{N-1}\left(S_{f^q}^{(t+1)}(k) + \frac{1}{N}\left|M_{f^q}^{(t+1)}(k)\right|^2\right)|Q(k)|^2\right]^{-1} \qquad (B.6)$$

$$\beta^{(t+1)} = \frac{1}{N}\sum_{k=0}^{N-1}\left(S_{f^q}^{(t+1)}(k) + \frac{1}{N}\left|M_{f^q}^{(t+1)}(k)\right|^2\right) \qquad (B.7)$$

$$\left(\sigma^2\right)^{(t+1)} =$$
$$\frac{1}{N}\sum_{k=0}^{N-1}\left(|G(k)|^2 - 2Re\left\{M_{f^q}^{(t+1)}(k)M_{h^q}^{(t+1)}(k)G^*(k)\right\}\right)$$
$$+ \sum_{k=0}^{N-1}\left(S_{f^q}^{(t+1)}(k) + \frac{1}{N}\left|M_{f^q}^{(t+1)}(k)\right|^2\right)\left(S_{h^q}^{(t+1)}(k) + \frac{1}{N}\left|M_{h^q}^{(t+1)}(k)\right|^2\right)$$
$$(B.8)$$

As for the VAR3 approach the updates for α and β remain the same. However, to obtain the updates for the noise variance we apply the same rules that were previously used to obtain the variational bounds $\mathbf{F}_f(q,\theta)$ and $\mathbf{F}_h(q,\theta)$ from the bound $\mathbf{F}(q,\theta)$ in Equation (4.5).

For the VAR1 approach, the update equations for the parameters θ_q of $q(s)$ (which are complex in the DFT domain) are easily obtained after by equating the corresponding gradient of $\mathbf{F}(q,\theta)$ to zero. This yields the following update equations for $k = 0,\ldots,N-1$:

E-step (VAR1 approach)

$$Re(M_{f^q}^{(t+1)}(k)) = \frac{Re(M_{h^q}^{(t)}(k))Re(G(k)) + Im(M_{h^q}^{(t)}(k))Im(G(k))}{\alpha\sigma^2|Q(k)|^2 + NS_{h^q}^{(t)}(k) + |M_{h^q}^{(t)}(k)|^2} \qquad (B.9)$$

$$Im(M_{f^q}^{(t+1)}(k)) = \frac{-Im(M_{h^q}^{(t)}(k))Re(G(k)) + Re(M_{h^q}^{(t)}(k))Im(G(k))}{\alpha\sigma^2|Q(k)|^2 + NS_{h^q}^{(t)}(k) + |M_{h^q}^{(t)}(k)|^2} \qquad (B.10)$$

$$S_{f^q}^{(t+1)}(k) = \frac{\sigma^2}{\alpha\sigma^2|Q(k)|^2 + NS_{h^q}^{(t)}(k) + |M_{h^q}^{(t)}(k)|^2} \qquad (B.11)$$

$$Re(M_{h^q}^{(t+1)}(k)) =$$
$$\frac{\beta\left[Re(M_{f^q}^{(t+1)}(k))Re(G(k)) + Im(M_{f^q}^{(t+1)}(k))Im(G(k))\right] + \sigma^2 Re(M_h^{(}k))}{\sigma^2 + \beta NS_{f^q}^{(t+1)}(k) + \beta|M_{f^q}^{(t+1)}(k)|^2}$$
$$(B.12)$$

$$Im(M_{h^q}^{(t+1)}(k)) =$$

$$\frac{\beta\left[Re(M_{f^q}^{(t+1)}(k))Im(G(k)) - Im(M_{f^q}^{(t+1)}(k))Re(G(k))\right] + \sigma^2 Im(M_h(k))}{\sigma^2 + \beta N S_{f^q}^{(t+1)}(k) + \beta|M_{f^q}^{(t+1)}(k)|^2}$$

$$\text{(B.13)}$$

$$S_{h^q}^{(t+1)}(k) = \frac{\beta\sigma^2}{\sigma^2 + \beta N S_{f^q}^{(t+1)}(k) + \beta|M_{f^q}^{(t+1)}(k)|^2} \qquad \text{(B.14)}$$

References

[1] D. Kundur and D. Hatzinakos, "Blind image deconvolution," *IEEE Signal Processing Magazine,* pp. 43-64, May 1996.

[2] R. L Lagendijk, J. Biemond, and D.E. Boekee, "Identification and restoration of noisy blurred images using the expectation-maximization algorithm," *IEEE Transactions on Acoustics, Speech, and Signal Processing,* Vol. 38, No. 7, pp. 1180-1191, July 1990.

[3] A. K. Katsaggelos and K. T. Lay, "Maximum likelihood identification and restoration of images using the expectation-maximization algorithm", A. K. Katsaggelos (editor), *Digital Image Restoration,* Ch. 6, New York: Springer-Verlag, 1991.

[4] A. K. Katsaggelos and K. T. Lay, "Maximum likelihood blur identification and image restoration using the EM algorithm," *IEEE Transactions on Signal Processing,* Vol. 39 No. 3, pp: 729-733, March 1991.

[5] Y. Yang, N. P. Galatsanos, and H. Stark, "Projection based blind deconvolution," *Journal of the Optical Society of America-A,* Vol. 11, No. 9, pp. 2401-2409, September 1994.

[6] Y. L. You and M. Kaveh, "A Regularization approach to joint blur identification and image restoration," *IEEE Transactions on Image Processing,* Vol. 5, pp. 416-428, March 1996.

[7] Y. L. You and M. Kaveh, "Blind image restoration by anisotropic regularization," *IEEE Transactions on Image Processing,* Vol. 8, pp. 396-407, March 1999.

[8] V. N. Mesarovic, N. P. Galatsanos, and M. N. Wernick, "Iterative LMMSE restoration of partially-known blurs," *Journal of the Optical Society of America-A,* Vol. 17, pp. 711-723, April 2000.

[9] N. P. Galatsanos, V. N. Mesarovic, R. M. Molina, and A. K. Katsaggelos, "Hierarchical Bayesian image restoration from partially-known blurs,"

IEEE Transactions on Image Processing, Vol. 9, No. 10, pp. 1784-1797, October 2000.

[10] N. P. Galatsanos, V. N. Mesarovic, R. M. Molina, J. Mateos, and A. K. Katsaggelos, "Hyper-parameter estimation using gamma hyper-priors in image restoration from partially-known blurs," *Optical Engineering*, 41(8), pp. 1845-1854, August 2002.

[11] A. D. Dempster, N. M. Laird, and D. B. Rubin, "Maximum likelihood from incomplete data via the EM algorithm," *Journal of the Royal Statistical Society*, Vol. B39, pp 1-37, 1977.

[12] C. Robert, *The Bayesian choice: from decision-theoretic foundations to computational implementation*, Springer-Verlag; 2nd edition, June 2001.

[13] B. Carlin and T. Louis, *Bayes and empirical Bayes methods for data analysis*, CRC Press; 2nd edition, 2000.

[14] A. M. Djafari, Editor, *Bayesian inference for inverse problems*, Proceedings of SPIE-The International Society for Optical Engineering, Vol. 3459, July 1998.

[15] K. H. Yap, L. Guan, and W. Liu, "A recursive soft decision approach to blind image deconvolution," *IEEE Transactions on Signal Processing*, Vol. 51, No. 2, pp. 515-526, February 2003.

[16] R. M. Neal and G. E. Hinton, "A view of the EM algorithm that justifies incremental, sparse and other variants," In M. I. Jordan (ed.) *Learning in Graphical Models*, pp. 355-368. Cambridge, MA: MIT Press, 1998.

[17] M. I. Jordan, Z. Ghahramani, T. S. Jaakola, and L. K. Saul, "An introduction to variational methods for graphical models," in *Learning in Graphical Models* (editor M.I. Jordan), pp. 105-162, Cambridge, MA: MIT Press 1998.

[18] R. Molina and B. D. Ripley, "Using spatial models as priors in astronomical images analysis," *Journal of Applied Statistics*, Vol. 16, pp. 193-206, 1989.

[19] T. S. Jaakkola, *Variational methods for inference and learning in graphical models*, Ph.D. Thesis, MIT, 1997.

[20] M. Cassidy and W. Penny, "Bayesian nonstationary autoregressive models for biomedical signal analysis," *IEEE Transactions on Biomedical Engineering*, Vol. 49, No. 10, pp. 1142-1152, October 2002.

[21] J. W. Miskin and D. J. C. MacKay, "Ensemble learning for blind image separation and deconvolution," in *Advances in Independent Component Analysis* (Ed. by Girolami, M). Springer-Verlag Scientific Publishers, July 2000.

[22] Z. Ghahramani and M. J. Beal, "Variational inference for Bayesian mixtures of factor analysers," in *Advances in Neural Information Processing Systems*, 12:449-455, MIT Press, 2000.

[23] M. J. Beal, "Variational algorithms for approximate Bayesian inference", PhD. Thesis, Gatsby Computational Neuroscience Unit, University College, London, 2003.

[24] K. T. Lay and A. K. Katsaggelos, "Image identification and restoration based on the expectation-maximization algorithm," *Optical Engineering*, Vol. 29, pp. 436-445, May 1990.

[25] J. Mateos, R. Molina, and A.K. Katsaggelos, "Approximations of posterior distributions in blind deconvolution using variational methods," *IEEE International Conference on Image Processing*, 2005. ICIP 2005, Volume 2, 11-14 Sept. 2005 Page(s):II - 770-3

[26] A. C. Likas and N. P. Galatsanos, "A variational approach for Bayesian blind image deconvolution," *IEEE Transactions on Signal Processing*, Vol. 52, No. 8, pp. 2222-2233, Aug. 2004.

5

Deconvolution of Medical Images from Microscopic to Whole Body Images

Oleg V. Michailovich

Department of Electrical and Computer Engineering
University of Alberta, Edmonton, AB T6G 2V4, Canada
email:olegm@ece.ualberta.ca

Dan R. Adam

Department of Biomedical Engineering
Technion - Israeli Institute of Technology, Haifa 32000, Israel
e-mail: dan@bm.technion.ac.il

Abstract

Medical images – images created for the purpose of diagnosing and providing treatment – are generally acquired at either microscopic (cellular, tissue) or macroscopic (organ, whole-body) levels by radiation interacting with living-tissue, where the radiation can be either ionizing or pertaining to waves. In many practical settings, this interaction can be approximated by a convolution model, in which the *point spread function* (PSF) of the imaging system blurs the data obtained from the tissue, thereby necessitating the procedure of image reconstruction by means of *deconvolution*. While in some cases, either an experimentally measured PSF or its analytical prediction may be used for the restoration, in most cases, only few *a priori* assumptions may be made regarding the characteristics of the PSF. As a result, the methods for deconvolving medical images with an unknown PSF have become of great practical relevance. Such methods of *blind deconvolution* have long been extensively used in medical imaging, and a myriad of their variations nowadays exist. This chapter provides an overview of a number of well-known blind deconvolution algorithms, with special attention paid to the methodologies that are distinctive to medical imagery. A substantial part of this chapter is therefore concerned with the problem of blindly deconvolving medical ultrasound imaging – a topic that has promoted the development of many generic methods of blind deconvolution, which are applicable to a wide spectrum of other imaging modalities. In addition, this chapter attempts to provide some assessment of future deconvolution tools that are required for processing biological data acquired by some of the new macro and micro systems used for imaging small organs and cells.

5.1   Introduction

5.1.1   Medical Imaging: Tendencies and Goals

Medical imaging is a scientific field concerned with reproducing and displaying anatomical structures and physiological functions of the human body. The clinical use of medical imaging, both for the purpose of diagnosis and guidance of treatment, has shown a constantly increasing scope and importance. Several imaging modalities have nowadays become a standard and indispensable part of quality healthcare available to large segments of the population in nearly all hospitals all over the world. At the same time, improving the quality and diagnostic reliability of medical imaging has always represented a continuous challenge for many scientists working in the field. Thus, before discussing *post-processing* methods designed to this end, it is imperative to understand the reasons for their development. These reasons turn out to be diverse and multilateral in several aspects. Below, however, only three of them are described, those which we recognize as the most important ones.

Unfortunately, the progress and proliferation of many lethal (predominatingly, oncological) diseases is surprisingly fast, and by the time a lesion can be detected by one of the many diagnostic imaging modalities, it is frequently too late for saving the life. This fact brings up the importance of *early detection* of the disease as the key to the survival of patients. The earlier detection implies an improved resolution of the scanners being used for diagnosis. The resolution, however, is usually subject to limitations imposed by the physical properties of the modality in use and by technical considerations underpinning the process of data acquisition. In such cases, in order to improve the imaging resolution so as to allow the clinicians to diagnose a disease at its early stages of development, efficient post-processing methods are needed.

Another major reason for the development of post-processing methods for medical imagery originates from some recent tendencies in the world market of medical equipment. Since this market tends to expand towards the health institutions of developing countries with large populations, the affordability of medical equipment there seems to have assumed the key role. Consequently, an overall tendency is created for selection of relatively inexpensive imaging devices as the modality of choice. Unfortunately, since in many cases such inexpensive imaging devices employ relatively simple physical phenomena, their image quality is often worse than those of the more sophisticated scanners. This fact creates the need for boosting the performance of cost-effective scanners via incorporation of post-processing methods into the process of image formation and display.

The third reason for the need to compensate for the reduced quality of medical imaging is related to applications. When either a transient phenomena or moving organs are being imaged, the amount of acquired information is often (deliberately) reduced so as to allow the data processing units to keep pace with the changes of the environment. Naturally, this information loss deteriorates the quality of the resulting images. Similarly, the miniaturization of imaging devices, which is intended to make the latter more portable (and less expensive), often causes deterioration of the image quality as well. In both cases mentioned above, post-processing can enhance the resolution and contrast of the medical diagnostic images.

All the reasons provided above advocate the necessity of developing efficient signal processing methods for improving the resolution and contrast in medical imagery, and we believe that these reasons are sufficient to justify the efforts invested in improving the quality of medical diagnosis as a crucial component of quality healthcare.

5.1.2 Linear Modeling of Image Formation

Most methods of image reconstruction are based on an *image formation model*, which relates properties of the interrogated biological tissue (e.g., its transmissivity, emissivity, reflectivity, conductivity, and so on) to the values of the corresponding observed image. Many important physical phenomena underlying such measurements can be reasonably accounted for by linear models. In many cases, however, the real interaction between the tissue and the probing field is nonlinear, and, hence, the linearization of models is a simplification intended to allow the images to be analyzed and reproduced using feasible computational resources.

Under the assumption of linearity, it is convenient to describe the relationship between the quantity of interest and its observation by means of *linear operators*. Specifically, let y be a function/vector representing the tissue property to be recovered and let $\mathbf{H}$ be the linear operator that accounts for the effect of the imaging system on y. Since ideal measurements are impossible in real life, the observed signal g is always contaminated by noise u, which we assume to be additive throughout the rest of this chapter. Consequently, the image formation model can now be simply defined as given by:

$$g = \mathbf{H}(y) + u. \tag{5.1}$$

In order to make the presentation formal, a number of functional specifications need to be made first. Since in the modern analysis using computers, the processed signals are discrete and finite, it is reasonable to assume all functions under consideration to be members of $\ell_2(\mathbb{Z}^d)$ (i.e., the space of square-summable sequences of finite ℓ_2-norm) with $d \in \{1, 2, 3\}$. Note that this class of functions is the most common to use for describing the discretized versions of scalar valued functions defined over a real line, plane, or volume, depending on the value of d. Moreover, although in reality the function y represents a physical quantity, for the sake of generality it will be referred to below as the *true* image.

The fundamental problem of estimating the values of y using the actual results of the measurements of g is termed the *inverse problem*. In the case when $\mathbf{H}$ is a convolution operator, the resulting inverse problem is known as *the problem of deconvolution* [1, 2], the solution of which is in the main focus of the material presented here.

For the case $d = 2$, the convolution model of image formation is defined as:

$$g[n, m] = \mathbf{H}(y[n, m]) + u[n, m] = \sum_k \sum_l h[n - k, m - l]\, y[k, l] + u[n, m], \tag{5.2}$$

where $h[n, m]$ is generally referred to as the *convolution kernel*, or (as long as imaging is concerned) the *point spread function* (PSF) of the imaging system. Note that setting $d = 2$ should not be seen as a restriction, as most of the considerations below will be valid for higher (as well as lower) dimensional cases as well.

A particularly challenging subset of deconvolution problems results when little (if at all) *a priori* information is available on the PSF $h[n, m]$. In this case, the solution of the deconvolution problem requires the estimation of both the true image $y[n, m]$ and the PSF $h[n, m]$ directly from an observation of $g[n, m]$, and, hence, the resultant problem is referred to as *blind*. Before presenting some key approaches to the solution of this problem in medical imaging, it is instructive to review several examples of medical imaging modalities, in which the problem of blind deconvolution naturally arises.

5.1.3 Blind Deconvolution in Medical Ultrasound Imaging

The use of ultrasound in medicine, which goes back to the 1950s, is the basis for several procedures that are widespread in today's clinical practice. Most of these procedures, including medical imaging, are presently used for diagnosis and supervision, but there is an ever-increasing number of therapeutical applications as well [3]. Among the most important advantages of medical ultrasound imaging are its noninvasive nature, availability, cost-effectiveness, its being practically harmless (as no ionizing radiation is used), and mobility. Unfortunately, all these advantages come at a price – the reduced resolution and contrast, as compared to those of some alternative imaging modalities, such as x-ray Computed Tomography (CT) or Magnetic Resonance Imaging (MRI). The most significant factors that limit the resolution of ultrasound imaging are the finite bandwidth of the signals produced by the ultrasound transducers and the nonnegligible width of the acoustic beams.

To better understand the origin of this resolution limitation, let us take a closer look at the mechanism of formation of ultrasound images. The present clinical ultrasound scanners process signals in a manner similar to that of sonar or radar units. To interrogate a tissue, the ultrasound probe produces a (pulsed) acoustic pressure field. The field propagates through the tissue and is partially reflected and scattered due to the inherent inhomogeneity of most tissues. The backscattered signal is received usually by the same probe, supplying useful information about the locations of tissue inhomogeneities and their relative "strengths."[1]

Ultrasound waves, reflected from biological tissues may, in general, be represented by a three-dimensional (3-D) function that describes the strengths of the acoustic reflectors and scatterers as a function of their spatial coordinates. Usually, this function is referred to as the spatial response of the insonified tissue, or *tissue reflectivity function* [4,5]. Employing the assumption of weak scattering (which permits applying the first Born approximation [6]), it can be shown that the reflectivity function and the backscattered signal can be related to each other by a simple Fourier transform relationship [7]. Formally, the acoustic pressure measured by the transducer can be described as given by [5,8]:

$$P(\mathbf{r}, t) = v(t) *_t y(\mathbf{r}) *_{\mathbf{r}} h(\mathbf{r}, t), \qquad (5.3)$$

where $\mathbf{r} \in \mathbb{R}^3$ and $t \in \mathbb{R}$ are the space and time variables. In (5.3), $*_{\mathbf{r}}$ and $*_t$ correspondingly denote the spatial and temporal convolution; $v(t)$ is the pulse-echo impulse response, which includes the transducer electrical excitation and the transducer electromechanical impulse response during emission and reception of the pulse; $h(\mathbf{r}, t)$ is the modified pulse-echo spatial impulse response that relates the transducer geometry to the spatial extent of the scattered field; and $y(\mathbf{r})$ is the tissue reflectivity function.

The model (5.3) is general, and it is usually reduced when 2-D imaging is considered. In order to acquire a 2-D image, a focused ultrasound pulse-pressure wave (frequently referred to as the *ultrasound beam*) is transmitted in a certain direction in the tissue being investigated. Due to the inhomogeneity of the tissue, part of the wave energy is backscattered or reflected towards the ultrasound transducer, where it is sensed as a composite return

---

[1]Typically, small changes in density, compressibility, and absorption give rise to a scattered wave radiating in all directions, whereas boundary reflections are encountered from the diaphragm, blood vessel walls, and organ boundaries.

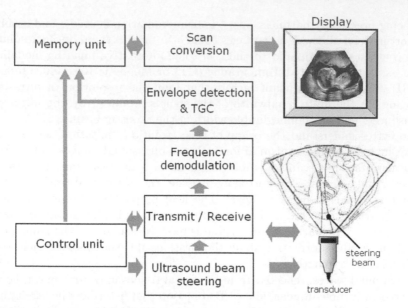

FIGURE 5.1:   Block diagram of a B-scan imaging system.

signal formed as a superposition of acoustic echoes. The piezoelectric elements of the transducer convert the return signal into a voltage trace, whose digitized version (commonly referred to as a *radio frequency line* [RF-line]) encodes the information about the tissue inhomogeneity pertaining to the chosen direction. Subsequently, the above process is repeated for a different propagation direction, resulting in another RF-line. By planar sweeping the ultrasound beam with a certain increment within a predefined plane, a set of RF-lines can be collected and, subsequently, used for visualization of the tissue inhomogeneity structure corresponding to the chosen crosssection. Note that this set of RF-lines can be organized and stored as a matrix, which is commonly referred to as a *radio frequency image* (RF-image). As only the low-frequency portion of the information contained in RF-images is useful for visualization of the tissue inhomogeneity structure, it is standard to subject each RF-line (composing a given RF-image) to the processes of frequency demodulation and sample-rate reduction. The demodulated RF-lines are usually referred to as *in-phase/quadrature lines* (IQ-lines), while the image they form is referred to as an *in-phase/quadrature image* (IQ-image). As a final point, the *envelope* (i.e., absolute value) of the IQ-image is computed, amplified by a time-gain compensation (TGC) amplifier, decimated, scan-converted, and displayed. The overall process of image formation described above forms the basis of operation of B-scan ultrasound systems, a standard blockdiagram of which is depicted in Figure 5.1.

The formation of an RF-image can be described by the convolution model of (5.2), when the tissue properties are assumed to be approximately uniform in the direction perpendicular to the scanning plane, within the extent of the ultrasound beam. In this case, the RF-image can be viewed as a result of the convolution of a 2-D tissue reflectivity function with the 2-D PSF of the imaging system. The convolution model implies that the received RF-image can be considered as a distorted version of the reflectivity function. Indeed, the convolution effect appears as a pronounced blurring of various tissue boundaries and fine structures, as is usually observed in ordinary ultrasound images. Moreover, the blurring tends to obscure very small structures, thereby deteriorating the reliability of many tissue characterization schemes. Thus, the task of recovery of the reflectivity function by means of deconvolution is of great clinical importance.

It should be noted, however, that even if the tissue under study is composed of weak scatterers alone, the stationary convolution model may not describe well the formation of a whole RF-image. The main reason is that, as long as *in vivo* imaging is of concern, the PSF may change considerably due to the presence of interrogated tissues between the transducer and the target. The *dispersive attenuation* and *phase aberrations* are among the main effects that contribute to the spatial variability of the PSF [9–11]. This variability considerably complicates the reconstruction procedure, making preliminary measurements or calibrations useless. In this case, the PSF should be estimated along with the tissue reflectivity function, which transforms the reconstruction problem at hand into a problem of blind deconvolution.

5.1.4 Blind Deconvolution in Single Photon Emission Computed Tomography

Single photon emission computed tomography (SPECT) images are obtained by measuring the radiation (viz., gamma rays) coming from radioactive isotopes injected into the human body. Specifically, a source of internal radiation is administered into the patient's blood system in the form of a radiopharmaceutical substance, i.e., pharmaceutics labeled with a radioactive isotope. The gamma rays emitted by the decaying radioactive isotope are registered, and the resulting information can be used for measuring and visualizing the level of activity within the studied tissues and organs. Thus, the SPECT imaging provides functionally based rather than anatomically based information, such as the metabolic behavior of organs (e.g., human brain, liver, lymph glands). One of the most important applications of SPECT is in neurology, where it is used for diagnosing some severe cerebrovascular diseases and brain disorders (e.g., Alzheimer's disease, Parkinson's disease, and so on) based on the metabolic activity in certain brain regions.

After a radiopharmaceutical (sometimes called a *tracer*) is administered into the patient's blood system, the gamma ray emission is detected by means of a *gamma camera*, which is rotated around the patient. In this manner, the emission distribution can be recorded in the form of *projections* (as obtained for multiple angles), which are subsequently used to recover the 3-D distribution of the radiopharmaceutical using a tomographic reconstruction algorithm.

A typical gamma camera consists of a collimator, a detector crystal, a photomultiplier tube array, position logic circuits, and a data analysis computer. Here, the collimator acts as the "front end" of the imaging system, projecting the gamma rays onto the detector crystal. Having passed through the collimator, the gamma ray photons interact with the iodide ions of the detector crystal by means of the *photoelectric effect* [12] that causes the release of electrons, which, in turn, interact with the crystal lattice to produce light. This process is known as *scintillation*. Afterward, the photomultiplier tubes (attached to the back of the detector crystal) transform the scintillation into a stream of electrons, which is converted into an electrical pulse at their cathode end. As a final stage, the resulting pulses are analyzed by the position logic circuits to determine where each scintillation event occurred in the detector crystal.

Due to the specific process of image formation, however, SPECT imaging is prone to errors caused by both poor statistics and poor spatial resolution. Because of the relatively low sensitivity of the collimator and the low dose of the injected radiopharmaceutical, the number of photons produced by the photoelectric effect is usually small, and, as a result, statistical errors are relatively large. Additionally, the intrinsic resolution of the camera is limited, which, along with the scattering of the emitted photons, constitute the main factors that result in blurred images, typical to SPECT cross sections. The blurring considerably impedes the interpretation of these images, making this process difficult, labor intensive, and subjective. The resolution of SPECT images can be improved by the application of an image reconstruction procedure, which can be formulated as a blind deconvolution problem [13].

In order to solve the deconvolution problem in a stable and reliable manner, an accurate estimate of the PSF must be provided. Some preliminary works on this subject used the Gaussian blur model with some success [14]. However, the results of such algorithms seem to be of limited usage, due to their high sensitivity to the assumptions made regarding the model of blur.

In theory, the PSF can be measured directly from the SPECT camera by visualizing the response of a point source against a uniform background. Unfortunately, in practice such experiments are generally difficult to realize, and, moreover (even if realized), their results are often imprecise due to limitations of the experimental setup. As a consequence, there comes a need for blind deconvolution methods, which are nowadays considered to be among the most useful post-processing tools in SPECT imagery.

5.1.5 Blind Deconvolution in Confocal Microscopy

Confocal microscopy is currently considered to be a substantial advancement over regular light microscopy, as it allows creation of 3-D images of cells and tissues as well as following specific cellular reactions over extended periods of time [15]. Although the original idea employed by this imaging modality dates back to the late 1950s, its further development and popularity seem to have received an additional significant impetus over the last two decades due to some recent advances in laser technologies and computing. These advances have allowed the true power of confocal microscopy to become available to wide segments of the research community, specifically to biologists that recognize this tool as the most powerful and versatile tool for visualizing living cells – an area of study where other forms of light microscopy have great difficulty in producing high-quality images. Some other fields where confocal microscopy is widely used include immunolabeling, organelle identification, protein trafficking, analyzing molecular ability, locating genes on chromosomes, as well as measuring membrane potentials and ion concentrations. When compared to conventional light microscopy, confocal microscopy has several advantages due to its exceptionally high resolution, extremely high sensitivity (that allows collecting single-molecule fluorescence), and the ability to efficiently coalign multichannel fluorescence. Moreover, the ability to perform optical sectioning, the 3-D nature of the image formation, the improved multiple labeling, the digitization, and the controllability by modern computational means are among many other useful features of this imaging modality.

There are many design aspects to confocal microscopes that make them much more versatile than conventional fluorescence microscopes. In its basic form, a confocal microscope consists of a conventional light microscope (typically equipped with epifluorescence optics), one or more lasers for irradiating the sample, a scan head for scanning the laser beam across the sample, a controller box containing much of the optics and some of the electronics needed for acquiring the images, and finally a computer for both controlling the microscope and collecting the images.

In *laser scanning confocal microscopy*, the image is created by scanning a diffraction-limited point of excitation (laser) light across the sample in a raster formation. The irradiating laser light is used to excite a suitable fluorophore that can be either applied to or naturally occurring within the sample. A small amount of the fluorescent light emitted from the sample passes back through the objective and gets separated from light of a second fluorophore, or from unwanted reflected excitation light, by means of partially reflecting dichroic mirrors.

The out-of-focus light, which originates from unwanted focal planes and has the potential of destroying the image quality, is rejected by focusing the fluorescent light through a small aperture, called a *"pinhole."* In this way, light rays from other focal planes are not correctly aligned with the confocal "pinhole" and are therefore eliminated from the image. At the same time, the "in-focus" light, after passing through the confocal "pinhole," is detected by a photomultiplier tube, which produces the corresponding analogue signal. The latter is subsequently converted to a digital format that usually contains both the information on the position of the laser in the image and the amount of light coming from the sample. Note that the digitization results in a grayscale image, which can be colored by a suitable computer algorithm. It should be noted that the confocal "pinhole" is a principal element of confocal microscopy that allows the creation of images of exceptional resolution and contrast. This ability to remove out-of-focus light has the effect of producing an image that consists of information from only a very thin focal plane and, for this reason, is often referred to as an "optical slice." A series of such slices can then be combined to visualize the 3-D structure of the original sample.

As stated above, a conventional microscope constitutes the heart of the confocal microscope. Just as in the conventional light microscopy, the objective lens forms the image in the confocal microscopy. Consequently, the quality of the final image, and particularly its resolution, are dependent on the quality of the objective lens used, as well as on the numerical aperture of the objective and the wavelength of the light. Although confocal microscopy pushes the limit of resolution to the theoretical limit of light microscopy, a confocal microscope does not resolve better than about 0.1 μm. Unfortunately, this limit in resolution is very important in biology, as many subcellular structures are at, or bellow, this size.

The resolution limitation in confocal microscopy can be modeled using the convolution model (5.2), with $y[n, m]$ being the true object (fluorescence) to be reconstructed and $h[n, m]$ being the PSF of the imaging system, which represents the combined effect of the optical blurring and electronic filtering. Although the functional form of the PSF can be predicted from the geometry of the objective and the shape of the "pinhole" aperture, this prediction is often inaccurate, as it may not take into account the effects imposed by the hardware as well as by the design imperfections. In this case, the reconstruction should be performed in a blind manner, which defines the importance of blind deconvolution to this imaging modality. It is worthwhile noting that this problem seems to have attracted a considerable interest of several software developers, and, as a result, a number of deconvolution packages for confocal microscopy are currently available on the market [16, 17].

5.1.6   Organization of the Chapter

Due to the space limitations, it is impossible to review all the medical imaging modalities that take advantage of blind deconvolution as the preferred method for enhancing the image resolution and contrast. Thus, several important medical applications, such as x-ray CT [18] and optical coherent tomography [19], have been left out. However, we believe that the blind deconvolution methods presented in this chapter (as well as those presented in this book) are general enough to be applicable to a wide range of medical imaging modalities. Furthermore, since in most cases, the methods designed for blindly deconvolving general (i.e., nonmedical) images can be applied to medical imaging as well, we decided to focus on those algorithms that seem to be distinctive to *medical* scanners. For this reason, the main part of the discussion below is dedicated to medical ultrasound, where the topic of blind deconvolution seems to have triggered the proposal of a number of generic approaches. Subsequently, this material is supplemented by a discussion on blind deconvolution methods used in some other medical imaging modalities. However, before turning to describing these methods, it is essential to review some basic approaches to solving the problem of *non*blind deconvolution, which we briefly review in the next section.

5.2 Nonblind Deconvolution

5.2.1 Regularization via Maximum *a Posteriori* Estimation

The problem of nonblind deconvolution arises when the true image $y[n, m]$ in (5.2) should be recovered from its observation $g[n, m]$ given the *full* knowledge about the PSF $h[n, m]$. In this case, the most convenient way to tackle this problem is a statistical estimation framework. In particular, the theory of *maximum likelihood* (ML) estimation [20] can be readily employed to this end, where the true image is considered as a parameter array, whose optimal value is computed as a *maximizer* of the *probability density function* (*PDF*) (commonly referred to as the *likelihood*) of $g[n, m]$. It is worthwhile noting that the optimality of the ML estimation is supported by the well-known Likelihood Principle [21], which states that in making either inferences or decisions about a parameter after data are observed, all relevant experimental information are contained in the likelihood function of the observed data.

Unfortunately, there are important practical cases in which the ML estimation framework may be of little use. For example, let the samples of the noise $u[n, m]$ in (5.2) be independent, identically distributed (*i.i.d.*) and obeying a Gaussian probability distribution with zero mean. It can then be easily shown that the ML estimate of the true image $y[n, m]$ is obtained as the solution to the following linear equation:

$$\mathbf{R}(y)[n, m] = \mathbf{H}^\star(g)[n, m], \qquad (5.4)$$

where $\mathbf{H}^\star$ denotes the *adjoint* of the convolution operator $\mathbf{H}$, and the operator $\mathbf{R}$ is defined as a composition of $\mathbf{H}$ and its adjoint, viz., $\mathbf{R}(\cdot) \equiv \mathbf{H}^\star(\mathbf{H}(\cdot))$. Let us recall that in ℓ_2, the adjoint $\mathbf{A}^\star$ of a linear operator $\mathbf{A}$ is defined by means of the equality $\langle x_1, \mathbf{A}(x_2) \rangle = \langle \mathbf{A}^\star(x_1), x_2 \rangle$, where $x_1, x_2 \in \ell_2$, and $\langle x_2, x_2 \rangle$ denotes the standard inner product, i.e., $\langle x_1, x_2 \rangle \equiv \sum_n \sum_m x_1[n, m] \, x_2[n, m]$ (where both x_1 and x_2 are assumed to be real valued). Using these definitions, it is straightforward to show that if the operator $\mathbf{H}$ convolutes its argument with $h[n, m]$, then the adjoint operator $\mathbf{H}^\star$ convolutes its argument with a space-reversed version of $h[n, m]$, viz., with $h[-n, -m]$. Consequently, the operator $\mathbf{R}$ is nothing else but the convolution with the autocorrelation function $r[n, m]$ of $h[n, m]$, defined simply as $r[n, m] \equiv h[n, m] * h[-n, -m]$.

Unfortunately, it can be rigorously proven that when the kernel $h[n, m]$ is bandlimited in the domain of its Fourier transformation, the system (5.4) cannot be solved in a unique and stable manner [1], and, hence, the problem of inversing $\mathbf{H}$ is *ill-posed*. It should be noted that this property is not related to a specific solution method being used, but is inherent to the problem itself. The bandlimitedness of $h[n, m]$ implies that the "out-of-band" spectral components of $y[n, m]$ are attenuated by $\mathbf{H}$, making the process of their recovery extremely unstable.

In order to render the inverse problem well-posed and reduce the sensitivity of its solution to noise, some additional information needs to be incorporated into the reconstruction procedure. This can be achieved by using a more general probabilistic point of view, in which the function $y[n, m]$ to be recovered is considered as a random variable characterized by its *prior* distribution. In this case, the *maximum a posteriori* (MAP) estimation [22] yields the most likely solution, given the observed data and a reasonable assumption regarding the statistical nature of the true image. Specifically, given $\mathbf{H}$ and an observed $g[n, m]$, the MAP estimator searchers for a maximizer of the *a posteriori* probability $P(y \mid g, \mathbf{A})$ that can be expressed using the Bayes rule as:

$$P(y \mid g, \mathbf{A}) \propto p(g \mid y, \mathbf{A}) \Pr(y), \tag{5.5}$$

where $p(g \mid y, \mathbf{H})$ is the data likelihood function and $\Pr(y)$ is the prior probability of the true image.

It turns out that, in numerous cases of practical interest, the problem of maximizing the posterior probability in (5.5) is equivalent to the problem of minimizing the following cost functional [1, 2]:

$$E(y) = \langle w_1, \rho_1(\mathbf{H}(y) - g) \rangle + \lambda \langle w_2, \rho_2(\mathbf{T}(y)) \rangle, \tag{5.6}$$

where $\rho_1 : \mathbb{R} \to \mathbb{R}$ and $\rho_2 : \mathbb{R} \to \mathbb{R}$ are convex, low-semicontinuous functions, $w_1[n, m]$ and $w_2[n, m]$ are two nonnegative *weighting* sequences, $\mathbf{T} : \ell_2 \to \ell_2$ is a predefined operator (usually of a differential type), and $\lambda > 0$ is a *regularization* (scalar) parameter, which is intended to balance the mutual influence of the two terms of (5.6).

It should be noted that the first term in (5.6) represents all the relevant experimental information defined by the data likelihood, while the second term comes from assumptions or guesses concerning the prior distribution of the true image $y[n, m]$. Note that this second term is intended to compensate for the information loss incurred as a result of the rank deficiency (i.e., ill-posedness) of $\mathbf{H}$, and, hence, it plays the role of a *regularization* term. Moreover, it can be rigorously proven that, whenever the weighting sequences $w_1[n, m]$ and $w_2[n, m]$ are positive and $\mathbf{T}$ is a bijection, the problem of minimizing $E(y)$ as given by (5.6) has a unique and stable solution [23]. Before describing some common methods of minimizing the cost functional $E(y)$, let us first present some special cases.

For the case of both the samples of the noise $u[n, m]$ and those of the true image $y[n, m]$ being assumed to be mutually independent *i.d.* Gaussian random variables, the functional (5.6) becomes:

$$E(y) = \|\mathbf{H}(y) - g\|_2^2 + \lambda \|y\|_2^2, \tag{5.7}$$

where $\| \cdot \|_2$ stands for the standard ℓ_2-norm.

Let $G(\omega_1, \omega_2)$ and $H(\omega_1, \omega_2)$ denote the Fourier transforms of $g[n, m]$ and $h[n, m]$, correspondingly[2]. Then, it can be shown that the Fourier transform $F(\omega_1, \omega_2)$ of the optimal solution of (5.7) is given by [1]:

$$F(\omega_1, \omega_2) = \frac{G(\omega_1, \omega_2) H^*(\omega_1, \omega_2)}{|H(\omega_1, \omega_2)|^2 + \lambda}, \tag{5.8}$$

where the asterisk denotes the complex conjugate. Therefore, for the case of (5.7), the deconvolution can be performed using analytical computations, by virtue of (5.8). This type of deconvolution is well known as *Wiener filtering* [24]. Moreover, for this case, the optimal value of the regularization parameter λ is equal to the ratio of the variance of the noise to that of the true image, and, for this reason, it is referred to as the *inverse signal-to-noise ratio* (ISNR).

The main shortcoming of this reconstruction method is the generation of Gibbs artifacts in the vicinity of the discontinuities of the true image $y[n, m]$. Besides, being a linear reconstruction method in nature, Wiener filtering is incapable of interpolating the information that is lost during the image formation process. Consequently, the Wiener solutions are frequently oversmoothed.

The disadvantages of the Wiener filtering mentioned above can be overcome by using *non*-Gaussian statistical priors for the true image. For example, when the samples of the true image are assumed to be *i.i.d.* Laplacian random variables and the noise is white and Gaussian, the cost functional becomes:

$$E(y) = \|\mathbf{H}(y) - g\|_2^2 + \lambda \sum_n \sum_m |y[n, m]| = \|\mathbf{H}(y) - g\|_2^2 + \lambda \|y\|_1, \tag{5.9}$$

where $\|y\|_1$ denotes the ℓ_1-norm of $y[n, m]$. It is interesting to note that Laplacian random variables are fairly similar to Gaussian ones, except that due to the larger density in the tails of the Laplacian *PDF*, there is a greater allowance for a few occasional large-amplitude samples to occur. Consequently, the Laplacian model is especially suitable for describing tissues, which are composed of diffusive scatterers superimposed with a few strong specular reflectors (e.g., liver arterioles) [25, 26].

In a similar way, when the true image is piecewise smooth, its gradient $\nabla y[n, m]$ is expected to be sparse. In this case, the *total variation* (TV) deconvolution [27] seems to be the most appropriate method to estimate the

---

[2]Here and hereinafter, by the Fourier transform of a discrete sequence $x[n, m]$ we mean the trigonometric series $X(\omega_1, \omega_2) = \sum_n \sum_m x[n, m] \exp\{-\imath(\omega_1 n + \omega_2 m)\}$.

true image. Specifically, the TV cost functional is defined by:

$$E(y) = \|\mathbf{H}(y) - g\|_2^2 + \lambda \sum_n \sum_m |\nabla y[n,m]| = \|\mathbf{H}(y) - g\|_2^2 + \lambda \|y\|_{TV}, \quad (5.10)$$

where $\| \cdot \|_{TV}$ denotes the TV-norm. It should be noted that the procedures of minimizing (5.9) and (5.10) are nonlinear. As a result, these procedures are often capable of recovering the sharp and miniature details of the true image $y[n,m]$, which have been blurred out by the convolution with $h[n,m]$. This feature makes the above estimation methods preferable over linear deconvolution, when the "sharpness" of the deconvolved images is of primary concern. Unfortunately, these methods require iterative computations, and, hence, in the situations when fast processing is required, the analytical solution by (5.8) could be a more realistic choice.

5.2.2 Numerical Optimization via Newton Method

A standard approach to minimizing the functional (5.6) is by using the *Newton method*, which is known to be among the most efficient tools for unconstrained optimization [28]. The standard Newton minimization of the functional $E(y)$ consists in iteratively finding its minimizer according to:

$$y^{k+1} = y^k + a_k \, d^k, \quad (5.11)$$

where a_k is determined via the *line search* $a_k = \arg\min_a E(y + ad^k)$ with d^k being the *Newton direction*, which is computed as a solution of the following system of equations:

$$\nabla^2 E_{y^k}(d^k) = -\nabla E_{y^k}. \quad (5.12)$$

Here, ∇E_{y^k} and $\nabla^2 E_{y^k}$ denote the gradient and the Hessian of the functional $E(y)$, evaluated at the iteration point y^k. It is important to note that due to the use of second-order information on $E(y)$, the Newton algorithm often provides a quadratic rate of convergence near the optimal point.

In order to complete the description of the procedure for minimizing the cost functional $E(y)$, its gradient and Hessian need to be specified. Let $\rho_1'(y)$ and $\rho_2'(y)$ denote the first derivatives of the functions $\rho_1(y)$ and $\rho_2(y)$, respectively. Then, the gradient $\nabla E_y \in \ell_2$ of $E(y)$ can be shown to be given by:

$$\nabla E_y = \mathbf{H}^\star \left(w_1 \cdot \rho_1'(\mathbf{H}(y) - g) \right) + \lambda \, \mathbf{T}^\star \left(w_2 \cdot \rho_2'(\mathbf{T}(y)) \right), \quad (5.13)$$

where the dot stands for the pointwise multiplication and $\mathbf{T}^\star$ is the adjoint of $\mathbf{T}$.

In an analogous manner it can be shown that if $\rho_1''(y)$ and $\rho_2''(y)$ denote the second derivatives of $\rho_1(y)$ and $\rho_2(y)$, respectively, then the Hessian $\nabla^2 E$: $\ell_2 \rightarrow \ell_2$ is an operator, which *at the coordinate* $x[n,m]$ works on $y[n,m]$ according to:

$$\nabla^2 E_x(y) = \mathbf{H}^\star \left(w_1 \cdot \rho_1''(\mathbf{H}(x) - g) \cdot \mathbf{H}(y) \right) + \lambda \, \mathbf{T}^\star \left(w_2 \cdot \rho_2''(\mathbf{T}(x)) \cdot \mathbf{T}(y) \right). \quad (5.14)$$

It should be noted that in many cases of practical interest, it is impossible to inverse the Hessian analytically, and hence the Newton direction d^k in (5.12) should be computed by means of a suitable iterative procedure. A standard approach here is to solve (5.12) by the *conjugate gradient* algorithm [28], in which case the resulting optimization procedure is known as the *truncated Newton* optimization [29]. Finally, we note that if either $\rho_1(y)$ or $\rho_2(y)$ is chosen to be the *absolute value* function (which is usually the case when the resulting norms are required to be robust), its derivatives are not well defined. In this case, it is common to use a smooth approximation of the absolute value function, which could be defined, e.g., as:

$$|x| \approx \eta(x) = |x/c| - \log(1 + |x/c|), \qquad (5.15)$$

with $c \in (0, 1]$ being a *proximity* parameter obeying $\eta(x) \to |x|$ as $c \to 0^+$.

5.2.3 Blind Deconvolution with Shift-Variant Blurs

Generally, there are three standard approaches to solving the problem of blind deconvolution. First, one can try to estimate the true image and the PSF simultaneously using a properly designed optimization procedure [17]. The second approach is to estimate the PSF first, followed by using the resultant estimate to solve the deconvolution problem in a *nonblind* manner [31]. It should be noted that though methods of the latter group seem to be less general as compared to those of the first group, they are often more computationally efficient. Finally, there is a group of "hybrid" approaches, where only *partial* information on the PSF is recovered first and, subsequently, used to estimate both the true image and the PSF simultaneously [32,33].

Unfortunately, all the methods mentioned above are not directly applicable for the cases when the PSF is spatially variant, and hence the resulting deconvolution problem is *nonstationary*. This situation is typical in ultrasound imaging, where the PSF constantly changes due to the processes of dispersive attenuation and phase aberration. In order to avoid the complexity of solving the problem of nonstationary deconvolution, a number of approximate approaches have been proposed (see, e.g., [34] and the references therein). Perhaps, the simplest approach is to assume that the spatial variability of the PSF is slow varying, and hence well approximable by a piecewise constant function. In this case, the data image can be divided into a number of (possibly overlapped) segments, whose size is chosen to be small enough to guarantee that each of the segments is formed by a *stationary* convolution with a different PSF.

In the case of the methods which estimate the PSF and the true image concurrently, the segments produced as above can be deconvolved independently, followed by combining the results thus obtained. On the other hand, in the case when the PSF is estimated first (either partially or in full), further division of the segments into blocks of a smaller size can be performed as shown by the illustration in Figure 5.2 for the case of ultrasound imaging.

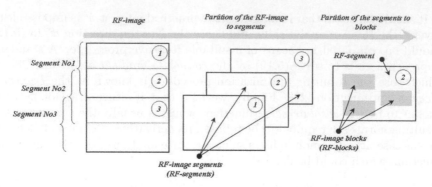

FIGURE 5.2:   Block diagram of the nonadaptive segmentation.

Subsequently, the estimation of the PSF can be performed independently for each block, followed by averaging all the estimates that pertain to the same segment. Thus, at the end of this process, the image segments are available together with their corresponding PSF. The latter can be used for either deconvolving the image segments separately or for recovering the true image as a whole using an approximation of the nonstationary convolution operator [34].

Once the tools have been defined, in the following we present an overview of a number of blind deconvolution methods used in the field of medical imaging, with the first modality to be considered being medical ultrasound imaging.

5.3   Blind Deconvolution in Ultrasound Imaging

As it was pointed out in the preceding subsection, in ultrasound imaging, due to the spatial variability of the PSF, a stationary convolution model can be applied only locally, and, as a result, a standard practice is to employ the segmentation approach of Section 5.2.3. Since usually the same blind deconvolution method is applied to all the image segments, we confine the discussion below to the case of processing a single segment that, with a slight abuse of notation, will be referred to below as an RF-image. Specifically, we assume that such an RF-image $g[n, m]$ is obtained as a result of the convolution of a tissue reflectivity function $y[n, m]$ with the PSF $h[n, m]$ of the ultrasound imaging system, viz.:

$$g[n, m] = (y * h)[n, m] + u[n, m]. \tag{5.16}$$

Here the term $u[n, m]$ is added to account for both experimental and model noises, and the pair $[n, m]$ designates the sample indices, corresponding to the axial and lateral (or radial and angular, for B-scan sector images) directions, respectively.

The various blind deconvolution algorithms that have been proposed so far for recovering $y[n, m]$ and $h[n, m]$ can be categorized into a number of principal groups. In this review, we follow the categorization scheme, in which the methods that are capable of deconvolving the RF-images along their axial and lateral dimensions in a *separable* manner are presented first. Subsequently, a number of *nonseparable* blind deconvolution techniques are reviewed.

5.3.1 Blind Deconvolution via Statistical Modeling

One of the classical methods of blind deconvolution used in medical ultrasound imaging originates from the theory of system identification [35, 36]. Although recent advances in this theory make it possible to deal with systems of virtually arbitrary dimensionality, in medical ultrasound, only 1-D techniques of this kind have been so far in use. In this case, the dimensionality reduction is achieved via a simplified representation of the 2-D PSF as a tensor product of two 1-D convolution kernels. Specifically, it is assumed that $h[n, m] = h_1[n] \, h_2[m]$, where the 1-D kernels $h_1[n]$ and $h_2[m]$ represent the axial (i.e., along the orientation of the transmitted ultrasound beam, into the depth of the tissue) and lateral (i.e., perpendicular to the orientation of the transmitted ultrasound beam, but usually within the scanned plane) PSF, respectively. Note that, while the axial kernel $h_1[n]$ includes the blurring effects due to the finite bandwidth of the transducer and the dispersive attenuation of the acoustical pressure pulse in tissue, the lateral kernel $h_2[n]$ represents the convolutional components of lateral blurring due to the complex beam pattern.

The separability of the PSF allows one to reduce the problem of a 2-D deconvolution to a sequence of 1-D deconvolution problems, applied along each direction independently. Hence, in order to facilitate the discussion, we resort to a simplified, 1-D version of (5.16). Specifically, denoting by $g[n]$, $y[n]$, $h[n]$, and $u[n]$ the 1-D versions of the functions in (5.16), the 1-D model is given by:

$$g[n] = (y * h)[n] + u[n]. \qquad (5.17)$$

In what follows, $h[n]$ may be either the axial or lateral convolution kernel, and, therefore, $g[n]$ may denote either an axial or lateral "slice" of a given RF-image. In ultrasound imaging, however, a common practice is to apply blind deconvolution methods of the type considered next, only along the axial direction [37, 38].

5.3.1.1 Modeling by ARMA Process

The system-based approach to the problem of blind deconvolution considers the PSF $h[n]$ to be the impulse response of a linear time-invariant system that is characterized by the generic difference equation as given by:

$$g[n] = \sum_{k=1}^{p} a[k]\, g[n-k] + \sum_{k=0}^{q} b[k]\, y[n-k]. \tag{5.18}$$

Hence, in this interpretation, the RF-signal $g[n]$ is considered to be the output sequence of a causal filter $h[n]$ (which is identified as the PSF of the ultrasound system in use), while the reflectivity function $y[n]$ is thereby considered to be an input "driving" sequence. The latter is commonly assumed to be a realization of an *i.i.d.* process of zero mean and variance σ_y^2. Moreover, $b[0]$ is normally constrained to be equal to 1, as the input $y[n]$ can always be rescaled so as to account for an arbitrary gain of the filter.

In the theory of system identification, the model of (5.18) is known as the *autoregressive moving average* (ARMA) model, and it constitutes a generalization of the *autoregressive* (AR) and *moving average* (MA) models, which are obtained by setting either $q = 0$ or $p = 0$, respectively. As a side note, it should be mentioned that while the AR model is useful for modeling random signals that possess "peaky" power spectral densities, the MA model is more appropriate for signals that have broad peaks or sharp nulls in their spectra. Needless to say, the ARMA model incorporates the properties of both these special cases, thereby having superior ability to generate diverse spectral shapes.

Adapting (5.18) for modeling RF-sequences implies a simple strategy for their blind deconvolution. First, the PSF $h[n]$ is recovered by estimating the ARMA parameters $\{a[k]\}_{k=1}^{p}$ and $\{b[k]\}_{k=1}^{q}$, since the z-transform $H(z) \equiv \sum_{n=0}^{\infty} h[n]\, z^{-n}$ of $h[n]$ is defined by:

$$H(z) = \frac{B(z)}{A(z)} = \frac{1 + \sum_{k=1}^{q} b[k]\, z^{-k}}{1 + \sum_{k=1}^{p} a[k]\, z^{-k}}. \tag{5.19}$$

Subsequently, the estimate of $h[n]$ can be used for deconvolving $g[n]$ in a nonblind manner, by means of, e.g., any of the methods mentioned in Section 5.2 of this chapter.

Unfortunately, due to their highly nonlinear nature, the number of methods which can be used for reliably and robustly recovering the ARMA parameters is quite limited. One of the standard approaches here is based on the concept of *maximum likelihood estimation* as detailed in [36]. However, the methods of this kind are computationally expensive and not guaranteed to converge to a desired global solution. Consequently, a number of suboptimal methods have been proposed for estimating the ARMA parameters, which are capable to compute the required estimate in a computationally efficient manner [35].

Some of these suboptimal methods take advantage of the equivalency between the ARMA, AR, and MA models. The equivalency suggests that, given any of the above models, it is possible to express that model in terms of the other two models (of, generally, infinite orders). Moreover, for the case of a "pure" AR process, many simple and efficient estimation methods are available [39]. In particular, if $g[n]$ was a real and stationary AR process of zero mean, then its autocovariances $\{r_{gg}[k]\}_{k=0}^{p}$ would be related to the AR parameters $\{a[k]\}_{k=1}^{p}$ by:

$$\sum_{k=1}^{p} r_{gg}[n-k]\,a[k] = -r_{gg}[n], \tag{5.20}$$

where $n = 1, 2, \ldots, p$ and $r_{gg}[k] = r_{gg}[-k]$, $\forall k$. The equation (5.20) is the well-known *Yule–Walker normal system* [39], which can be efficiently solved by the fast Levinson–Durbin algorithm [40]. In practice, of course, the auto-covariance sequence is replaced by its sample estimate.

At this point, a reasonable question comes to mind: is it worthwhile to proceed with the ARMA identification with all of its inherent problems, while having available the efficient AR modeling? Indeed, the principle of equivalency assures that for any ARMA model of a finite dimension, there always exists an AR system, which will produce the output $g[n]$ whose power spectral density is identical to that produced by the ARMA system. For the case of ultrasound imaging, however, the choice of an ARMA model is well justified. This is because the spectra of the RF-lines are normally wide-peaked, and, hence, their accurate approximation would require relatively high orders of the AR model. On the other hand, since the requirement of stationarity imposes limitations on the range of sampling intervals, stationary RF-lines are usually short in duration. Consequently, apart from worsening the computational efficiency, increasing the number of parameters would unavoidably deteriorate the robustness of the estimation as well as its well-posedness, since in a well-posed estimation problem there should be fewer parameters than observations. For these reasons, although using the AR models was considered in the ultrasound literature [41, 42], the ARMA models seem to be preferable, in view of their superior modeling capability for relatively low orders of p and q and for short data segments.

5.3.1.2 Estimation of the ARMA Parameters

The model equivalency principle mentioned above can be used for estimating the ARMA parameters via *separately* estimating the AR and MA constituents of the model. This approach has the advantage of being computationally efficient, as it only requires the solution of few linear systems [35]. Unfortunately, its performance noticeably deteriorates when the number of available data points becomes relatively small, and, since this is usually the case for ultrasound imaging, alternative solutions to the problem of estimating the ARMA parameters have been suggested. Among such alternative methods are those based on minimizing the sum of squares of the *one-step-forward prediction error* within the sampling range. The resulting estimates are commonly referred to as the *least-square* (LS) estimates [43], and they can be proven to converge to the maximum likelihood estimates [36] as the sample size increases. Moreover, when correctly constructed, the LS estimates are guaranteed to fulfill the conditions of stationarity and invertibility. As the methods of this kind seem to have been the most commonly used in medical ultrasound, in the rest of this subsection we linger round some essential details regarding their implementation.

Denoting by $G(z)$ and $Y(z)$ the z-transforms of $g[n]$ and $y[n]$, respectively, the LS cost functional for the ARMA estimation is given by:

$$E_{LS} = \frac{1}{2\pi\imath} \oint Y(z)Y(z^{-1})\frac{dz}{z}, \qquad (5.21)$$

where $Y(z) = \frac{A(z)}{B(z)}G(z)$. Alternatively, E_{LS} can be defined as:

$$E_{LS} = N\frac{\sigma_y^2}{2\pi\imath} \oint \frac{C(z)}{\Gamma(z)}\frac{dz}{z}, \qquad (5.22)$$

where $\Gamma(z) \equiv \sigma_y^2 \frac{B(z)B(z^{-1})}{A(z)A(z^{-1})}$ and $C(z) \equiv N^{-1}G(z)G(z^{-1})$ are the theoretical and empirical *autocovariance generating functions* of the ARMA process, correspondingly, and N denotes the number of sample points composing the RF-line $g[n]$. As the empirical autocovariance generating function $C(z)$ converges to its theoretical counterpart $G(z)$ as N increases, it can be proven that the parameters $\{a[k]\}$ and $\{b[k]\}$, obtained through minimization of E_{LS}, will converge to their true values (provided that $g[n]$ is generated by an ARMA process, of course) [43].

In order to formulate a practical method for computing the LS estimates of the ARMA parameters, we define L to be an $N \times N$ matrix whose *first* $N-1$ columns are equal to the *last* $N-1$ columns of an $N \times N$ identity matrix $I_{N \times N}$, while the last column of L is composed of zeros. Additionally, let us define the following $N \times N$ matrices:

$$A \equiv \sum_{k=0}^{p} a[k]\,L^k,\ B \equiv \sum_{k=0}^{q} b[k]\,L^k,\ G \equiv \sum_{n=0}^{N-1} g[k]\,L^n,\ Y \equiv \sum_{n=0}^{N-1} y[k]\,L^n. \quad (5.23)$$

Assuming $L^0 = I_{N \times N}$ and setting $g[n] = 0$ and $y[m] = 0$ for $n = -p, \ldots, -1$ and $m = -q, \ldots, -1$, respectively, it is straightforward to show that the model (5.18) implies that $A\,G = B\,Y$, and hence a discrete version E_{LS}^N of (5.21) can now be defined as:

$$E_{LS}^N = \mathbf{Tr}[Y^T Y] = \mathbf{Tr}[G^T A^T B^{-T} B^{-1} A G], \qquad (5.24)$$

where T stands for the matrix transpose operation and $\mathbf{Tr}$ denotes the trace of a square matrix. It should be noted that by "artificially" increasing N via zero-padding the data, the cost functional (5.24) can be made to approximate the "ideal" cost functional (5.21) to any degree of accuracy.

In order to find the LS-optimal parameters of the ARMA model, the discrete functional E_{LS}^N should be minimized as a function of the combined vector of parameters $\theta = [a_1, a_2, \ldots, a_p, b_1, b_2, \ldots, b_q]^T$. Traditionally, the minimization is performed using the Gauss–Newton minimization procedure:

$$\theta_{t+1} = \theta_t + \lambda_t \left[\frac{\partial Y^T}{\partial \theta} \frac{\partial Y}{\partial \theta} \right]_t \left[\frac{\partial Y^T}{\partial \theta} Y \right]_t, \qquad (5.25)$$

with Y depending on θ via $Y = B^{-1}AG$ and λ_t being a "step-adjustment" scalar, which is usually set to be equal to 1. The subscript t in (5.25) designates the iteration index, so that, e.g., $\left[\frac{\partial Y}{\partial \theta} \right]_t$ denotes the derivative of Y evaluated at $\theta = \theta_t$.

The Gauss–Newton iterations can be implemented using the `lsqnonlin` routine of MATLAB®. However, in order to use this routine in an efficient manner the derivative $\left[\frac{\partial Y}{\partial \theta} \right]_t$ needs to be known. Denoting by $(B^{-1}G)_{[1:p]}$ and $(B^{-1}Y)_{[1:q]}$ the matrices composed of the first p columns of the matrix $B^{-1}G$ and the first q columns of the matrix $B^{-1}Y$, respectively, it is easy to show that [43]:

$$\frac{\partial Y}{\partial \theta} = [(B^{-1}G)_{[1:p]} \quad - (B^{-1}Y)_{[1:q]}], \qquad (5.26)$$

where $Y = B^{-1}AG$.

The parameters $a[k]$ and $b[k]$ obtained via minimizing the functional (5.24) can be described as *conditional* LS estimators of the ARMA parameters, in the sense specified in [44]. Unfortunately, these estimators are not guaranteed to satisfy the conditions of stationarity and invertibility. Moreover, due to both nonconvexity and nonlinearity of (5.24), the Gauss–Newton procedure is not guaranteed to converge to a desired global minimum. However, it was demonstrated in [37] that the convergence characteristics of (3.10) can be significantly improved if a number of independent measurements of $g[n]$ are incorporated into the estimation process. Such measurements can be obtained as segments of RF-sequences, which compose the same RF-image and correspond to the same penetration depth.

It should be noted that the Gauss–Newton procedure specified above is not the only way to compute the LS-optimal estimates of the ARMA model. There exists an alternative approach as detailed in [35], which is based on a slightly different definition of the cost functional (5.24) and seems to have some computational advantages over (5.25). Yet, neither this method nor the one discussed above is immune to the problem of multiple minima. Moreover, neither of them guarantees that the zeros and poles of the resulting ARMA model will reside inside the unit circle, and hence some stability issues may arise. These problematic aspects of the ARMA modeling become even more pronounced when the estimation is applied to higher-dimensional RF-data. This seems to be the main reason for which 2-D ARMA modeling has never been attempted in medical ultrasound imaging.

Finally, we note that the *square* nature of the LS criterion (5.24) makes the whole estimation procedure prone to errors caused by occasional, large-amplitude samples in the input sequence $y[n]$. Consequently, the LS-based procedure is not robust. The estimation performance, however, can be improved by using only those segments of the data which correspond to reflectivity functions, whose samples can be considered to be *i.i.d.* and Gaussian. (Such data segments can be identified using, e.g., the procedure reported in [45].) The above property of the LS estimation is interesting to note, in view of the fact that the algorithm presented next requites the reflectivity function to be essentially different in nature.

5.3.2   Blind Deconvolution via Higher-Order Spectra Analysis

While the methods for estimating the 1-D PSF $h[n]$ presented in the preceding subsection can be viewed as parametric techniques related to the theory of system identification, the method we present next is a nonparametric representative of the very same theory. This method reconstructs the PSF using the *high-order spectra* (HOS) of RF-signals [8, 46]. As before, the resulting estimate of the PSF can be subsequently used to deconvolve the RF-sequences in a nonblind way.

The HOS-based approach to estimating the PSF has proven to be a powerful alternative to the model-based approaches. As compared to the latter, the HOS-based approach is less sensitive to measurement noises, the properties of which are now explicitly taken into consideration. Moreover, this approach offers some computational advantages, as it is noniterative. A more important fact is that the HOS-based approach has been used for both axial and lateral deconvolution of RF-images, as opposed to the methods based on the ARMA modeling.

The HOS-based approach is applied under the following statistical assumption: 1) the PSF $h[n]$ is a deterministic and mixed-phase sequence; 2) the reflectivity sequence $y[n]$ is stationary, white, zero-mean, and *non-Gaussian*; 3) the noise $u[n]$ is zero-mean Gaussian and independent of $y[n]$. Note that the second assumption is the most crucial here and it defines a conceptual difference between this approach and the model-based method of Section 5.3.1.

In order to formally introduce the HOS-based estimation of the PSF, let $c_g[\tau_1, \tau_2] \equiv \mathcal{E}\{g[n]g[n + \tau_1]g[n + \tau_2]\}$ be the third-order cumulant of a zero-mean observation of the RF-signal $g[n]$. Then, under the above assumptions, the *bispectrum* $C_g(\omega_1, \omega_2)$ of $g[n]$ (which is nothing but the Fourier transform of $c_g[\tau_1, \tau_2]$) is given by [46]:

$$C_g(\omega_1, \omega_2) = C_y(\omega_1, \omega_2)H(\omega_1)H(\omega_2)H^*(\omega_1 + \omega_2) + C_u(\omega_1, \omega_2), \qquad (5.27)$$

where $C_y(\omega_1, \omega_2)$ and $C_u(\omega_1, \omega_2)$ denote the bispectra of $y[n]$ and $u[n]$, respectively, and $H(\omega)$ is the spectrum of $h[n]$. If the additive noise $u[n]$ is zero-mean and Gaussian, then $C_u(\omega_1, \omega_2) = 0$, and (5.27) becomes:

$$C_g(\omega_1, \omega_2) = C_y(\omega_1, \omega_2)H(\omega_1)H(\omega_2)H^*(\omega_1 + \omega_2). \qquad (5.28)$$

The inverse Fourier transform of the logarithm of the bispectrum of a random signal is known as its *bicepstrum*. It follows directly from (5.28) that the bicepstrum $b_g[k, l]$ of $g[n]$ is related to those of $y[n]$ and $h[n]$ (which we denote by $b_y[k, l]$ and $b_h[k, l]$, respectively) as given by:

$$b_g[k, l] = b_y[k, l] + b_h[k, l]. \tag{5.29}$$

Moreover, if the samples of $y[n]$ are *i.i.d.* random variables obeying a *non*-Gaussian probability distribution, then $C_y(\omega_1, \omega_2)$ is a constant function with its amplitude equal to the skewness of the distribution. As a result, the bicepstrum $b_g[k, l]$ will be an impulse located at the origin and, therefore, a scaled and shifted version of $h[n]$ can be recovered from values of $b_g[k, l]$ taken along the main axes *except* at the origin. Specifically, denoting by $\mathcal{F}$ and $\mathcal{F}^{-1}$ the operators of direct and inverse Fourier transform, respectively, an estimate $h[n]$ can be computed according to:

$$h[n] \simeq \mathcal{F}^{-1}\{\exp\{\mathcal{F}\{\tilde{b}_h[n]\}\}\}, \tag{5.30}$$

where

$$\tilde{b}_h[n] = \begin{cases} b_g[n, 0], & \text{if } n > 0; \\ 0, & \text{if } n = 0; \\ b_g[-n, 0], & \text{if } n < 0. \end{cases} \tag{5.31}$$

It should be reemphasized that the HOS-based approach to estimating $h[n]$ is only useful in the case when $g[n]$ is "produced" by a non-Gaussian reflectivity sequence $y[n]$.

In practice, the estimation of the bicepstrum in (5.29) is performed according to the following steps. First, the RF-sequence $g[n]$ is segmented into K records of length N, followed by subtracting the mean values of the resulting subsegments. As the next step, each subsegment $g_j[n]$, $j = 1, 2, \ldots, K$ is used to compute its third-order cumulant $\hat{c}_{g_j}[\tau_1, \tau_2]$ according to:

$$\hat{c}_{g_j}[\tau_1, \tau_2] = N^{-1} \sum_{k=0}^{N-1} g_j[k]\, g_j[k + \tau_1]\, g_j[k + \tau_2], \tag{5.32}$$

where $|\tau_1| \leq M$ and $|\tau_2| \leq M$ with M being the length of the third-order correlation lags (e.g., $M = 10$, [8]). Subsequently, the estimates $\{\hat{c}_{g_j}\}_{j=1}^{K}$ are averaged over j to compute the cumulant estimate $\hat{c}_g[\tau_1, \tau_2]$ corresponding to the whole RF-line. Finally, an estimate $\hat{b}_g[k, l]$ of the bicepstrum $b_g[k, l]$ in (5.29) is computed as given by [8]:

$$\hat{b}_g[k, l] = \frac{1}{k}\mathcal{F}_{2D}^{-1}\left(\frac{\mathcal{F}_{2D}(k\,\hat{c}_g[k, l])}{\mathcal{F}_{2D}(\hat{c}_g[k, l])}\right), \quad k \neq 0; \tag{5.33}$$

$$\hat{b}_g[0, l] = \hat{b}_g[l, 0],$$

where $\mathcal{F}_{2D}$ and $\mathcal{F}_{2D}^{-1}$ denote the operation of the 2-D discrete Fourier transform and its inverse, respectively. Note that, at this point, the Equation (5.33) may appear somewhat perplexing, since, according to the definition of bicepstrum, we should have computed $\hat{b}_g[k, l]$ by applying the inverse Fourier transform to the (complex) logarithm of the Fourier transform of $\hat{c}_g[\tau_1, \tau_2]$. This "formal" way of computing the bicepstrum $\hat{b}_g[k, l]$, however, can be shown to be equivalent to that given by (5.33) [46]. (Moreover, the latter allows one to avoid solving the problem of *phase unwrapping*, which is discussed in the next subsection.) In the given case, we have decided to use (5.33) for the sake of consistency with the original reference [8], where the HOS-based approach was first proposed for blind deconvolution in medical ultrasound.

As a conclusion of this subsection, let us make a few remarks regarding some limitations of the estimation procedure just presented. First, since in practice, the bispectra in (5.27) are estimated from finite data samples, the condition $C_u(\omega_1, \omega_2) = 0$ is achievable only asymptotically. As a result, the noise cannot be guaranteed to be completely rejected, and hence its influence on the estimation should be expected to be not negligible. For the very same reason, using the finite data implies that the estimates in (5.29) through (5.31) are "noisy" due to their being random quantities. In order to decrease the variance of the estimation, a relatively large number of independent estimates should be computed, and subsequently averaged. Unfortunately, from the practical point of view, this is not always acceptable, as the amount of data is usually limited.

Finally, although the HOS-based approach to estimating the PSF can be theoretically extended to higher-dimensional cases, this extension does not seem to be practical. Indeed, in the 1-D case, the estimation was performed based on 2-D statistics (e.g., bispectra, bicepstra, and so on), which would have been four-dimensional, had the recovery of a 2-D PSF been attempted. This fact certainly raises some issues regarding the computational efficiency of the estimation as well as its feasibility for real-time processing, and this seems to be the main reason for which only the 1-D version of this estimation method has been applied so far in medical ultrasound.

5.3.3   Homomorphic Deconvolution: 1-D Case

Although the blind deconvolution approach we are about to present next can be directly formulated for the 2-D (and even-higher dimensional [47]) case, its 1-D version will be reviewed first, as it allows a closer look at some aspects of the estimation which are not that obvious and intuitive in 2-D.

The idea of this approach is based on concepts of *homomorphic signal processing* [48], whose application to medical ultrasound seems to have been first reported in [49], followed by some substantial developments in [50]. Basically, this idea consists in using the concept of *homomorphic mapping* to transform the product of two functions into the sum of the derived functions. Subsequently, if the derived functions possess different smoothness properties (or, equivalently, occupy different "bands" in the Fourier transform domain), they can be separated by means of *linear* filtering.

In particular, let $G(\omega)$, $Y(\omega)$, and $H(\omega)$ denote the Fourier transforms of their lowercase counterparts in (5.17)[3]. Then, temporarily ignoring the noise $u[n]$, the convolution model can be alternatively specified in the Fourier domain as given by:

$$G(\omega) = Y(\omega)\,H(\omega). \tag{5.34}$$

In the case of (5.34), the homomorphic mapping is performed by applying the complex logarithmical transformation to both its sides, which results in:

$$\hat{G}(\omega) \equiv \log|G(\omega)| + \imath\angle G(\omega) = \tag{5.35}$$
$$= (\log|Y(\omega)| + \imath\angle Y(\omega)) + (\log|H(\omega)| + \imath\angle H(\omega)),$$

where the symbols $|\cdot|$ and $\angle$ stand for the magnitude and phase of a complex function. It should be noted that, although in (5.34) and (5.35) the Fourier transforms are continuously defined and periodic, in practice, only their values at the discrete set of points $\Omega_N \equiv \{\omega_n = 2\pi n/N\}_{n=0}^{N-1}$ are usually computed and stored. For the sake of simplicity of notation, however, the continuous variable ω is used below, while its discrete counterpart is implied.

From (5.35), one can see that the log-spectrum $\hat{H}(\omega) \equiv \log|H(\omega)| + \imath\angle H(\omega)$ of the PSF can be estimated from $\hat{G}(\omega)$ by filtering out the tissue-related term $\hat{Y}(\omega) \equiv \log|Y(\omega)| + \imath\angle Y(\omega)$. In other words, (5.35) suggests that the log-magnitude $\log|H(\omega)|$ and phase $\angle H(\omega)$ of the PSF can be estimated from measurements of $\log|G(\omega)|$ and $\angle G(\omega)$, correspondingly, by filtering out the respective terms related to the reflectivity function. However, the above filtering makes sense only for the case when the functions $\log|H(\omega)|$ and $\log|Y(\omega)|$, as well as $\angle H(\omega)$ and $\angle Y(\omega)$, have considerably different smoothness properties. Fortunately, evidence exists that this is indeed the case [49–51]. In particular, it has been observed in numerous studies that the log-spectrum $\log|H(\omega)|$ is a regular and slow-varying function, while $\log|Y(\omega)|$ appears to be broadband and "spiky," thereby expressing the extremely intricate and random nature of the reflectivity functions of biological tissues.

5.3.3.1 Smoothness Properties of Fourier Phases

In order to demonstrate that the functions $\angle H(\omega)$ and $\angle Y(\omega)$ can also be considered to be of different nature regarding their smoothness properties, it is imperative to take a closer look at their analytical form. Given an RF-signal $g[n]$ composed of N sample points, its z-transform can be factorized as given below:

$$G(z) = \sum_{n=0}^{N-1} g[n]z^{-1} = \prod_{k=1}^{N-1}(1 - z_k z^{-1}), \tag{5.36}$$

---

[3]In the 1-D case, by the Fourier transform of a sequence $x[n]$ we mean the trigonometric series $X(\omega) = \sum_n x[n]\exp\{\imath n\omega\}$.

where it was assumed, without loss of generality, that the leading coefficient of $G(z)$ is equal to one, i.e., $g[N-1] = 1$. The polynomials $Z_k(z) = (1 - z_k z^{-1})$ are the irreducible factors of $G(z)$ and the complex constants $z_k = r_k e^{i\theta_k}$ are the zeros of $G(z)$.

The factorization (5.36) implies that the phase $\angle G(\omega)$ of the Fourier transform $G(\omega)$ of $g[n]$ is given by the sum of all the "elementary" phases $\angle Z_k(z)$ evaluated at $z = e^{i\omega}$ [52], viz.:

$$\angle G(\omega) = \sum_{k=1}^{N-1} \angle Z_k(e^{i\omega}) = \sum_{k=1}^{N-1} \arctan\left(\frac{r_k \sin(\omega - \theta_k)}{1 - r_k \cos(\omega - \theta_k)} \right). \qquad (5.37)$$

The properties of $\angle G(\omega)$ are uniquely defined by the distribution of the complex zeros of $G(z)$, while this distribution depends on the statistics of $g[n]$. In order to infer about the overall behavior of $\angle G(\omega)$, it would be instructive to scrutinize the behavior of a single term of (5.37). The dependence of the phase response $\angle Z_k(z)$ on the radius r_k is shown in Figure 5.3. The upper subplot of the figure shows the phase responses for zeros within the unit circle with $r_k = \{0.75, 0.85, 0.95, 0.9999\}$, whilst the lower subplot shows the phase responses for zeros outside the unit circle with $r_k = \{1.25, 1.15, 1.05, 1.0001\}$. The angle θ_k is fixed to be equal to π for all cases. One can see that the phase remains a continuous function until r_k approaches unity. Moreover, one can easily verify that the derivative of $\angle Z_k(z)$ evaluated at $z = e^{i\theta_k}$ is equal to $r_k/(r_k - 1)$, and therefore it tends to infinity as r_k approaches 1. Thus, for $r_k = 1$, the phase is discontinuous at $\omega = \theta_k$ with a jump of π radians. On the other hand, when the corresponding zero moves away from the unit circle, the overall regularity of the phase response increases.

In the noise-free case, the convolution model implies that the zeros of $G(z)$ are given as a combination of the zeros of both $H(z)$ and $Y(z)$. Formally, let Γ_h be a subset of the indices k, such that the zeros $\{z_k\}_{k \in \Gamma_h}$ correspond to the PSF. In an analogous manner, let Γ_y be a (complimentary) subset of indices for the reflectivity function $y[n]$. Then, one can decompose (5.37) into two terms as given by:

$$\angle G(\omega) = \angle Y(\omega) + \angle H(\omega) = \sum_{k \in \Gamma_y} \angle Z_k(e^{i\omega}) + \sum_{k \in \Gamma_h} \angle Z_k(e^{i\omega}). \qquad (5.38)$$

Although there is no unified agreement regarding the exact distribution of the samples of the tissue reflectivity function, most of the existing models assume these samples to be generated by an *i.i.d.* random process. The z-transform of such a sequence can be viewed as a polynomial in z^{-1} with random coefficients, i.e., a *random polynomial.* The properties of random polynomials have been extensively studied [53]. In the context of the present study, the most important property is stated as follows: the zeros of a polynomial, having *i.i.d.* random coefficients, are uniformly distributed at different angles with a vertex at the origin (i.e., at $|z| = 0$) and concentrated at the annulus

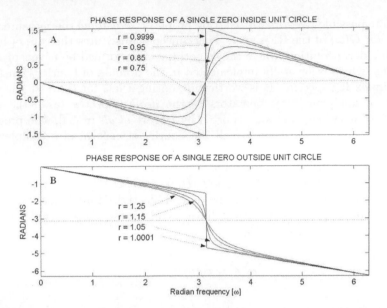

FIGURE 5.3: (Subplot A): phase responses of zeros with $r_k = \{0.75, 0.85, 0.95, 0.9999\}$ and $\theta_k = \pi$. (Subplot B): phase responses of zeros with $r_k = \{1.25, 1.15, 1.05, 1.0001\}$ and $\theta_k = \pi$.

$1 - \epsilon \le |z| \le 1 + \epsilon$, where $\epsilon > 0$ decreases exponentially, when $N \to \infty$. As a result, the phase $\angle Y(\omega)$ is comprised of the "elementary" phases, which have relatively large derivatives at $\omega = \theta_k$, with θ_k distributed uniformly within the interval $[0, 2\pi)$. Such a phase will be necessarily an irregular, "jagged" function of ω.

On the other hand, the PSF can be considered to be a band-pass filter. Hence, its zeros should be distant from the unit circle for those θ_k which lie *within* the transducer passband. This implies that within this frequency range, the PSF phase $\angle H(\omega)$ is dominated by smooth, slowly varying "elementary" phases, and therefore, it should be a smooth and regular function there. In regions outside the passband, the zeros can approach the unit circle and, as a result, the phase might be considerably less regular there. However, these "out-of-band" values of the phase are of little importance, as they hardly influence the overall shape of the corresponding PSF. Hence, for all practical purposes, the behavior of $\angle H(\omega)$ outside the pass-band can be safely ignored, and it can be assumed that the overall phase behavior is the same as that within the passband.

5.3.3.2 Cepstrum-Based Estimation of the PSF

From the discussion above, it follows that the log-spectrum $\hat{H}(\omega)$ is likely to be an essentially smoother function as compared with $\hat{Y}(\omega)$, and this fact makes it possible to estimate $\hat{H}(\omega)$ by properly *smoothing* the measured log-spectrum $\hat{G}(\omega)$ of the RF-signal. This concept constitutes the core of homomorphic deconvolution, where the smoothing is performed by means of *linear filtering*, which is generally implemented in the domain of Fourier transforms of complex log-spectra. It is worthwhile noting that, for the first time, the utility of applying linear operators to the *decibel* spectra (as if the latter were "ordinary" signals) was reported by Bogert *et al.* in [54], who proposed the term *complex cepstrum* to call the Fourier transform of a complex log-spectrum.

For the case under consideration, applying the Fourier transform to the complex log-spectra $\hat{G}(\omega)$, $\hat{Y}(\omega)$, and $\hat{H}(\omega)$ yields the complex cepstra $c_g[n]$, $c_y[n]$, and $c_h[n]$ of the RF-signal, the tissue reflectivity function, and the PSF, correspondingly. By virtue of the linearity of (5.35), these complex cepstra are related as given by [48, 55]:

$$c_g[n] = c_y[n] + c_h[n]. \tag{5.39}$$

Thus, in the noise-free case, the complex cepstrum $c_g[n]$ of $g[n]$ can be viewed as a superposition of the complex cepstra of the PSF and of the reflectivity function. Moreover, due to the differences in the smoothness properties of corresponding log-spectra, the cepstra $c_y[n]$ and $c_h[n]$ have different rates of convergence. In particular, $c_h[n]$ is a fast converging sequence, with its energy concentrated about the origin of the cepstrum domain, while the energy of $c_y[n]$ is spread over the entire cepstrum domain due to the nonregularity of $\hat{Y}(\omega)$. Although the sequences $c_h[n]$ and $c_y[n]$ do have a certain overlap, the latter is usually neglected, and the complex cepstrum of the PSF (and, consequently, the PSF itself) is recovered via truncating the cepstrum $c_g[n]$ of the RF-signal $g[n]$.

The truncation is usually performed by setting to zero all samples of $c_g[n]$, except for N^+ and N^- samples taken in the causal and anticausal directions, correspondingly. The parameters N^+ and N^- are commonly referred to as the *minimum-phase* and *maximum-phase* cutoff values [48, 50]. Knowing the approximate support of the PSF, and using the general rule that N^+ and N^- should be about one third of the PSF half-width with no zero-padding [50], an estimate of the cutoff values can be computed. In practice, however, these values have to be adjusted systematically because of the varying properties of the PSF caused by the dispersive attenuation and phase abberations.

Finally, we note that the truncation (which amounts to filtering $\hat{G}(\omega)$ with an ideal "box" filter) is likely to cause discontinuities in the estimate of the complex cepstrum $c_h[n]$. As a result, the corresponding estimate of the spectrum of the PSF can be contaminated by undesirable ringing artifacts (also known as the Gibbs phenomenon [56]). To overcome this problem, one can replace the truncation by multiplication with the frequency response of a smooth filter (e.g., the Butterworth filter was used to this end in [57, 58]).

5.3.3.3 Phase Unwrapping in 1-D

Computing the complex cepstrum $c_g[n]$ requires availability of both the log-amplitude $\log |G(\omega)|$ and the phase $\angle G(\omega)$ of the RF-signal. Unfortunately, the latter is generally unobservable in its *original* form. To provide uniqueness and analyticity of the definition of the complex logarithm, the phase $\angle G(\omega)$ can only be computed *modulo* 2π. This *wrapped* phase (also known as the *principal* phase) can be formally described as a result of applying the wrapping operator $\mathbf{W}$ to the original phase $\angle G(\omega)$. The wrapping operator $\mathbf{W}$ adds to $\angle G(\omega)$ a piecewise constant function $K(\omega) : \mathbb{R} \to \{2\pi k\}_{k\in\mathbb{Z}}$, resulting in $\mathbf{W}\{\angle G(\omega)\} = \angle G(\omega) + K(\omega)$ that obeys:

$$-\pi \leq \mathbf{W}\{\angle G(\omega)\} \leq \pi. \tag{5.40}$$

The wrapping introduces spurious discontinuities into the principal phase, thereby making the direct execution of any smoothing procedure of no use. In this case, a standard solution is to reconstruct the original phase values via *phase unwrapping* before the smoothing is applied [59]. Phase unwrapping, however, is known to be a nontrivial reconstruction problem. We shall allude to a number of possible solutions to this problem in the next subsections of the chapter, where 2-D blind deconvolution techniques are discussed. Meanwhile, it should be noted that for the 1-D case, the problem of phase unwrapping is not as acute as for the 2-D case. This is because of the property of 1-D polynomials of being factorizable, which allows one to recover the original phase $\angle G(\omega)$ directly from the zeros $\{z_k\}$ of $G(z)$ according to (5.39). Unfortunately, it is impossible to perform analogous computations in 2-D, as two-(and higher-) dimensional polynomials are, in general, not factorizable.

Finally, it is important to note that having recovered the original phase $\angle G(\omega)$ (either by a direct computation or by means of phase unwrapping), it is essential to estimate and subtract its linear component before the cepstrum is computed. If not subtracted, the linear slope can induce a discontinuity in the phase $\angle G(\omega)$ at $\omega = \pi$ (which, in theory, is supposed to be a continuous and odd function for $\omega < |\pi|$). As a result, $\hat{G}(\omega)$ would cease to be an analytic function on the unit circle and numerical problems would arise [48].

5.3.3.4 Computation of Complex Cepstra

In this subsection, we shall review a few approaches (as well as the assumptions they are based on) which can be used to estimate the complex cepstrum $c_h[n]$ of the PSF with no need to recover the original phase $\angle G(\omega)$ of the corresponding RF-data. One such approach is based on assuming the PSF $h[n]$ to be a *minimum-phase* sequence [48, 49]. In this case, the complex cepstrum of $h[n]$ can be uniquely recovered from the Fourier transform of the *real* part of $\hat{H}(\omega)$, i.e., $\log |H(\omega)|$, thereby leaving aside the necessity to deal with the phase $\angle H(\omega)$. Note that the sequence obtained by applying the Fourier transform to $\log |H(\omega)|$ is commonly referred to as the *real cepstrum* of $h[n]$. Provided $h[n]$ is a minimum-phase sequence, its real cepstrum $c_h^r[n]$

can be used for computing the corresponding complex cepstrum according
to [48, Chapter 10.5]:

$$c_h[n] = \begin{cases} c_h^r[n], & n = 0, N/2; \\ 2c_h^r[n], & 1 \le n < N/2; \\ 0, & N/2 < n < N. \end{cases} \tag{5.41}$$

Accordingly, (5.41) can be used for estimating the complex cepstrum of
a minimum-phase $h[n]$ from the real cepstrum $c_g^r[n] \equiv \mathcal{F}(\log|G(\omega)|)$ of the
RF-signal as given by [49]:

$$c_h[n] \simeq \begin{cases} c_g^r[n], & n = 0; \\ 2c_g^r[n], & 1 \le n \le N^+; \\ 0, & N^+ < n < N, \end{cases} \tag{5.42}$$

where $N^+ \ll N$ denotes a minimum-phase cutoff. Note that in order to reduce
the ringing artifacts, the truncation in (5.42) can be replaced by multiplication
with a smooth window.

Since neither phase unwrapping nor linear phase estimation are necessary
when computing the real cepstrum $c_g^r[n]$, assuming the PSF to be minimum
phase offers significant computational advantages. Yet, unfortunately, this
assumption is generally not true [50].

The phase unwrapping can also be avoided when the complex cepstrum
of $g[n]$ is evaluated by means of the logarithmic derivatives. The idea of
this approach is quite simple, and it is based on the following fact (which
follows directly from the basic properties of the Fourier transform). While the
Fourier transform of the complex log-spectrum $\log(G(\omega))$ yields the complex
cepstrum $c_g[n]$, the Fourier transform of the *derivative* of $\log(G(\omega))$ results in
the polynomially weighted sequence $n\,c_g[n]$. Now, denoting the derivative of
$G(\omega)$ by $G'(\omega)$ and using the fact that $d\log(G(\omega))/d\omega = G'(\omega)/G(\omega)$, it is
rather simple to show [48, 50] that, in the discrete case:

$$c_g[n] = \begin{cases} \frac{1}{inN} \sum_{k=0}^{N-1} \frac{G'[2\pi k/N]}{G[2\pi k/N]} e^{i(2\pi/N)nk}, & 1 \le n \le N - 1; \\ -\frac{1}{N} \sum_{k=0}^{N-1} \log G[2\pi k/N], & n = 0. \end{cases} \tag{5.43}$$

The main disadvantage of (5.43) is that, in this case, the aliasing error in the
complex cepstrum is usually much more severe in comparison with the phase
unwrapping method.

There exist several alternative techniques for computing the complex spectrum, which have been reviewed in [50], and will not be described here, since
all these methods seem to be similar to those that have already been presented
above. It should be noted, however, that all the cepstrum-based techniques
seem to share a number of common drawbacks. One such drawback stems
from the fact that even if the spectrum $G(\omega)$ is not aliased (as the RF-signals

are normally sampled in compliance with the Shannon–Nyquist theorem [60]), there is no *a priori* guarantee that the corresponding complex cepstrum will be free of the aliasing error. For this reason, a common practice is to extensively zero-pad the RF-signals before their spectra are computed. It is needless to say that such zero-padding significantly increases the computation cost of the overall estimation procedure.

Among other disadvantages of the cepstrum-based estimation of the PSF is its excessive sensitivity to noise and instability caused by the complex zeros close to the unit circle. However, the most important disadvantage is the one inherent to the basic concept of homomorphic processing: the log-spectrum of the PSF is recovered by means of a *linear* smoothing. We shall demonstrate below that a considerable improvement of the estimation quality can be achieved by extending the scope of all possible smoothing procedures to include *non*linear methods as well. However, before turning to describe such methods, we would like to supplement the review of the cepstrum-based techniques by pointing out some peculiarities of their 2-D formulation.

5.3.4 Homomorphic Deconvolution: 2-D Case

The applicability of homomorphic signal processing to the problem of 2-D *nonseparable* deconvolution in medical ultrasound seems to have been first reported in [50]. At that time, it had been among the few methods which were capable of improving both the axial and lateral resolutions of ultrasound images while allowing a fairly direct real-time implementation.

The formulation of homomorphic estimation of a 2-D PSF is identical to the 1-D case. Specifically, denoting by $G(\omega_1, \omega_2)$, $Y(\omega_1, \omega_2)$, and $H(\omega_1, \omega_2)$ the 2-D Fourier transforms of the RF-image, the tissue reflectivity function and the PSF, respectively, and ignoring for the moment the influence of the additive noise term $u[n, m]$, the convolution model (5.16) implies the following relations in the log-Fourier domain:

$$\log |G(\omega_1, \omega_2)| = \log |Y(\omega_1, \omega_2)| + \log |H(\omega_1, \omega_2)|, \qquad (5.44)$$

$$\angle G(\omega_1, \omega_2) = \angle Y(\omega_1, \omega_2) + \angle H(\omega_1, \omega_2). \qquad (5.45)$$

Just like in the 1-D case, the relations (5.44) and (5.45) suggest that the log-spectrum $\hat{H}(\omega_1, \omega_2)$ of the PSF (and, hence, the PSF itself) could be recovered by filtering out the tissue-related quantities $\log |Y(\omega_1, \omega_2)|$ and $\angle Y(\omega_1, \omega_2)$ from the acquired log-magnitude $\log |G(\omega_1, \omega_2)|$ and phase $\angle G(\omega_1, \omega_2)$ of the RF-image, respectively. Moreover, considering the fact that the overall behavior of the log-magnitudes $\log |H(\omega_1, \omega_2)|$ and $\log |Y(\omega_1, \omega_2)|$, as well as those of the phases $\angle H(\omega_1, \omega_2)$ and $\angle Y(\omega_1, \omega_2)$, is likely to be analogous to the behavior of their 1-D counterparts, an extension of the results of the preceding section to the 2-D case seems to be quite straightforward.

Indeed, if the unwrapped phase $\angle G(\omega_1, \omega_2)$ was known, it would then be possible to compute the 2-D cepstrum $c_g[n, m]$ of $g[n, m]$. Consequently, due to the differences in smoothness properties of the log-spectra $\log(H(\omega_1, \omega_2))$ and $\log(Y(\omega_1, \omega_2))$, the sequence $c_g[n, m]$ could be considered as a sum of the fast converging cepstrum $c_h[n, m]$ of the PSF and the slow converging, broad-spread cepstrum $c_y[n, m]$ of the reflectivity function. Subsequently, the cepstrum $c_h[n, m]$ could be estimated from $c_g[n, m]$ by multiplying the latter with the rectangular window defined by: $|n| \leq N_1$ and $|m| \leq N_2$, with N_1 and N_2 being the axial and lateral cutoff parameters. Once again, we note that using the rectangular window can cause the ringing artifacts, which could be reduced by using smoother window functions [57, 58]. It should also be noted that, irrespective of its choice, the window function should be constantly adjusted in compliance with the changes in the shape of the PSF.

5.3.4.1   Phase Unwrapping in 2-D

The discussion on 2-D homomorphic deconvolution is not complete without mentioning some key methods of phase unwrapping, which can be used for recovering the original values of the Fourier phase $\angle G(\omega_1, \omega_2)$ of the RF-image. The problem of phase unwrapping is known to be common to a number of applications of applied science, with a multitude of solutions nowadays available, many of which can be readily applied to recover $\angle G(\omega_1, \omega_2)$ [59]. A standard approach is to estimate the unwrapped phase through numerically integrating its *partial differences* (i.e., algebraic differences of the phase values at adjacent points on a chosen lattice), whilst the the latter is estimated from the available wrapped phase.

Specifically, let the discrete values of the wrapped phase $\mathbf{W}\{\angle G(\omega_1, \omega_2)\}$ be arranged in an $N \times M$ matrix Ψ (with $\mathbf{W}$ denoting the wrapping operator, as before). Here the discretization is assumed to be uniform, viz., it is performed by sampling the phase $\mathbf{W}\{\angle G(\omega_1, \omega_2)\}$ at the points of the set $\Omega_{N \times M} \equiv \{(\omega_n, \omega_m) \mid \omega_n = 2\pi n/N, \omega_m = 2\pi m/M\}$, with $n = 0, 1, \ldots, N-1$ and $m = 0, 1, \ldots, M-1$. In the same way, the discretized values of the unwrapped phase $\angle G(\omega_1, \omega_2)$ are assumed to be collected in another $N \times M$ matrix Φ.

Given the discrete phase Ψ, let Ψ'_n and Ψ'_m denote the matrices of its partial differences, which are obtained by subtracting the successive elements of Ψ in the column and row directions, respectively. (Note that the boundary element of the difference matrices can be computed using periodic boundary conditions due to the property of Fourier phases of being periodic functions.) Analogously, let Φ'_n and Φ'_m be the matrices of partial differences corresponding to the unwrapped phase Φ. Then, it can be rigorously proven [59] that conditioned:

$$|\Phi'_n[l, k]| < \pi \quad \text{and} \quad |\Phi'_m[l, k]| < \pi, \quad \forall l, k, \tag{5.46}$$

the following equalities hold:

$$\Phi'_n = \mathbf{W}\{\Psi'_n\}, \tag{5.47}$$
$$\Phi'_m = \mathbf{W}\{\Psi'_m\}.$$

Thus, if it was *a priori* guaranteed that the partial differences of the unwrapped phase do not exceed π in their absolute values, then it would be possible to compute their values *precisely* (by virtue of (5.47)) from the samples of the observed wrapped phase. Subsequently, the unwrapped phase Φ could be recovered by numerically integrating these differences. It is important to note that the uniqueness of this integration is assured through the property of Fourier phases to be odd (and hence zero mean) functions over the domain of their fundamental period.

Unfortunately, in practice, the condition of (5.46) is often violated. In this case, the resulting gradient of Φ (as defined by its components Φ_n' and Φ_m') is no longer a conservative vector field, and, as a result, its integral over an arbitrary closed pass will not, in general, be equal to zero (as would have been the case had the condition (5.46) held true), but rather to $2\pi k$, with $k \in \mathbb{Z}$. Note that such nonzero values of the integral are known as *phase residua* [59], and their presence, in turn, indicates that the phase differences estimated according to (5.47) are perturbed by sizable errors.

The potential inaccuracy in estimating the partial differences Φ_n' and Φ_m' using (5.47) can be alleviated by several methods, with different degrees of success [59, 61, 63]. Most of these methods are implemented in two steps. First, the partial differences of the unwrapped phase are estimated from the available samples of its wrapped counterpart according to (5.47). Second, the *optimal* unwrapped phase is found by minimizing a distance between its partial differences and the differences provided by the estimation. It should be noted that, in this case, the optimal solution only approximates the true phase, in the sense defined by the nature of the "fitting" procedure. However, if it was possible to recover the *exact* values of the phase differences (which would be the case if the condition (5.46) held true), this approximation would have coincided with the true unwrapped phase for most of the existing methods [59].

Among some other works on phase unwrapping is [64], where it was argued that most of the unwrapping algorithms do not recover the original phase by adding to its wrapped measurements integer multipliers of 2π. This can be viewed as a serious limitation, since it can destroy the integrity (i.e., the congruence modulo 2π) of the unwrapped phase values. To overcome this deficiency, an adaptive integration scheme based on the Block Least Square method was proposed in [64]. However, since in homomorphic deconvolution the unwrapped phases are eventually subjected to linear filtering, the method presented in [64] does not seem to be essential for the problem at hand. A nice comparison of some alternative unwrapping methods can be found in [65].

5.3.4.2 Phase Unwrapping via Smoothing Integration

All unwrapping methods mentioned above seem to share a common drawback: their performance inevitably deteriorates as the number of inaccurate estimations of the partial phase differences increases. Thus, in these methods, the unwrapping error is strongly dependent on the functional class of the phases being considered. On the other hand, a considerable improvement in the quality of the phase estimation can be achieved by taking advantage of some specific properties of the Fourier phases. Among the most important attributes of the latter are their property of being odd periodic functions over $\Omega \equiv [-\pi, \pi) \times [-\pi, \pi)$ and the possibility to compute their partial *derivatives* (not differences) *precisely* at arbitrary points of Ω. These properties allow the problem of phase unwrapping to be approached from a very different direction, as shown below.

First of all, let us note that, given the 2-D Fourier transform $F(\omega_1, \omega_2)$ of a discrete sequence $f[n, m]$, and denoting its phase $\angle F(\omega_1, \omega_2)$ by $\varphi(\omega_1, \omega_2)$, it is rarely a problem to compute the partial derivatives of the latter. It can be easily shown that for $k \in \{1, 2\}$ [66]:

$$\frac{\partial^n \varphi(\omega_1, \omega_2)}{\partial \omega_k^n} = \frac{\partial^n \angle F(\omega_1, \omega_2)}{\partial \omega_k^n} = \operatorname{Im}\left\{\frac{\partial^n \log(F(\omega_1, \omega_2))}{\partial \omega_k^n}\right\}, \tag{5.48}$$

where Im stands for the imagery part. In particular,

$$\frac{\partial \varphi(\omega_1, \omega_2)}{\partial \omega_k} = \operatorname{Im}\left\{\frac{1}{F(\omega_1, \omega_2)} \frac{\partial F(\omega_1, \omega_2)}{\partial \omega_k}\right\}, \tag{5.49}$$

$$\frac{\partial \varphi^2(\omega_1, \omega_2)}{\partial \omega_k^2} = \operatorname{Im}\left\{\frac{1}{F(\omega_1, \omega_2)} \frac{\partial^2 F(\omega_1, \omega_2)}{\partial \omega_k^2} - \left(\frac{1}{F(\omega_1, \omega_2)} \frac{\partial F(\omega_1, \omega_2)}{\partial \omega_k}\right)^2\right\}. \tag{5.50}$$

Note that the derivatives of $F(\omega_1, \omega_2)$ can be computed by applying the Fourier transform to a polynomial-weighted signal $f[n, m]$ in the spatial domain [48].

The availability of precise values of the partial derivatives of a Fourier phase implies that $\varphi(\omega_1, \omega_2)$ can be unwrapped by integrating these derivatives rather than their partial differences, as estimated by using (5.47). This possibility was first advocated in [67] for the 1-D case, where Fourier phases are recovered by numerically integrating their first-order derivatives using the trapezoidal integration rule. An extension of this idea to the 2-D case was later proposed in [66]. This extension was based on formulating the problem of reconstruction of the unwrapped phase $\varphi(\omega_1, \omega_2)$ in terms of the Neumann problem for the Poisson equation. It should be noted that as opposed to using the discrete Poisson equation in [63], the formulation in [66] is continuously defined and based on the *computed* derivatives of the phase. Specifically, in [66], the (unwrapped) phase $\varphi(\omega_1, \omega_2)$ is estimated as a solution to the following problem:

$$\Delta \varphi = -p \text{ in } \Omega \tag{5.51}$$

$$\vec{n} \nabla \varphi = r \text{ on } \partial\Omega$$

where $\Delta\varphi(\omega_1, \omega_2) \equiv \partial^2\varphi(\omega_1, \omega_2)/\omega_1^2 + \partial^2\varphi(\omega_1, \omega_2)/\omega_2^2$ and $\nabla\varphi(\omega_1, \omega_2)$ denote the Laplacian and gradient of the phase $\varphi(\omega_1, \omega_2)$, respectively, $\partial\Omega$ is the boundary of Ω and $\vec{n}$ is the outward normal to Ω.

It can be rigorously proven that given the measurements $p(\omega_1, \omega_2) \in \mathbb{L}_2(\Omega)$ and $r(s) \in \mathbb{L}_2(\partial\Omega)^4$, there exists a weak form solution of (5.51) in $\{\varphi(\omega_1, \omega_2) \in \mathbb{W}_2^1 \mid \int\int \varphi(\omega_1, \omega_2)\partial\omega_1\partial\omega_2 = 0\}$, where $\mathbb{W}_2^1$ is the Sobolev space of the first order [68]. Moreover, this solution is unique up to an additive constant provided $\int\int \varphi(\omega_1, \omega_2)\partial\omega_1\partial\omega_2 + \int\int r(s)\partial s = 0$ [69]. Fortunately, due to the oddness and periodicity of the Fourier phases this condition is always guaranteed to hold true and, moreover, the additive constant can be unambiguously defined from the constraint $\varphi(0,0) = 0$.

It should be noted that the formulation in (5.51) is continuous, while in practice the values of the partial derivatives of Fourier phases can only be computed on a predefined lattice in the Fourier domain. The discrepancies between the continuous formulation and the discrete measurements can be resolved by using an efficient interpolation scheme, which was also proposed in [66].

In addition to extension of the method of [67] to the 2-D case, the study in [66] proposed another, rather more advanced idea, based on the fact that for homomorphic deconvolution, the phase $\angle G(\omega_1, \omega_2)$ *does not* need to be recovered, but rather its *smoothed* version. Hence, by restricting the solution of (5.51) to belong to a properly defined subspace of smooth, low-varying function, one can combine the problems of unwrapping of $\angle G(\omega_1, \omega_2)$ and estimation of $\angle H(\omega_1, \omega_2)$. In particular, it was proposed to consider the phase $\angle H(\omega_1, \omega_2)$ to be a member of the subspace defined by:

$$\mathbb{V} = \text{Span}\left\{\Phi_{k_1,k_2}^{J_1,J_2}(\omega_1, \omega_2) = \phi_{J_1,k_1}(\omega_1)\,\phi_{J_2,k_2}(\omega_2)\right\}, \qquad (5.52)$$

where

$$\phi_{J,k}(\omega) = \left(2^{J+1}\pi\right)^{-1} \sum_{l\in\mathbb{Z}} \phi\left(2^{-J}(\omega/2\pi + l) - k\right), \qquad (5.53)$$

and $k_1 = 0, 1, \ldots, 2^{-J_1} - 1$, $k_2 = 0, 1, \ldots, 2^{-J_2} - 1$, while $J_1, J_2 \in \mathbb{Z}^-$.

The subspace $\mathbb{V}$ defined by (5.52) is known as a periodic *principal shift-invariant* subspace generated by $\phi(\omega)$ [70]. The *generator* $\phi(\omega)$ is usually chosen to be a smooth function with compact support, whose Fourier transform does not possess 2π-periodic zeros. The latter property of $\phi(\omega)$ assures that the functions $\{\Phi_{k_1,k_2}^{J_1,J_2}(\omega_1, \omega_2)\}$ are linearly independent, and, hence, can be used for stable approximation in $\mathbb{V}$ [71]. Note that the subspace $\mathbb{V}$ is finite-dimensional, since for any $J_1, J_2 < 0$, there exist exactly $2^{-(J_1+J_2)}$ basis functions. Moreover, $\mathbb{V}$ is typically considered as a subspace of all the "low-resolution" approximations of the functions from $\mathbb{L}_2(\Omega)$, with its "resolution" being controlled by J (where a finer resolution is achieved for smaller J).

Having defined the "target" space $\mathbb{V}$, an estimate of $\angle H(\omega_1, \omega_2)$ can be computed by projecting the phase $\angle G(\omega_1, \omega_2)$ onto $\mathbb{V}$. (Note that with $\mathbb{V}$ defined as a low-resolution subspace, this projection can be viewed as a smoothing operation.) However, in order to estimate $\angle H(\omega_1, \omega_2)$ as a projection of

---

[4]Note that both these functions are assumed to be computed by virtue of (5.49) and (5.50)

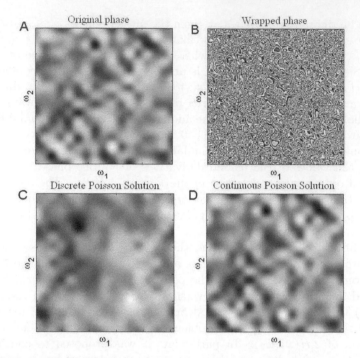

FIGURE 5.4: (Subplot A) Original phase; (Subplot B) wrapped phase; (Subplot C) estimation obtained using the discrete Poisson solver, based on estimated partial differences; (Subplot D) estimation obtained using the continuous Poisson equation, based on computed partial derivatives.

$\angle G(\omega_1, \omega_2)$ onto $\mathbb{V}$, it is not necessary to first recover the unwrapped phase $\angle G(\omega_1, \omega_2)$, as this projection can be computed *directly* as a solution of (5.51) using the Riesz method [69] as shown in [66].

The difference in performances of a standard unwrapping method (based on estimated phase differences) and the method of [66] is demonstrated in Figure 5.4. As the standard method, the method specified in [63] was chosen, since it seems to reasonably represent a large group of approaches based on integrating the partial differences. Specifically, the method of [63] estimates the unwrapped phase by minimizing the *least square* error between the partial differences of the optimal solution and their estimates, computed from the observed (wrapped) phase. It is worthwhile noting that this approach solves a discrete version of (5.51), as opposed to the continuous formulation proposed in [66].

Subplots A and B of Figure 5.4 show a simulated Fourier phase of an image and its wrapped version, respectively. Note that the simulated phase was generated so that condition (5.46) was violated. The solution obtained using the standard method (which is referred to as the *discrete* Poison estimation) is shown in Subplot C of Figure 5.4, while the solution obtained by the method of [66] is shown in Subplot D of the same figure[5]. One can see that the standard estimation hardly resembles the original phase due to the violations of the condition (5.46). At the same time, for the case of the continuous estimation, there is no clear difference between the origin and the reconstruction. This result demonstrates that integrating the *exact* partial derivatives (rather than the estimated partial differences) allows one to perform the phase unwrapping with considerably smaller errors, for a wider class of phases under consideration.

In conclusion, a number of important remarks should be made. First, even though it seems to be possible to improve the performance of homomorphic deconvolution by using more advanced methods for either filtering in the cepstrum domain or for phase unwrapping, we believe that the main weakness of this methodology stems from its being linear and indifferent to the properties of the "noise" $\log(F(\omega_1, \omega_2))$. This suggests the possibility of improving the performance of homomorphic deconvolution by carefully taking into consideration both functional and statistical properties of the Fourier transforms of the PSF and tissue reflectivity function. In particular, such an improvement can be achieved using the *generalized* homomorphic deconvolution that is presented next.

5.3.5 Generalized Homomorphic Deconvolution

The generalization of the main concepts of homomorphic deconvolution was recently proposed [72]. Similar to the case of the classical homomorphic deconvolution, the generalized approach separates the functions $\log|H(\omega_1, \omega_2)|$ and $\log|Y(\omega_1, \omega_2)|$, as well as $\angle H(\omega_1, \omega_2)$ and $\angle Y(\omega_1, \omega_2)$, based on the differences of their smoothness properties. However, in this case, the design of the smoothing procedure is based on a rigorous analysis of the statistical properties of the noise to be rejected.

5.3.5.1 Estimation of the Fourier Magnitude of the PSF

In order to explain the generalized approach, let us simplify the notation first. Since, in practice, only discrete versions of the functions in (5.44) and (5.45) are dealt with, we assume that the Fourier transform of an acquired RF-image is sampled at the discrete points $\omega_n = 2\pi n/N$ and $\omega_m = 2\pi n/M$ with $n = 0, 1, \ldots, N-1$ and $m = 0, 1, \ldots, M-1$. Then, defining $\hat{i}[n, m] \equiv \log|i(\omega_n, \omega_m)|$, with $i \in \{G, H, Y\}$, the model equation (5.44) can be rewritten as given by:

$$\hat{G}[n, m] = \hat{H}[n, m] + \hat{Y}[n, m]. \tag{5.54}$$

---

[5]In the latter method, the subspace $\mathbb{V}$ was generated by a cubic B-spline for $J_1 = J_2 = -4$.

Note that, throughout the rest of this subsection, the pair $[n, m]$ is used to index the frequency (rather than spatial domain) samples.

The above expression suggests that the signal $\hat{H}[n, m]$ can be recovered from its "measurements" $\hat{G}[n, m]$, while considering $\hat{Y}[n, m]$ as noise that needs to be rejected. Thus, the problem of estimating $\hat{H}[n, m]$ is essentially a *denoising* problem [73–75].

Although $\log|H(\omega_1, \omega_2)|$ is generally a much smoother function as compared to $\log|Y(\omega_1, \omega_2)|$, it may not be uniformly regular (e.g., in the sense of Sobolev [68]), and some isolated singularities are to be expected as well. In this case, the *denoising by wavelet shrinkage* (or, simply, wavelet denoising), as first introduced in [73], can be effectively used for estimating its sampled values $\hat{H}[n, m]$. In its basic form, wavelet denoising consists of the following three steps: (1) the data signal $\hat{G}[n, m]$ is transformed into a wavelet orthogonal domain [76, 77]; (2) the resulting wavelet coefficients are subjected to the *soft-thresholding procedure* $\eta_t(x) = \text{sign}(x)(|x| - t)_+$ with threshold t (where the operator $(x)_+$ returns x if $x > 0$ and zero otherwise); (3) the thresholded wavelet coefficients are inversely transformed into the original domain, supplying an estimation of the noise-free signal.

Among the most significant properties of the wavelet denoising is its ability to minimize the estimation error, subject to an additional constraint that requires the estimate to be at least as smooth as the original function. Moreover, the resulting estimates achieve almost the *minimax* error rate for a wide range of smoothness classes. These properties demonstrate the main advantage of this algorithm over linear smoothing: it allows performing the estimation of the PSF log-spectrum *adaptively*, according to its smoothness, producing an estimate that is nearly optimal over a very wide range of functional classes [73].

It should be noted that the uniform thresholding is not the only possible way to suppress the wavelet coefficients of the noise, and many other methods have been proposed, based, for example, on the principles of Bayesian estimation [78]. In most cases, these methods were shown to outperform the soft-thresholding, and hence it is reasonable to assume that some of them can be used to further increase the accuracy of estimating $\hat{H}[n, m]$.

5.3.5.2   Outlier Resistant Denoising

The denoising algorithm is known to perform optimally when the noise to be rejected is white Gaussian noise. Consequently, the statistical nature of the "noise" $\hat{Y}[n, m]$ should be analyzed first to determine whether the denoising algorithm can reject it in an optimal way. In [75], it was shown that whenever the samples of the tissue reflectivity function behave as white Gaussian noise with zero-mean and the variance σ^2, the samples of $\hat{Y}[n, m]$ are *i.i.d.* random variables as well and they obey the *Fisher–Tippett* distribution, defined by its *PDF* as:

$$p_{\hat{Y}}(x) = 2\exp\left\{(2x - \log(2\sigma^2)) - \exp\left\{2x - \log(2\sigma^2)\right\}\right\}. \qquad (5.55)$$

It should be emphasized that this statistical model is valid for the cases when the reflectivity function can be assumed to behave as white Gaussian noise. As a result, the model (5.55) (as well as the estimation procedure based on it) should be applied to such segments of RF-images that comply with this assumption. However, even though this assumption seems to be limiting, it was observed in [72] that the performance of the estimation method detailed below is fairly insensitive to its violation.

When replacing the second exponential in (5.55) by the first three terms of its Taylor expansion, one obtains the following *Gaussian* approximation of the Fisher–Tippett *PDF*:

$$p_{\hat{Y}}(x) \simeq 2e^{-1} \exp\left\{ -\frac{1}{2}\frac{(x - \log(\sqrt{2}\sigma))^2}{0.25} \right\}. \tag{5.56}$$

The form of (5.56) suggests that $\hat{Y}[n, m]$ can be approximately considered as white Gaussian noise, having a mean value of $\log(\sqrt{2}\sigma)$ and a *constant* standard deviation of 0.5.

The Gaussian approximation closely approaches the original Fisher–Tippett *PDF* (5.55) in the vicinity of its mean value, while failing to account for the heavy "tail" on the left side of the distribution. In fact, in contrast to the Gaussian distribution, the *PDF* of the Fisher–Tippett distribution is *asymmetric* and *leptokurtic*, with its kurtosis and skewness being equal to 2.4 and $12\sqrt{6}\zeta/\pi^3$, respectively, with ζ being the Apery's constant. This difference between the two *PDF* causes the Fisher–Tippett "noise" to appear as a white Gaussian noise contaminated by occasional "outliers," i.e., a relatively small number of spiky noise samples of relatively large amplitudes. In such a case, the classical denoising algorithm may "recognize" the outliers as useful features of the signal to be recovered, thereby allowing them to elude the thresholding.

It is generally known that spiky noises are difficult to deal with, and many methods, which exploit the concept of $\mathbb{L}_2$-projections (including the wavelet de-noising), often fail to reject such noises in a satisfactory way. To reduce the sensitivity of the classical denoising to the outliers, it was proposed in [75] to *gaussianize* the noise $\hat{Y}[n, m]$ using the procedure of outlier shrinkage. The latter can be performed by subtracting from $\hat{G}[n, m]$ its robust residuals $R[n, m]$ computed according to:

$$R[n, m] = \text{sign}(\Delta\hat{G}[n, m])(|\Delta\hat{G}[n, m]| - \lambda)_+, \tag{5.57}$$

where $\Delta\hat{G}[n, m]$ is the difference between $\hat{G}[n, m]$ and its median-filtered version, and $\lambda > 0$ is a predefined threshold. It was observed that, in most cases, the robust residuals $R[n, m]$ correspond to the outliers induced by the spiky component of $\hat{Y}[n, m]$, when the size of the median filter is set to be 3×3 (or 5×5) and when the threshold λ is set to such a level that 93-95% of the samples $\Delta\hat{G}[n, m]$ are preserved. Moreover, it was demonstrated in [48] that

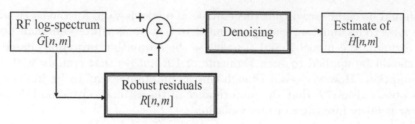

FIGURE 5.5: Block diagram of the algorithm for estimating the log-magnitude of the PSF using the outlier resistant denoising.

the noise contaminating the difference signal $\hat{G}[n, m] - R[n, m]$ behaves very similarly to white Gaussian noise, while the desired signal $\hat{H}[n, m]$ remains practically intact.

Once the spiky component of $\hat{Y}[n, m]$ has been estimated as $R[n, m]$, the signal $\hat{G}[n, m] - R[n, m]$ can be denoised in order to estimate $\hat{H}[n, m]$. The block diagram of Figure 5.5 shows the overall procedure for estimating $\hat{H}[n, m]$. In [72, 75], the wavelet denoising was performed using the classical dyadic wavelet transform, based on the maximally symmetric wavelets having six vanishing moments [76]. The value of the universal threshold t was set to be equal to $\sqrt{2 \log(NM)} \sigma$ with the standard deviation σ of the noise defined as the theoretically predicted standard deviation of the Gaussian approximation of the Fisher–Tippett *PDF*, viz., $\sigma = 0.5$.

5.3.5.3   Estimation of the Fourier Phase of the PSF

The above method for estimating the log-magnitude $\hat{H}[n, m]$ of the Fourier transform of the PSF can be shown to be related to the Bayesian estimation framework [22], since the wavelet denoising is a Bayesian estimator, in nature [73]. It is worthwhile noting that an often superior performance of Bayesian estimators (as compared, e.g., with linear filtering) is a result of their being optimized with respect to *both* the likelihood of the observed data and the expected behavior of the signal to be recovered. Therefore, to derive a Bayesian method for estimating the Fourier phase of the PSF would require one to specify the properties of $\angle H(\omega_1, \omega_2)$ and $\angle Y(\omega_1, \omega_2)$.

While, based on the arguments analogous to those presented in Section 5.3.3.1, one could reasonably assume the function $\angle H(\omega_1, \omega_2)$ to be piecewise smooth, little could be said about the properties of $\angle Y(\omega_1, \omega_2)$ without a deeper analysis. The latter would require a rigorous description of the distribution of the zero sheets of the 2-D z-transform $Y(z_1, z_2)$ around the unit bicircle. Performing such an analysis, however, is a complicated task, whose fulfillment is hard to achieve because of the absence of some relevant results in the theory of high-dimensional random polynomials.

In order to overcome the above difficultly, one can resort to a suboptimal approach to estimating the Fourier phase of the PSF. In particular, it was

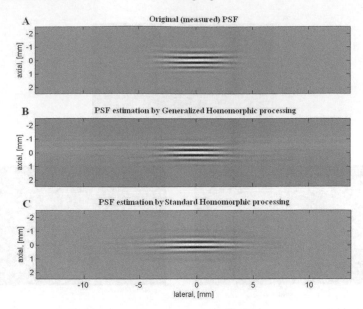

FIGURE 5.6: (Upper panel) Original PSF; (Central panel) Reconstruction by the generalized homomorphic method; (Lower panel) Cepstrum-based estimation.

proposed in [72] to combine the estimation of $\log|H(\omega_1, \omega_2)|$ by means of the wavelet denoising with the estimation of $\angle H(\omega_1, \omega_2)$ by means of the smoothing integration procedure, described in Section 5.3.4.2. (To be precise, the smooth integration in [72] was implemented via linear filtering, based on the method proposed in [79]).

5.3.5.4 Generalized Homomorphic Estimation vs. Standard Homomorphic Estimation

In a series of both *in silico* and *in vitro* experiments conducted in [72], it was demonstrated that the generalized homomorphic approach results in estimates having a mean squared error that is more than 2 times smaller than the error obtained for the standard cepstrum-based approach. An example of such estimation (obtained for SNR=10 dB) is shown in Figure 5.6, where the original (i.e., measured) PSF and its estimates obtained using the generalized and the standard homomorphic approaches are shown by Subplots A, B, and C of the figure, respectively. The parameters of both reconstruction methods were chosen by trial and error, so as to obtain the most accurate results. Note that all the PSFs are shown as grayscale images, with the black and white colors corresponding to their smallest negative and the largest positive values, respectively. One can see that whereas the generalized homomorphic processing succeeds well in recovering the PSF shape and support information,

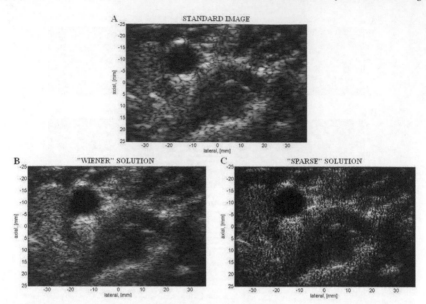

FIGURE 5.7: (Subplot A) Original abdominal ultrasound image; (Subplot B) Deconvolution under Gaussian priors using the analytical Wiener filter; (Subplot C) Sparse deconvolution under Laplacian priors.

the standard cepstrum-based reconstruction suffers from errors, which appear as "warping" and "smearing" of the estimated shape of the PSF.

Some deconvolution results, which were obtained by using the PSF estimated by means of the generalized homomorphic approach, are shown in Figure 5.7. Subplot A of the figure shows an abdominal image recorded from an adult volunteer with a VIVID-3 (GE, Medical System, Inc.) commercial ultrasound scanner. The image acquisition was performed using a linear array probe, with a central frequency of about 3.5 MHz. Subsequently, the image was deconvolved using the *maximum a posteriori* estimation of Section 5.2.2 under the Gaussian and Laplacian priors, which amounts to minimizing of the cost functionals (5.7) and (5.9), respectively. The resulting deconvolved images are shown in Subplots B and C of Figure 5.7, where they are referred to as the "Wiener" and "Sparse" solutions. Note that all the images have been normalized and, subsequently, log-compressed for visualization in 8-bit resolution, and thus all images in Figure 5.7 have the same dynamic range.

One can see that all structures within the deconvolved images appear considerably less blurred than in the original image. Moreover, the speckle pattern in the deconvolved images is much "finer" when compared to that of the original image, and as a result, homogeneous regions of the underlying tissue appear more uniform in the deconvolved images. Most of the "edge-like" structures of the tissue, however, are much better represented by the

"Sparse" solution. Moreover, this reconstruction has obviously the best contrast, thereby better representing the overall tissue structure. The gains in the mean resolution, obtained for the "Wiener" and "Sparse" reconstructions, were 2.1 and 5.6, respectively. Note that a numerical measure of the resolution gain was defined as the ratio between the number of pixels of the autocorrelation function with values higher than 0.75, computed for the standard image, and that for the deconvolved image [51].

5.3.6 Blind Deconvolution via Inverse Filtering

In all the blind deconvolution algorithms discussed in the preceding subsections, estimating the PSF preceded the recovery of the tissue reflectivity function. In this subsection, we provide a review of a number of alternative methods, in which the reflectivity function $y[n, m]$ and the PSF $h[n, m]$ are recovered *simultaneously*. One such method was recently presented in [32], where $y[n, m]$ and $h[n, m]$ are estimated iteratively by means of the Van Cittert algorithm. However, for the reason of space and to avoid repetitions, we do not describe this algorithm here, as it is based on the methodology very similar to that we shall present later in Section 5.4.2.

Another class of approaches to the problem of simultaneous recovery of the PSF and the reflectivity function is based on *linear inverse filtering* [80]. In this case, the blind deconvolution algorithm does not recover the PSF but rather its inverse, i.e., an *inverse convolution kernel* $s[n, m]$ that results in:

$$y[n, m] \simeq s[n, m] * g[n, m]. \tag{5.58}$$

It should be noted that because of the *scale* and *linear-phase* ambiguity problem[6], which is inherent in deconvolution methods of this type, (5.58) is not general and should be replaced by $\alpha y[n - n_0, m - m_0] \simeq s[n, m] * g[n, m]$ with α being an arbitrary scalar and $[n_0, m_0]$ being a pair of integers, designating an arbitrary translation of the estimate. However, in order to simplify the notation, the relation (5.58) is used throughout the text, while the more general case is tacitly implied.

The inverse filter $s[n, m]$ is usually estimated assuming the samples of the tissue reflectivity function to be *i.i.d. non-Gaussian* random variables. This assumption is essential, since if the samples were Gaussian, then the estimation methods presented here could not be applied. (It is quite interesting to note that this is the very same assumption, which was used in Section 5.3.2, where the HOS-based methods for estimating the PSF were presented.)

The significance of the "non-Gaussianity assumption" is quite intuitive and easy to understand. Indeed, let us assume that the samples of $y[n, m]$ are

---

[6]The scale ambiguity stems from the fact that $g[n] = y[n] * h[n] = (a^{-1}y[n]) * (ah[n]) = (ay[n]) * (a^{-1}h[n])$, where a is an arbitrary scalar. The linear-phase ambiguity results from $g[n] = y[n] * h[n] = y[n + n_0] * h[n - n_0] = y[n - n_0] * h[n + n_0]$ with n_0 being an arbitrary integer designating a signal's shift.

i.i.d. and non-Gaussian. In the course of convolution with the PSF, these samples are "mixed up" so that every sample of the resulting RF-image is, actually, a *linear combination* of some samples of the reflectivity function. Thus, by virtue of the central limit theory, one can conclude that samples of the RF-image are always *more* Gaussian than those of the related reflectivity function. This fact suggests that an optimal inverse filter should be able to restore the non-Gaussianity of the data, which is equivalent to minimizing the entropy of the deconvolved result. This was the original idea underpinning the *minimum entropy deconvolution*, which was first proposed by Wiggins in [81], and later developed by Donoho in [82].

5.3.6.1   Moment-Based Estimation of the Inverse Filter

The discussion above suggests that, provided the samples of the reflectivity function are "produced" by a non-Gaussian random process, the optimal inverse filter can be found by maximizing a measure of non-Gaussianity of the deconvolved result. The question that now arises is what measure one should use. In [81], this measure was defined as the *fourth* moment of the deconvolved output. Specifically, the optimal inverse filter $s_{opt}[n, m]$ was found as a solution of the following problem:

$$s_{opt}[n, m] = \arg\max_s \left[\frac{\sum_n \sum_m |(s * g)[n, m]|^4}{\sum_n \sum_m |(s * g)[n, m]|^2} \right]. \tag{5.59}$$

Though the fourth moment has long been known as an effective measure of non-Gaussianity, its practical use may be quite problematic, due to the fact that the resultant measure is not *robust*. As a result, a small number of erroneously large, unsuccessful data samples can completely destroy the estimation based on (5.59).

In order to improve the robustness of (5.59), it was proposed in [83] to generalize this measure and to find the optimal inverse filter as the solution of the following optimization problem:

$$s_{opt}[n, m] = \arg\max_s \left[\frac{\sum_n \sum_m |(s * g)[n, m]|^p}{[\sum_n \sum_m |(s * g)[n, m]|^2]^{p/2}} \right]. \tag{5.60}$$

An alternative candidate for the measure may be the Claerbout's measure, and the resulted Claerbout's parsimonious deconvolution [84], based on inverse filtering with s_{opt} defined as:

$$s_{opt}[n, m] = \arg\min_s \left[\frac{1}{NM} \sum_n \sum_m \left(\frac{NM|(s * g)[n, m]|^2}{\sum_n \sum_m |(s * g)[n, m]|^2} \times \right. \right. \tag{5.61}$$

$$\left. \left. \times \log\left(\frac{NM|(s * g)[n, m]|^2}{\sum_n \sum_m |(s * g)[n, m]|^2} \right) \right) \right],$$

where it was assumed that the RF-image $g[n, m]$ is represented by an $N \times M$ matrix.

In all the cases above, the optimal inverse filter is supposed to be found by means of an optimization procedure, which is typically performed using the *steepest decent* algorithm [28]. The reader is kindly referred to [83], where a number of additional measures of non-Gaussianity (as well as those presented above) are given together with their gradients, and some numerical aspects of the optimization procedures are discussed as well.

5.3.6.2 Sparsity and Non-Gaussianity

All the measures of non-Gaussianity mentioned above are nonspecific to the form of the probability distribution characterizing the samples of the tissue reflectivity function $y[n, m]$. Although there is no universal agreed-upon form of this distribution, evidence exists [25,26] that when acoustic scattering takes place in conjunction with specular reflections (so that the corresponding reflectivity function has a *sparse* structure), this distribution is likely to be Laplacian. For this case, it can be shown [82] that minimizing the entropy of deconvolved RF-images is equivalent to minimizing their ℓ_1-norm.

The above equivalency between minimization of the entropy and of the ℓ_1-norm was utilized in [85] where the Laplacian probability model was used to recover the optimal inverse filter as a maximizer of the likelihood function of the observed RF-image. Specifically, in [85], the Fourier transform $S_{opt}(\omega_1, \omega_2)$ of the optimal inverse filter was found as the solution of the following minimization problem:

$$S_{opt}(\omega_1, \omega_2) = \arg \min_S \left[-\int\limits_{-\pi}^{\pi} \int\limits_{-\pi}^{\pi} \log |S(\omega_1, \omega_2)| d\omega_1 d\omega_2 + \right. \tag{5.62}$$

$$\left. + \lambda \, \|\mathcal{F}^{-1}(S(\omega_1, \omega_2) \, G(\omega_1, \omega_2))\|_1 \right],$$

where $\mathcal{F}^{-1}$ denotes the inverse Fourier transform, $G(\omega_1, \omega_2)$ is the Fourier transform of the RF-image, and $\lambda > 0$ is a regularization parameter controlling the mutual influence of the two terms in (5.62).

5.3.6.3 "Hybrid" Deconvolution

Unfortunately, the maximum likelihood approach of [85] seems to have a flaw in the case of *bandlimited* PSF, as it explicitly requires the convolution with the PSF to be invertible. Indeed, while minimizing the Jacobian term $-\int_{-\pi}^{\pi} \int_{-\pi}^{\pi} \log |S(\omega_1, \omega_2)| d\omega_1 d\omega_2$ would prevent the inverse filter from converging to a trivial solution, it also seems to "encourage" the amplitude of $S(\omega_1, \omega_2)$ to become arbitrarily large. Moreover, such an increase of the amplitude of $S(\omega_1, \omega_2)$ is likely to take place outside of the transducer passband, where the values of $H(\omega_1, \omega_2)$ are relatively small and, as a result, $G(\omega_1, \omega_2)$ is dominated by noises. Consequently, the resulted estimates of the reflectivity function may not be reliable.

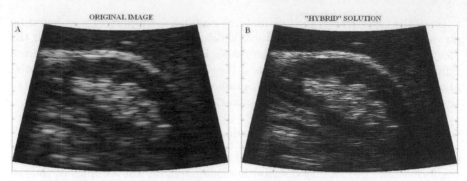

FIGURE 5.8: (Subplot A) Original ultrasound image of a kidney; (Subplot B) Reconstruction by the "hybrid" approach.

In order to overcome the above deficiency of [85], it was proposed in [33] to replace the logarithmic term in (5.62) by a functional that explicitly constrains the amplitude of the combined response $S(\omega_1, \omega_2)G(\omega_1, \omega_2)$ to be as close to unity as possible. In this modified formulation, the inverse filter is found as the solution to the following optimization problem:

$$S_{opt}(\omega_1, \omega_2) = \arg\min_{S} \left[-\int_{-\pi}^{\pi}\int_{-\pi}^{\pi} \left| |S(\omega_1, \omega_2)|^2 |\hat{H}(\omega_1, \omega_2)|^2 - 1 \right| d\omega_1 d\omega_2 + \right.$$

$$\left. + \lambda \left\| \mathcal{F}^{-1}(S(\omega_1, \omega_2)G(\omega_1, \omega_2)) \right\|_1 \right],$$

(5.63)

where $|\hat{H}(\omega_1, \omega_2)|$ is an estimate of the Fourier magnitude of the PSF, as computed, e.g., using the method presented in Section 5.3.5.2[7].

In conclusion, two interesting properties of (5.63) should be mentioned. First, by arguments similar to those in [82], it can be proven that if the samples of $y[n, m]$ were *i.i.d.* and Laplacian, and the PSF $h[n, m]$ was invertible, then the global minimizer of (5.63) would be the optimal inverse filter (i.e., the filter that inverts the PSF up to an arbitrary scalar and a shift) [86]. Second, the method can be thought of a "hybridization" of the two standard approaches to the problem of blind deconvolution, which are based on either *successive* or *concurrent* estimation of the PSF and the image of interest. This is because the functional in (5.63) incorporates some *partial* information on the PSF, viz., its power spectrum. It should be noted that this methodology is conceptually different from that of the methods in which an initial estimate of

---

[7]As opposed to estimating the Fourier phase of the PSF, which is quite problematic in view of the reasons mentioned in Section 5.3.5.3, the estimation of the Fourier magnitude $|H(\omega_1, \omega_2)|$ can be performed in a reliable and computationally efficient manner.

the PSF is computed first, following by its update though a properly contrived optimization procedure [32].

An *in vivo* example that demonstrates the performance of the "hybrid" deconvolution is shown in Figure 5.8. Subplot A of the figure shows an ultrasound image of the right kidney recorded from an adult volunteer by using the VIVID-3 (GE Medical Systems, Inc.) commercial ultrasound scanner. A phased array probe with a fundamental frequency of 3.3 MHz was used for image acquisition. On the other hand, the reconstruction computed by the "hybrid" method is shown by Subplot B of Figure 5.8. One can see that the reconstructed image has a better resolution and enhanced contrast, as compared to the original image. Moreover, many structural tissue details, which are blurred out in the original image, become visible after the blind deconvolution. This result defines the "hybrid" reconstruction as a prospective alternative to the currently available methods for blind deconvolution in medical ultrasound.

5.4 Blind Deconvolution in SPECT

5.4.1 Origins of the Blurring Artifact in SPECT

In Section 5.1.4 it was briefly mentioned that the image degradation phenomenon in SPECT imagery can be accounted for by a convolution with a PSF of the imaging system. In this section, however, before reviewing some blind deconvolution, methods which have been used to improve the quality of this imaging modality, a slightly deeper insight should be provided into the mechanism of formation of SPECT images. Specifically, during the data acquisition process, the head of the SPECT camera is incrementally positioned at multiple angles around the patient's body, and for each position of the camera, a 2-D *projection* of the 3-D spatial distribution of a radioactive tracer is registered. Subsequently, the acquired projections are processed to reconstruct the corresponding 3-D image (by means of a tomographic reconstruction algorithm), 2-D slices of which are what is typically analyzed by radiologists.

Having in mind the general idea of how a SPECT image is formed, one could probably ask: where exactly does the blurring artifact occur? In other words, is it the projections that are blurred or is it the tomographically reconstructed 3-D data? It turns out that both are blurred. As a result, image restoration by means of deconvolution has been applied to improve the quality of the projection data (pre-reconstruction) [13, 14] as well as of the final 3-D data (post-reconstruction) [87, 88], either exclusively or in combination [89, 90]. It should be noted that, although in the case of post-reconstruction, deconvolution could be directly applied to 3-D data sets [91], it is more common to process single 2-D slices of the latter. Hence, the 2-D convolution model (5.2) is usually used for both the pre- and the post-reconstruction stages, and, for this reason, the discussion below is confined to the 2-D case. However, all the algorithms reviewed in this section are also applicable to higher-dimensional cases, with no need for substantial modifications.

Before presenting these algorithms, a few remarks should be made concerning the applicability of the stationary convolution model for describing the blurring artifact in SPECT. As to the projection data, the main origins of this artifact are simple to figure out. First, the blur is caused by the resolution limitations of the gamma camera and collimation, as well as by the scattering and absorption of the radiation by the tissue. Motion of the patient and of his or her organs (e.g., heart, lungs) is another source of blurring. It should be noted that few of the above causes of blur can be considered as shift-invariant phenomena, since attenuation and motion are, in general, inhomogeneous over the patient's body. For this reason, approximating the blur in projections by assuming the convolution with a single PSF is a simplification whose limitations should be taken into account.

The blurring artifact in the final (i.e., tomographically reconstructed) images is caused mainly by the blur in each of the projections. However, in the post-reconstruction setting, the convolution can be shown to be approximately shift-invariant, provided the size of an object of interest is small in comparison to its distance to the collimator [92]. In such a case, the stationary convolution model (5.2) can be used with greater confidence, though motion artifacts are still the main factor reducing the validity of the stationary assumption.

It is interesting to note that the tomographic reconstruction is, by itself, an *inverse* problem, in which the true image (i.e., the 3-D spatial distribution of a tracer) is recovered as the "best fit" to the observed projections. Moreover, the relation between the true image and its projections is described by a linear operator (commonly referred to as the *system matrix*). Considering the fact that the stationary blurring is a linear operation as well, it seems possible to modify the system matrix so that it takes into account both the data formation and the blurring artifact. In this case, it would be possible to carry out the deconvolution *implicitly* in the course of the tomographic reconstruction. Although this reconstruction approach seems to be a feasible and even a preferable way of combining the advantages of both the pre- and the post-reconstruction [93], it still remains relatively unexplored, and hence its further investigation is needed.

Considering the similarity between the deconvolution methods used at the pre- and the post-reconstruction stages in SPECT, the following subsections will consider the more fundamental problem of estimating the true (nuclear medicine) image $y[n, m]$ from its blurred and noisy observation $g[n, m]$, provided no *a priori* information exists on the PSF $h[n, m]$ of the SPECT scanner in use.

5.4.2 Blind Deconvolution via Alternative Minimization

In blind deconvolution methods of this kind, the optimal estimates of the true image and the PSF are computed as (in general, local) minimizers of a bivariate functional $F(y, h)$. Typically, this functional is defined so that for a fixed $y = y^*$ (respectively, $h = h^*$) the resulting functional $F(y^*, h)$ (respectively, $F(y, h^*)$) is convex and, hence, has a unique global minimizer. The *alternative minimization* (AM) approach to minimizing the functional $F(y, h)$ consists in iteratively solving the following sequence of optimization problems:

$$\text{given } y_{k-1}[n, m], \text{ solve } h_k[n, m] = \arg\min_h F(y_{k-1}, h); \qquad (5.64)$$

$$\text{given } h_{k-1}[n, m], \text{ solve } y_k[n, m] = \arg\min_y F(y, h_{k-1}),$$

where $y_{k-1}[n, m]$ and $h_{k-1}[n, m]$ denote the estimates of the true image and of the PSF, respectively, that are computed in iteration $k - 1$, and $y_k[n, m]$ and $h_k[n, m]$ are their updates in iteration k. Note that when computing the estimate $y_k[n, m]$ in (5.64), the updated estimate $h_k[n, m]$ can be used instead of $h_{k-1}[n, m]$. Commonly, the initial estimate of the true image is set to be equal to the observed image, i.e., $y_0[n, m] = g[n, m]$, and the initial estimate of the PSF is set to be the Kronecker delta function, i.e., $h_0[n, m] = \delta[n, m]$ (though an experimentally measured PSF, if available, would be a much better choice).

In the case of the AM approach, the optimal estimates of $y[n, m]$ and $h[n, m]$ are obtained as limit elements of the sequences of intermediate solutions $\{y_k[n, m]\}$ and $\{h_k[n, m]\}$, correspondingly. In practice, however, it is common to terminate the iterations after certain convergence criteria are fulfilled (e.g., the norms $\|y_k - y_{k-1}\|_2$ and $\|h_k - h_{k-1}\|_2$ become smaller than a predefined positive constant). Unfortunately, the functional $F(y, h)$ in most cases is *non*convex in two variables. For this reason, the convergence of (5.64) to a global minimum is not *a priori* guaranteed, and virtually all AM methods are prone to getting stuck in local minima.

In order to facilitate the convergence of the AM algorithms to useful estimates, the intermediate solutions can be made to possess some known properties of the true image and the PSF. In the case of SPECT, for example, it is reasonable to assume that both the true image and the PSF are positive valued functions. Consequently, it makes sense to force the intermediate solutions $y_k[n, m]$ and $h_k[n, m]$ to be positive by using the following adjustments:

$$h_k[n, m] = \{h_k[n, m]\}_+; \qquad (5.65)$$

$$y_k[n, m] = \{y_k[n, m]\}_+,$$

where the operator $\{x\}_+$ returns x if $x > 0$ and zero otherwise.

For the case when it is *a priori* known that the true image and the PSF are supported in $\Omega_y \subseteq \mathbb{Z}^2$ and $\Omega_h \subseteq \mathbb{Z}^2$, respectively, the adjustment could be performed according to:

$$h_k[n, m] = h_k[n, m] \chi_h[n, m]; \tag{5.66}$$
$$y_k[n, m] = y_k[n, m] \chi_y[n, m],$$

where $\chi_h[n, m]$ and $\chi_y[n, m]$ denote the characteristic functions of the subsets Ω_h and Ω_y, correspondingly. (Note that the values of a characteristic function are equal to unity within the subset it represents, while they are equal to zero everywhere else).

In order to prevent convergence of the PSF to a trivial solution, it is common to enforce the following normalization constraint:

$$h_k[n, m] = \frac{h_k[n, m]}{\sum_n \sum_m h_k[n, m]}, \tag{5.67}$$

which requires the sum of the values of the PSF to be equal to unity. Moreover, the intermediate estimates of a *centrosymmetric* PSF can be additionally adjusted using:

$$h_k[n, m] = \frac{h_k[n, m] + h_k[-n, -m]}{2}. \tag{5.68}$$

It is worthwhile noting that the constraints enforced via (5.65) through (5.68) define *convex subsets* in the space of possible solutions. Moreover, enforcing these constraints is, in fact, equivalent to projecting the intermediate solutions onto the corresponding convex sets. Note that the technique of projection onto convex sets (POCS) by itself is known to be a powerful tool for solving diverse inverse problems [25]. For the case at hand, however, POCS is used as a supporting stage, which is intended to prevent the sequences of intermediate estimates from converging to useless local minima.

In order to complete the review of the AM methods, a number of possible choices of the functional $F(y, h)$ in (5.64) should be considered next. In situations when the computational efficiency is of utmost importance, it is common to define the functional in such a way that the corresponding first-order optimality conditions are represented by *linear* equations (or systems thereof). In such a case, a standard choice is to minimize $F(y, h)$ given by [94]:

$$F(y, h) = \|h * y - g\|_2^2 + \alpha_1 \|v_1 * y\|_2^2 + \alpha_2 \|v_2 * h\|_2^2, \tag{5.69}$$

where $\| \cdot \|_2$, as before, denotes the ℓ_2-norm, and $v_1[n, m]$ and $v_2[n, m]$ are the kernels of the regularization operators, which are typically defined to be high-pass filters (e.g., a discrete Laplacian operator). The regularization parameters α_1 and α_2 in (5.69) are intended to control the tradeoff between the fidelity to the observations (as accounted for by the first term in (5.69)) and the smoothness of both the estimated image and the estimated PSF.

In order to minimize the functional $F(y, h)$ in (5.69), its first derivatives in $y[n, m]$ and $h[n, m]$ should be computed and equated to zero. This will give us the first-order optimality condition which, for the case at hand, results in:

$$(\partial F / \partial h)[n, m] = y[-n, -m] * e[n, m] + \alpha_2 (r_2 * h)[n, m] = 0, \tag{5.70}$$

$$(\partial F/\partial y)[n,m] = h[-n,-m] * e[n,m] + \alpha_1(r_1 * y)[n,m] = 0, \qquad (5.71)$$

where $r_1[n,m] = v_1[n,m] * v_1[-n,-m]$, $r_2[n,m] = v_2[n,m] * v_2[-n,-m]$, and $e[n,m] = (h * y)[n,m] - g[n,m]$.

By applying the discrete Fourier transform to both parts of (5.70) and (5.71), it is straightforward to show that the first-order optimality conditions can be equivalently represented in the Fourier domain by:

$$(|Y(\omega_1,\omega_2)|^2 + \alpha_2|V_2(\omega_1,\omega_2)|^2)H(\omega_1,\omega_2) - Y^*(\omega_1,\omega_2)G(\omega_1,\omega_2) = 0, \quad (5.72)$$

$$(|H(\omega_1,\omega_2)|^2 + \alpha_1|V_1(\omega_1,\omega_2)|^2)Y(\omega_1,\omega_2) - H^*(\omega_1,\omega_2)G(\omega_1,\omega_2) = 0, \quad (5.73)$$

where $V_1(\omega_1,\omega_2)$ and $V_2(\omega_1,\omega_2)$ denote the Fourier transforms of $v_1[n,m]$ and $v_2[n,m]$, respectively, and the asterisk stands for the complex conjugate. Consequently, the AM method can be performed in the Fourier domain using the following update equations:

$$H_k(\omega_1,\omega_2) = \frac{Y_{k-1}^*(\omega_1,\omega_2)G(\omega_1,\omega_2)}{|Y_{k-1}^*(\omega_1,\omega_2)|^2 + \alpha_2|V_2(\omega_1,\omega_2)|^2}, \qquad (5.74)$$

$$Y_k(\omega_1,\omega_2) = \frac{H_{k-1}^*(\omega_1,\omega_2)G(\omega_1,\omega_2)}{|H_{k-1}^*(\omega_1,\omega_2)|^2 + \alpha_1|V_1(\omega_1,\omega_2)|^2}. \qquad (5.75)$$

In order to facilitate the convergence, every time $H_k(\omega_1,\omega_2)$ and $Y_k(\omega_1,\omega_2)$ are computed, they need to be back-transformed to the spatial domain, followed by application of the corrections according to (5.65) through (5.68). It should also be noted that as long as the estimate of the PSF is computed first, it seems to be better if one uses $H_k(\omega_1,\omega_2)$ instead of $H_{k-1}(\omega_1,\omega_2)$ for updating $Y_k(\omega_1,\omega_2)$ in (5.75).

One can see that (5.74) and (5.75) have the structure of the classical Wiener filter [24]. Furthermore, these expressions would have the precise analytical form of the Wiener filter, if α_1 and α_2 were replaced by the variance σ_u^2 of the additive noise $u[n,m]$ in (5.1), and $|V_1(\omega_1,\omega_2)|^{-2}$ and $|V_2(\omega_1,\omega_2)|^{-2}$ were replaced by the *power spectra* of the true image and of the PSF, respectively. Unfortunately, in the case when little *a priori* information is available on both the true image and the PSF, their power spectra cannot be generally considered as known. In order to overcome this difficulty, it was proposed in [17] to estimate the required spectra at iteration k, using the intermediate solutions computed at the previous iteration step $k-1$. Specifically, it was proposed to use the following AM updates:

$$H_k(\omega_1,\omega_2) = \frac{Y_{k-1}^*(\omega_1,\omega_2)G(\omega_1,\omega_2)}{|Y_{k-1}^*(\omega_1,\omega_2)|^2 + \sigma_u^2|H_{k-1}(\omega_1,\omega_2)|^{-2}}, \qquad (5.76)$$

$$Y_k(\omega_1,\omega_2) = \frac{H_{k-1}^*(\omega_1,\omega_2)G(\omega_1,\omega_2)}{|H_{k-1}^*(\omega_1,\omega_2)|^2 + \sigma_u^2|Y_{k-1}(\omega_1,\omega_2)|^{-2}}. \qquad (5.77)$$

In the equations above, the power spectra of $y_k[n, m]$ and $h_k[n, m]$ are approximated by the squared magnitudes of their respective Fourier transforms. Considering the random nature of the true images in SPECT, however, this approximation approach does not seem to be sufficiently robust. This fact suggests the possibility of further improving the performance of the blind deconvolution method of [17] via employing more sophisticated and robust techniques for power spectrum estimation [95].

The AM algorithms presented above are particularly useful for cases when the true images are smooth or fast processing is required. However, whenever the true image is discontinuous (e.g., possessing sharp edges or isolated singularities), its estimate will be oversmoothed owing to the fact that the update equations (5.74) and (5.75) (or, equivalently, (5.76) and (5.77)) are linear and, hence, incapable of interpolating the harmonic components of the true images, which were destroyed by the convolution with the PSF. In order to overcome this deficiency, a new definition of the functional $F(y, h)$ is needed. A possible form of such an "edge preserving" functional $F(y, h)$ was proposed in [96], where it was defined as:

$$F(y, h) = \|h * y - g\|_2^2 + \alpha_1 \sum_n \sum_m |\nabla y[n, m]| + \alpha_2 \sum_n \sum_m |\nabla h[n, m]|, \quad (5.78)$$

with ∇ denoting the (discrete) gradient operator.

Though the blind deconvolution methods based on minimizing (5.78) have not been applied to the SPECT imagery as yet, we believe that they represent a powerful alternative to the methods discussed above.

As in the previous cases, the functional (5.78) is minimized using the AM approach, in which each update step is followed by the corrections (5.65) through (5.68). As opposed to the methods discussed earlier, the functional (5.78) is particularly advantageous for recovering *piecewise smooth* functions. Indeed, if the true image $y[n, m]$ could be considered to be a piecewise smooth function (which is a reasonable assumption in SPECT, where the true image represents the distribution of a radioactive tracer in a patient's organ), then its gradient would be necessarily sparse. In this case, the deconvolution problem can be formulated as a problem of finding the "best fit" to the measurements that possess the sparsest gradient. The sparseness of the latter, in turn, is regularly assessed by evaluating its ℓ_1-norm, which is referred to as the *total variation* (TV) norm of the image. (Note that in the SPECT imagery, the PSF is expected to be a smooth function, and, thus, it seems to be more reasonable to regularize the estimation of $h[n, m]$ by constraining its ℓ_2-norm.)

The first-order optimality conditions for minimizing (5.78) can be shown to be given by:

$$(\partial F / \partial h)[n, m] = y[-n, -m] * e[n, m] - \alpha_2 \, \mathbf{div} \left(\frac{\nabla h[n, m]}{|\nabla h[n, m]|} \right) = 0, \quad (5.79)$$

$$(\partial F / \partial y)[n, m] = h[-n, -m] * e[n, m] - \alpha_1 \, \mathbf{div} \left(\frac{\nabla y[n, m]}{|\nabla y[n, m]|} \right) = 0, \quad (5.80)$$

where **div** denotes a numerical approximation of the divergence operator. Consequently, the corresponding update equations have the following form:

solve for $h_k[n, m]$: $\hfill$ (5.81)

$$y_{k-1}[-n, -m] * ((h_k * y_{k-1})[n, m] - g[n, m]) - \alpha_2 \, \mathbf{div} \left(\frac{\nabla h_k[n, m]}{|\nabla h_k[n, m]|} \right) = 0,$$

solve for $y_k[n, m]$: $\hfill$ (5.82)

$$h_k[-n, -m] * ((h_k * y_k)[n, m] - g[n, m]) - \alpha_2 \, \mathbf{div} \left(\frac{\nabla y_k[n, m]}{|\nabla y_k[n, m]|} \right) = 0.$$

As in the previous case, the functional $F(y, h)$ in (5.78) is not convex simultaneously in both y and h, and, therefore, the blind deconvolution algorithm is prone to converge to useless local minima. Moreover, the solution of (5.81) and (5.82) are quite hard to find due to the nonlinearity of these equations. This fact demonstrates the main disadvantage of the TV deconvolution as compared to the previously discussed methods.

To solve equations (5.81) and (5.82), the *fixed point algorithm* can be used, as detailed in [96]. This algorithm is based on the fact that if $|\nabla h_k[n, m]|^{-1}$ in (5.81) (respectively, $|\nabla y_k[n, m]|^{-1}$ in (5.82)) was fixed, then the resulting equation would be *linear*. This linearization can be performed by lagging the diffusive coefficients $|\nabla h_k[n, m]|^{-1}$ and $|\nabla y_k[n, m]|^{-1}$ by one (fixed point) iteration, followed by solving the resulting linear problems iteratively. The AM update equations then become:

solve for $h_k[n, m]$ (by iterating i until convergence): $\hfill$ (5.83)

$$y_{k-1}[-n, -m] * ((h_k^i * y_{k-1})[n, m] - g[n, m]) - \alpha_2 \, \mathbf{div} \left(\frac{\nabla h_k^i[n, m]}{|\nabla h_k^{i-1}[n, m]|} \right) = 0,$$

solve for $y_k[n, m]$ (by iterating i until convergence): $\hfill$ (5.84)

$$h_k[-n, -m] * ((h_k * y_k^i)[n, m] - g[n, m]) - \alpha_2 \, \mathbf{div} \left(\frac{\nabla y_k^i[n, m]}{|\nabla y_k^{i-1}[n, m]|} \right) = 0.$$

Note that at iteration i, the terms $|\nabla h_k^{i-1}[n, m]|^{-1}$ and $|\nabla y_k^{i-1}[n, m]|^{-1}$ are constant and, consequently, the resulting equations are linear. Unfortunately, there are no closed form solutions available for solving (5.83) and (5.84), that necessitates the use of iterative techniques. Although the *preconditioned conjugate gradient* method [28] can be efficiently used to this end, the overall complexity of the TV blind deconvolution still remains relatively high.

Finally, we note that the optimal values of the regularization parameters α_1 and α_2 are usually found by trial and error, though there are some consistent approaches to their automatic determination (e.g., using the theory of L-curves [97]). In general, the parameter α_1 can be shown to be inversely proportional to the SNR [96], and thus its value should be increased, when the observed image become more noisy (so as to prevent the solution from exploding as a result of noise amplification). On the other hand, the value of the parameters α_2 should be relatively large for more severe blurs [96].

5.4.3   Blind Deconvolution via Nonnegativity and Support Constrains Recursive Inverse Filtering

The last method we would like to present in this section is known as the *nonnegativity and support constraints recursive inverse filtering* (NAS-RIF). This method was originally proposed in [98] and applied to SPECT imagery data in [88]. The NAS-RIF method recovers the true image $y[n, m]$ via convolving the data image $g[n, m]$ with an inverse filter $s[n, m]$ according to (5.58), and, thus, this method is conceptually similar to the blind deconvolution algorithms reviewed in Section 5.3.6. However, as opposed to this previous case, where the optimal inverse filter was found based on some *statistical* properties of the deconvolved images, the NAS-RIF algorithms optimize the inverse filter using *deterministic* properties of the deconvolved images. In particular, the optimal inverse filter is found to be such that the deconvolved image is positive valued and constant (usually, equal to zero) outside a predefined domain $D \subseteq \mathbb{Z}_2$. From the algorithmic point of view, such an inverse filter can be found as the solution of the following optimization problem:

$$s_{opt} = \arg\min_s \left[\frac{1}{2} \sum_{[n,m]\in D} \hat{y}^2[n,m]\big(1 - \text{sign}(\hat{y}[n,m])\big) + \right. \tag{5.85}$$

$$\left. + \alpha_1 \sum_{[n,m]\nsubseteq D} (\hat{y}[n,m] - y_0)^2 + \alpha_2 \Big(\sum_{[n,m]} s[n,m] - 1\Big)^2 \right],$$

where $\hat{y}[n, m] = (s * g)[n, m]$, and y_0 is a (known) value of the true image outside of its domain of definition D. The first two terms force the deconvolved image $\hat{y}[n, m]$ to be positive and equal to a predefined value (i.e., y_0) outside of D, respectively, while the last term normalizes the inverse filter, thereby preventing it from converging to a trivial solution.

The most important feature of the NAS-RIF method stems from the property of the cost functional of being *strictly convex*. It implies that there always exists a global unique minimizer, and therefore the algorithm will not converge to a local minimum of little interest. Moreover, the derivatives of this cost functional are readily computable [98], which makes it possible to perform the minimization in a computationally efficient manner (using, for example, the Newton algorithm detailed in Section 5.2 of this chapter.

5.5 Blind Deconvolution in Confocal Microscopy

5.5.1 Maximum Likelihood Deconvolution in Fluorescence Microscopy

In this section of the chapter, we are going to present a blind deconvolution method which, similarly to the approach described in Section 5.3.6.1, is also based on the principles of *maximum likelihood* estimation. However, this time, the estimation is performed using a rather different technique which, by itself, deserves independent attention. In estimation theory, this technique is known as the *expectation maximization* (EM) algorithm [99]. It is interesting to note that the applicability of the EM algorithm to the maximum likelihood estimation seems to be first reported in [100], where it was used for reconstructions in positron emission tomography (PET). Later on, this approach was extended to reconstructions in fluorescence microscopy, both conventional and confocal [101–103].

Although the typical resolution of a confocal microscope is much improved over that of a conventional wide-field microscope, it may still contain severe anisotropy, which might be compensated for via the usage of suitable postprocessing methods. In particular, there exist studies in which the problem of enhancing the resolution of confocal microscopy is addressed by means of nonblind deconvolution methods (see, e.g., [101] and references therein). In such cases, the PSF of a microscope is assumed to be measured through a proper calibration routine. The calibration typically involves imaging of fluorescence beads of relatively small size (0.1–0.2 μm). Unfortunately, such a measurement often turns out to be a tedious and esoteric process, depending on the know-how and training of the operator. Besides, the measurement is further complicated by severe noises due to low light levels, photobleaching of beads, necessity to exactly match the refraction index of the immersion oil, necessity to isolate and identify individual beads, as well as by the fact that the measured PSFs are usually oversmoothed due to the finite size of the beads [103]. Moreover, the usage of a theoretically derived PSF may not be sufficiently accurate, as the actual PSF is a function of the refractive index of the embedding medium and the sample itself. In addition, situations are possible in which the scientist may want to retrospectively restore images, for which the PSF calibration was not carried out or was imprecise for some reason.

All these arguments emphasize the importance of blind deconvolution methods as a useful alternative to the nonblind restoration. Below, we elaborate on the methods that make use of the maximum likelihood estimation as a quantitative optimization criterion to solve the blind deconvolution problem in fluorescence microscope imagery. It is interesting to note that, in these methods, the true and observed images, as well as the PSF, are viewed as *probability distributions.*

For the sake of notational consistency with the references, in what follows, we use continuous coordinates, e.g., $x \in \mathbb{R}^2$, rather than their discrete counterpart $[n, m] \in \mathbb{Z}^2$. It should be noted that although the discussion here is confined to the 2-D case, all the derivations below can be straightforward extended to a higher-dimensional setting. (However, it is not done here so as to avoid unnecessary abstraction of the notations.)

Let $\lambda[x]$ denote the distribution of the fluorescence concentration, and let X_i denote the location of the i-th photon emission. Note that the latter would coincide with the i-th position detected by the CCD camera if the diffraction-induced blurring was *not* present. For this reason, the emission points X_i are typically referred to as the *true emission points,* and they can be shown to form an inhomogeneous Poisson random-point process, whose intensity function is given by $\lambda[x]$ [101–103]. Moreover, the true emission points X_i (the total number N of which is Poisson distributed with a mean value $\Lambda = \int_{\mathbb{R}^2} \lambda[x] dx$) are usually assumed to be *i.i.d.* and obeying the PDF $y_X[x] = \lambda[x]/\Lambda$ for each $X = X_i$.

Due to the diffraction, each of the true emission points is contaminated by a random translation vector B_i, that is supposed to be statistically independent of X_i and assumed to be *i.i.d.* and obeying the PDF $h[b]$, which is nothing else but the PSF of the optical system. The resulting positional measurement U_i is then expressed by:

$$U_i = X_i + B_i. \tag{5.86}$$

Due to the statistical independence between the true emission points and the random displacements, the *PDF* $g_U[u]$ of the positional measurement U_i is equal to:

$$g_U[u] = (y_X * h)[u]. \tag{5.87}$$

The expression (5.87) represents the model equation, which is actually analogous to the basic model (5.2). However, in the current case, the PDF $g_U[u]$ of the data is unknown, though "observable" through the collected data $N_u[du]$, which is the number of photons having measurements lying within the region defined by $du = [u_1, u_1+du_1) \times [u_2, u_2+du_2)$ with $u = (u_1, u_2)$. Consequently, the problem of blind deconvolution is now formulated as the problem of recovering the true concentration $\lambda[x]$ of the fluorescence from the measured $N_u[du]$.

A possible solution to this problem can be based on the EM procedure [102] which basically consists of two steps: the expectation step and the maximization step. To implement these steps involves identifying an *incomplete* data set which, in the present case, is defined to be the set of photon measurements $\mathbf{U} = \{U_i\}$. As typical for this kind of methods, it is difficult to perform the maximum likelihood estimation based on the incomplete data alone. However, the desired estimates would be straightforward computable, if a *complete* data set was available. For the case at hand, such complete data can be defined as the set of true emission vectors $\mathbf{X} = \{X_i\}$ *in union* with the set of error (displacement) vectors $\mathbf{B} = \{B_i\}$.

Let $N_x[dx]$ denote the number of true emission points lying within $dx = [x_1, x_1 + dx_1) \times [x_2, x_2 + dx_2)$ with $x = (x_1, x_2)$. Then, the log-likelihood of $\mathbf{X}$ can be expressed as [104]:

$$L_1\{\mathbf{X}; \lambda\} = -\int_{\mathbb{R}^2} \lambda[x]dx + \int_{\mathbb{R}^2} \log(\lambda[x])N_x[dx]. \qquad (5.88)$$

In a similar manner, assuming that the set of error vectors $\mathbf{B}$ constitutes an inhomogeneous Poisson random point with an intensity function $\gamma[b] = \Lambda h[b]$, it can be shown that the log-likelihood of $\mathbf{B}$ can be expressed as:

$$L_2\{\mathbf{B}; h, \Lambda\} = -\int_{\mathbb{R}^2} \gamma[b]db + \int_{\mathbb{R}^2} \log(\gamma[b])N_b[db], \qquad (5.89)$$

where $N_b[db]$ is the number of photons having error vectors within the region $db = [b_1, b_1 + db) \times [b_2, b_2 + db_2)$ with $b = (b_1, b_2)$.

In order to characterize the complete data set, its log-likelihood function should be computed next. It turns out that the latter can be expressed in two equivalent ways. Specifically, it follows from (5.88) and (5.89), as well as from the statistical independence of the elements of $\mathbf{X}$ and $\mathbf{B}$, that the log-likelihood function of the complete data can be expressed either as:

$$L_3\{\mathbf{X}, \mathbf{B}; \lambda, h, \Lambda\} = -\int_{\mathbb{R}^2} \lambda[x]dx + \int_{\mathbb{R}^2} \log(\lambda[x])N_x[dx] + \sum_{i=1}^{N} \log(h[B_i]), \quad (5.90)$$

or as:

$$L_4\{\mathbf{X}, \mathbf{B}; \lambda, h, \Lambda\} = -\int_{\mathbb{R}^2} \gamma[b]db + \int_{\mathbb{R}^2} \log(\gamma[b])N_b[db] + \sum_{i=1}^{N} \log(\gamma[X_i]/\Lambda).$$
$$(5.91)$$

Given all the definitions above, it is now possible to specify the EM algorithm. In the estimation step (E-step), the terms λ^{k+1} and h^{k+1} are substituted for λ and h, respectively, in the complete data likelihood expressions (5.90) and (5.91). Subsequently, the expectations of these expressions are evaluated, conditioned on both the observed (incomplete) data and the *event* C^k, which denotes that $\lambda = \hat{\lambda}^k$, $h = \hat{h}^k$, and $\Lambda = \hat{\Lambda}^k$. Note that here the superscripts k and $k + 1$ refer to the k-th and the $(k + 1)$-th iteration, and the hat indicates that the corresponding terms are iterative estimates, based on the maximization step (M-step) of the k-th iteration. Finally, in the M-step, the expectation of the complete data likelihood L_3 (that results from the E-step) is maximized over λ^{k+1}, h^{k+1}, and Λ. This yields the solutions $\lambda = \hat{\lambda}^{k+1}$, $h = \hat{h}^{k+1}$, and $\hat{\Lambda}$, which are the next iterative estimates of λ, h, and Λ, respectively.

Following the derivations detailed in [102], one can show that the complete EM algorithm can be tersely expressed as:

$$\hat{\lambda}^{k+1}[x] = \hat{\lambda}^k[x] \int_{\mathbb{R}^2} \frac{\hat{h}^k[u-x]N_u[du]}{\hat{h}^k[u-z]\hat{\lambda}^k[z]dz}, \qquad (5.92)$$

$$\hat{h}^{k+1}[b] = \frac{\hat{h}^k[b]}{N} \int_{\mathbb{R}^2} \frac{\hat{\lambda}^k[u-b]N_u[du]}{\hat{\lambda}^k[u-z]\hat{h}^k[z]dz}, \qquad (5.93)$$

with $\hat{\Lambda}$ estimated as the total photon count according to:

$$\hat{\Lambda} = \int_{\mathbb{R}^2} N_u[du]. \qquad (5.94)$$

Note that the equations (5.92) and (5.93) represent a full iteration of the EM algorithm, whose structure very much resembles that of the alternative minimization methods discussed in Section 5.4. Moreover, as discussed earlier in this section, it is known that the convergence of the methods of this type might be problematic in the case when the intermediate solutions are not corrected by enforcing some constraints, defined by known properties of the optimal solutions. It is interesting to note, however, that the update equations (5.92) and (5.93) are capable of automatically maintaining a number of implicit constraints. First of all, it can be shown that the nonnegativity of $\hat{\lambda}^{k+1}$ and $\hat{h}^{k+1}$ is preserved, as long as the initial guesses for these quantities are nonnegative. Second, by integrating both sides of (5.92) it can be shown that each iteration preserves the total photon count. As well, by integrating both sides of (5.93) it can be shown that the unit volume of the estimated PSF is preserved in each iteration (owing to the normalization via division by N).

It is definitely possible to incorporate some additional prior knowledge about the PSF into the algorithm. For instant, in the cases when PSF is known to be circularly symmetric, only a subtle correction to (5.93) needs to be done in order to take this constraint into account [102]. Another possible constraint, considered in [102], follows from the property of the PSF of being a bandlimited function. Specifically, if it was known that the Fourier transform of the PSF should be zero outside a predefined passband, then it would be possible to correct the intermediate estimate of $h[b]$ by computing its Fourier transform, followed by setting to zero all its values corresponding to the stopband. Note that this operation can be recognized as a projection onto a convex set, since the set of bandlimited functions having the same passband support is obviously convex [25].

5.5.2   Refinements of the EM Algorithms

It was observed in [103] that the EM algorithm presented above is prone to errors caused by the background intensities originating from the dark current and electron-traps of the cooled CCD camera. To overcome this deficiency, it was proposed to constrain the PSF background intensity with Lagrange multipliers. In particular, the method proposed in [103] defines a background region in the domain of the PSF and constrains the summation of the PSF values to be less than a predefined upper bound.

Specifically, the parasitic background intensities can be described by an additional inhomogeneous Poisson random-point process with an intensity function $\mu_0[u]$. The latter may be calibrated via collecting an image with the camera shutter closed and with a relatively large number of total photons counted (so as to minimize the effects of quantum-photon noise). Consequently, following the derivations in [103], the resulting EM equations can be shown to be given by:

$$\hat{\lambda}^{k+1}[x] = \hat{\lambda}^k[x] \int_{\mathbb{R}^2} \frac{\hat{h}^k[u-x]N_u[du]}{\left[\hat{h}^k[u-z]\hat{\lambda}^k[z]dz + \mu_0[u]\right]}, \tag{5.95}$$

$$\hat{h}^{k+1}[b] = \frac{\hat{h}^k[b]}{N} \int_{\mathbb{R}^2} \frac{\hat{\lambda}^k[u-b]N_u[du]}{\left[\hat{\lambda}^k[u-z]\hat{h}^k[z]dz + \mu_0[u]\right]}. \tag{5.96}$$

The refinement expressed by (5.95) and (5.96) was demonstrated to result in substantial improvement of the quality of the final reconstruction, and hence it may be a useful tool in the settings when background intensity corrections are needed.

5.5.3   Blind Deconvolution in 3-D Transmitted Light Brightfield Microscopy

In finalizing this section, we would like to say a few words on the applicability of post-processing by means of blind deconvolution for reconstruction in *transmitted light brightfield* (TLB) microscopy. This application seems to be very important, since the TLB microscope is probably the most widely used modality in the life-sciences laboratory. Moreover, it was demonstrated in [105] that blindly deconvolved TLB images of fine spinal structures provide better representation (in terms of resolution and distortion correction) than images acquired using a reflected light confocal microscope. Thus, the availability of blind deconvolution methods makes it possible to obtain high-resolution images of 3-D structures by light microscopy of absorbing stains.

In order to formalize the image formation model in TLB microscopy, let $y[x, y, z]$ denote the 3-D distribution of the optical density and $h[x, y, z]$ denote the PSF of the optical system. Consequently, denoting by B the background image intensity (which is a characteristic of TLB images) and by $u[x, y, z]$

a noise component in the data, the observed TLB image $g_u[x, y, z]$ can be represented as given by [105]:

$$g_u[x, y, z] = B\, y[x, y, z] * h[x, y, z] + u[x, y, z]. \tag{5.97}$$

Note that the pair $[x, y]$ usually stands for the in-plane coordinates, while z denotes the axial coordinate.

The blind reconstruction in TLB microscopy consists in recovering the true 3-D distribution of the optical density $y[x, y, z]$ given no *a priori* information on the PSF $h[x, y, z]$. We note that, though the latter can be measured via a properly designed calibration procedure, this measurement is often imprecise and difficult to implement, for the very same reasons that were mentioned in the beginning of Section 5.5.1.

The blind reconstruction procedure for the reconstruction of TLB images is usually started by detecting the background level B by searching for the maximum value in the data[8]. Subsequently, $g_u[x, y, z]$ can be subtracted from B, resulting in:

$$g_p[x, y, z] = y[x, y, z] * h[x, y, z] - u[x, y, z]. \tag{5.98}$$

From here on, the convolution model (5.98) is defined in its canonical form, which allows us to tackle the problem of recovering the true image $y[x, y, z]$ using some standard approaches. In particular, it was proposed in [105] to solve the blind deconvolution problem using the AM-type algorithm, which was derived by properly modifying the maximum likelihood approach presented in the preceding subsection. The resulting update equations given by [105] are

$$\lambda^{k+1}[x, y, z] = \lambda^k[x, y, z] \left[\frac{y_p[x, y, z]}{(\lambda^k * h^k)[x, y, z]} * h^k[-x, -y, -z] \right], \tag{5.99}$$

$$h^{k+1}[x, y, z] = \frac{h^k[x, y, z]}{N} \left[\frac{y_p[x, y, z]}{(\lambda^k * h^k)[x, y, z]} * \lambda^k[-x, -y, -z] \right], \tag{5.100}$$

where N denotes the total photon count in the 3-D data set. Note that this algorithm can also be shown to be optimal for the case when the Poisson statistics are assumed. More details regarding the performance of this method can be found in [105] (see also the references therein).

5.6   Summary

In this chapter we have reviewed a number of key methods of blind deconvolution which are commonly used for post-processing of medical images.

---

[8]This method, however, is reliable as long as the data are properly precorrected for photodetector sensitivity, bias and shutter-speed flicker, and no saturated pixels are present in the observations.

Considering the fact that the modern arsenal of tools for solving blind deconvolution problems contains tens (if not hundreds) of diverse approaches, an effort has been made to review only those methods that are specific to medical imaging. For this reason, most of this chapter was dedicated to describing the methods for blindly deconvolving medical ultrasound images – images, the diagnostic value of which is believed to be substantially improvable via proper post-processing. While quite a few of these methods have been adopted from other nonmedical applications (e.g., seismic prospecting), there also exists a large group of approaches, which have originated from ultrasound image processing and have the potential to be applicable to a wider range of reconstruction tasks. For this reason, the problem of reconstruction by means of blind deconvolution in optical coherent tomography (OCT) was not addressed in this chapter, since the mechanism of image formation and the statistical nature of the signals in OCT are very similar to those of ultrasound, thereby implying the similarity of possible reconstruction approaches.

Additionally, an attempt was made to associate each of the reviewed imaging modalities with the blind deconvolution methods that work best for this modality, and may not be that efficient if applied to the others. For example, the alternative minimization algorithms presented in Section 5.4 take advantage of the fact that both the true image and the PSF are expected to be real, positive-valued functions (of, probably, known support). These assumptions seem to be natural when applied to the SPECT imagery, as well as to many other medical imaging modalities (e.g., PET). At the same time, the methods described in Section 5.4 do not seem to be applicable to ultrasound images, where the samples of both the true image and the PSF can generally be of an arbitrary sign and even complex valued. On the other hand, no attempt has been made to apply the "ultrasound-oriented" methods of blind deconvolution to either SPECT or microscopy imagery, as these methods were derived under the assumption that the true object (i.e., the reflectivity function) is a random, zero-mean sequence – an assumption which seems to be reasonable for the ultrasound images, but senseless for both SPECT and optical microscopy images.

Another goal pursued in this chapter was to demonstrate some different approaches to modeling of the true image and of the PSF. In particular, in Sections 5.2 through 5.4 these quantities were considered to be members of either functional or probabilistic subspaces. In the latter case, the functions to be recovered were characterized by their corresponding probability densities, and a number of optimality criteria were derived based on this probabilistic description. On the other hand, the physics underpinning the process of data acquisition in fluorescence microscopy suggested a different approach to the modeling of the true image (i.e., the fluorescence distribution) and of the PSF, where they were regarded as probability distributions by themselves. This modeling has resulted in a different form of the maximum likelihood optimality, and, as a result, in a blind deconvolution algorithm with distinctive performance characteristic.

It is obvious that due to space limitations, this review includes only a relatively narrow spectrum of existing medical imaging modalities, together with their associated blind deconvolution algorithms. However, we believe that the material of this chapter represents a reasonable and important group of methodologies, which are general enough for being applicable to a broader variety of medical imaging modalities, where reconstruction by means of blind deconvolution is needed to make a difference in the quality of healthcare.

References

[1] P. C. Hansen, *Rank-deficient and discrete ill-posed problems.* Numerical Aspects of Linear Inversion, SIAM, 1998.

[2] C. Vogel, *Computational Methods for Inverse Problems.* SIAM, 2002.

[3] B. Angelsen, *Ultrasound Imaging: Waves, Signals, and Signal Processing.* Trondhejm, Norway: Emantec, 2000.

[4] P. R. Stepanishen, "An approach to computing time-dependent interaction forces and mutual radiation impedances between pistons in a rigid planar baffle," *Journal of the Acoustical Society of America*, vol. 49, pp. 283–292, Jan. 1971.

[5] J. A. Jensen, "A model for the propagation and scattering of ultrasound in tissue," *Journal of the Acoustical Society of America*, vol. 89, pp. 182–191, Jan. 1991.

[6] D. Colton, *Inverse Acoustic and Electromagnetic Scattering Theory.* Springer-Verlag, 1998.

[7] S. J. Norton and M. Linzer, "Ultrasonic reflectivity imaging in three dimensions: Exact inverse scattering solutions for plane, cylindrical and spherical apertures," *IEEE Transactions on Biomedical Engineering*, vol. BME-28, pp. 202–220, Feb. 1981.

[8] U. R. Abeyratne, A. P. Petropulu, and J. M. Reid, "Higher order spectra based deconvolution of ultrasound images," *IEEE Transactions on Ultrasonics, Ferroelectrics and Frequency Control*, vol. 42, pp. 1064–1075, Nov. 1995.

[9] M. Cloostermans and J. Thijssen, "A beam corrected estimation of the frequency dependent attenuation of biological tissues from backscattered ultrasound," *Ultrasonic Imaging*, vol. 5, pp. 136–147, April 1983.

[10] P. Narayana, J. Ophir, and N. Maklad, "The attenuation of ultrasound in biological fluids," *Journal of the Acoustical Society of America*, vol. 76, pp. 1–4, July 1984.

[11] K. A. Wear, "The effects of frequency-dependent attenuation and dispersion on sound speed measurements: Application in human trabecular bone," *IEEE Transactions on Ultrasonics, Ferroelectrics and Frequency Control*, vol. 47, pp. 265–273, January 2000.

[12] G. B. Saha, *Physics and radiobiology of nuclear medicine.* Springer-Verlag, 2001.

[13] D. Boulfelfel, R. M. Rangayyan, L. J. Han, and R. Kloiber, "Prereconstruction restoration of myocardial single photon emission computed tomography images," *IEEE Transactions on Medical Imaging*, vol. 11, no. 3, pp. 336–341, 1992.

[14] M. T. Madsen and C. H. Park, "Enhancement of spect images by Fourier filtering the projection set," *Journal of Nuclear Science and Technology*, vol. 26, pp. 2687–2690, 1979.

[15] A. R. Hibbs, *Confocal microscopy for biologists.* Kluwer Academic/Plenum Publishers, 2004.

[16] www.api.com/products/bio/deltavision.html, "Deconvolution software produced by DeltaVision."

[17] www.aqi.com, "Deconvolution software produced by AutoQuant."

[18] J. Ming, W. Ge, M. W. Skinner, J. T. Rubinstein, and M. Vannier, "Blind deblurring of spiral CT images-comparative studies on edge-to-noise ratios," *Medical Physics*, vol. 29, pp. 821–829, May 2002.

[19] T. S. Ralston, D. L. Marks, F. Kamalabadi, and S. A. Boppart, "Deconvolution methods for mitigation of transverse blurring in optical coherence tomography," *IEEE Transactions on Image Processing*, vol. 14, pp. 1254–1264, Sept. 2005.

[20] S. R. Eliason, *Maximum Likelihood Estimation: Logic and Practice.* Newbury Park, 1993.

[21] A. Birnbaum, "On the foundations of statistical inference (with discussion)," *Journal of the American Statistical Association*, vol. 57, pp. 269–326, 1962.

[22] B. P. Carlin and T. A. Louis, *Bayes and empirical Bayes methods for data analysis*, vol. 69 of *Monographs on Statistics and Applied Probability.* Chapman and Hall, 1996.

[23] E. D. Micheli, N. Magnoli, and G. A. Viano, "On the regularization of Fredholm integral equations of the first kind," *SIAM Journal on Mathematical Analysis* , vol. 29, pp. 855–877, July 1998.

[24] E. Sekko, G. Thomas, and A. Boukrouche, "A deconvolution technique using optimal Wiener filtering and regularization," *Signal processing*, vol. 72, pp. 23–32, Jan. 1999.

[25] H. Stark, *Image Recovery: Theory and Applications*. Academic Press, 1987.

[26] O. Michailovich and D. Adam, "Blind deconvolution of ultrasound images using partial spectral information and sparsity constraints," in *Proceedings IEEE ISBI*, 2002.

[27] P. L. Combettes and J. C. Pesquet, "Image restoration subject to a total variation constraint," *IEEE Transactions on Image Processing*, vol. 13, pp. 1213–1222, Sept. 2004.

[28] D. Bertsekas, *Nonlinear Programming*. Athena Scientific, 1999.

[29] R. Dembo and T. Steihaug, "Truncated-Newton algorithms for large-scale unconstrained optimization," *Mathematical Programming*, vol. 26, pp. 190–212, June 1983.

[30] G. Ayers and J. Dainty, "Iterative blind deconvolution methods and its applications," *Optics Letters*, vol. 13, pp. 547–549, July 1988.

[31] D. Kundur and D. Hatzinakos, "Blind image deconvolution," *IEEE Signal Processing Magazine*, vol. 13, pp. 43–64, May 1996.

[32] R. Jirik and T. Taxt, "Two-dimensional blind iterative deconvolution of medical ultrasound images," in *In 2004 IEEE Ultrasonics Symposium*, vol. 2, pp. 1262–1265, 2004.

[33] O. Michailovich and A. Tannenbaum, "Deconvolution of medical ultrasound images via parametrized inverse filtering," in *Proceedings IEEE ISBI*, 2006.

[34] J. Nagy and D. O'Leary, "Restoring images degraded by spatially variant blur," *SIAM Journal on Scientific Computing*, vol. 19, pp. 1063–1082, July 1998.

[35] S. L. Marple Jr., *Digital Spectral Analysis with Applications*. Prentice-Hall, 1987.

[36] S. M. Kay, *Modern Spectral Estimation*. Prentice-Hall, 1987.

[37] J. A. Jensen, "Estimation of in vivo pulses in medical ultrasound," *Ultrasonic Imaging*, vol. 16, pp. 190–203, 1994.

[38] K. B. Rasmussen, "Maximum likelihood estimation of the attenuated ultrasound pulse," *IEEE Transactions on Signal Processing*, vol. 42, pp. 220–222, Jan. 1994.

[39] J. Makhoul, "Linear prediction: A tutorial review," *Proceedings of IEEE*, vol. 63, pp. 561–580, Apr. 1975.

[40] W. F. Trench, "An algorithm for the inversion of finite Toeplitz matrices," *Journal of the Society for Industrial and Applied Mathematics*, vol. 12, pp. 515–522, 1964.

[41] F. Towfig, C. W. Barnes, and E. J. Pisa, "Tissue classification based on autoregressive models for ultrasound pulse echo data," *Acta Electronica*, vol. 26, pp. 95–110, 1984.

[42] L. Y. Shih, C. W. Barnes, and L. A. Ferrari, "Estimation of attenuation coefficient for ultrasonic tissue characterization using time-varying state-space model," *Ultrasonic Imaging*, vol. 10, pp. 90–109, 1988.

[43] D. S. G. Pollock, *A Handbook of Time-Series Analysis, Signal Processing and Dynamics*. Academic Press, 1999.

[44] G. E. P. Box and G. M. Jenkins, *Time Series Analysis Forecasting and Control*. Holden-Day, 1976.

[45] G. Georgiou and F. Cohen, "Statistical characterization of diffuse scattering in ultrasound images," *IEEE Transactions on Ultrasonics, Ferroelectrics and Frequency Control*, vol. 45, pp. 57–64, Jan. 1998.

[46] C. L. Nikias and A. P. Petropulu, *Higher-order spectral analysis: A nonlinear signal processing framework*. Englewood Cliffs, NJ: Prentice Hall, 1993.

[47] T. Taxt, "Three-dimensional blind deconvolution of ultrasound images," *IEEE Transactions on Ultrasonics, Ferroelectrics and Frequency Control*, vol. 48, pp. 867–871, July 2001.

[48] A. V. Oppenheim and R. W. Schafer, *Discrete Time Signal Processing*. Prentice Hall, 1989.

[49] J. A. Jensen and S. Leeman, "Nonparametric estimation of ultrasound pulses," *IEEE Transactions on Biomedical Engineering*, vol. 41, pp. 929–936, Nov. 1994.

[50] T. Taxt, "Comparison of cepstrum-based methods for radial blind deconvolution of ultrasound images," *IEEE Transactions on Ultrasonics, Ferroelectrics and Frequency Control*, vol. 44, pp. 666–674, May 1997.

[51] T. Taxt, "Restoration of medical ultrasound images using two-dimensional homomorphic deconvolution," *IEEE Transactions on Ultrasonics, Ferroelectrics and Frequency Control*, vol. 42, pp. 543–554, July 1995.

[52] K. Steiglitz and B. Dickinson, "Phase unwrapping by factorization," *IEEE Transactions on Acoustic, Speech, and Signal Processing*, vol. ASSP-30, pp. 984–991, Dec. 1982.

[53] A. T. Bharucha-Reid and M. Sambandham, *Random polynomials*. Probability and Mathematical Statistics, New York: Academic Press, 1986.

[54] B. P. Bogert, M. J. R. Healy, and J. W. Tukey, "The quefrency alanysis of time series for echoes: cepstrum, pseudo-autocovariance, cross-

cepstrum, and saphe cracking," in *Proceedings of the Symposium on Time Series Analysis*, pp. 209–243, New York: Wiley, 1963.

[55] D. G. Childers, D. P. Skinner, and R. C. Kemerait, "The cepstrum: A guide to processing," *Proceedings of IEEE*, vol. 65, pp. 1428–1443, 1977.

[56] R. W. Hamming, *Digital Filters*. Courier Dover Publications, 1989.

[57] T. Taxt and J. Strand, "Two-dimensional noise-robust blind deconvolution of ultrasound images," *IEEE Transactions on Ultrasonics, Ferroelectrics and Frequency Control*, vol. 48, pp. 861–866, July 2001.

[58] J. R and T. Taxt, "High-resolution ultrasonic imaging using two-dimensional homomorphic filtering," *IEEE Transactions on Ultrasonics, Ferroelectrics and Frequency Control*, vol. 53, pp. 1440–1448, August 2006.

[59] D. C. Ghiglia and M. D. Pritt, *Two-Dimensional Phase Unwrapping: Theory, Algorithms, and Software*. New York: Wiley-Interscience, 1998.

[60] H. Nyquist, "Certain topics in telegraph transmission theory," *Trans. AIEE*, vol. 47, pp. 617–644, April 1928.

[61] R. M. Goldstein, H. A. Zebker, and C. L. Werner, "Satellite radar interferometry: Two-dimensional phase unwrapping," *Radio Science*, vol. 23, pp. 713–720, July-Aug. 1988.

[62] M. D. Pritt and J. S. Shipman, "Least-square two-dimensional phase unwrapping using FFTs," *IEEE Transactions on Geoscience and Remote Sensing*, vol. 32, pp. 706–708, May 1994.

[63] S. M. Song, S. Napel, N. J. Pelc, and G. H. Glover, "Phase unwrapping of MR phase images using Poisson equation," *IEEE Transactions on Image Processing*, vol. 4, pp. 667–676, May 1995.

[64] J. Strand and T. Taxt, "Two-dimensional phase unwrapping using robust derivative estimation and adaptive integration," *IEEE Transactions on Image Processing*, vol. 11, pp. 1192–1200, Oct. 2002.

[65] J. Strand and T. Taxt, "Performance evaluation of two-dimensional phase unwrapping algorithms," *Applied Optics*, vol. 38, pp. 4333–4344, July 1999.

[66] O. Michailovich and D. Adam, "Phase unwrapping for 2-D blind deconvolution of ultrasound images," *IEEE Transactions on Medical Imaging*, vol. 23, pp. 7–25, Jan. 2004.

[67] J. M. Tribolet, "A new phase unwrapping algorithm," *IEEE Transactions on Acoustic, Speech, and Signal Processing*, vol. ASSP-25, pp. 170–177, April 1977.

[68] R. A. Adams, *Sobolev spaces.* Pure and Applied Mathematics Series, Academic Press, 1978.

[69] K. Rektorys, *Variational Methods in Mathematics, Science, and Engineering.* Dordrecht: Reidel, 1980.

[70] C. D. Boor, R. DeVore, and A. Ron, "Approximation from shift-invariant subspaces in $\mathbb{L}_2(\mathbb{R}^d)$," *Transactions of the American Mathematical Society*, vol. 341, pp. 787–806, 1994.

[71] G. Strang and G. Fix, "A Fourier analysis of the finite element variational method," *Constructive Aspects of Functional Analysis*, Edizioni Cremonese, pp. 795–840, 1973.

[72] O. Michailovich and D. Adam, "A novel approach to the 2-D blind deconvolution problem in medical ultrasound," *IEEE Transactions on Medical Imaging*, vol. 24, pp. 86–104, Jan. 2005.

[73] D. L. Donoho, "De-noising by soft-thresholding," *IEEE Transactions on Information Theory*, vol. 41, pp. 613–627, May 1995.

[74] D. L. Donoho and I. Johnstone, "Adapting to unknown smoothness via wavelet shrinkage," *Journal of The American Statistical Association*, vol. 90, pp. 1200–1224, 1995.

[75] O. Michailovich and D. Adam, "Robust estimation of ultrasound pulses using outlier-resistant de-noising," *IEEE Transactions on Medical Imaging*, vol. 22, pp. 368–392, March 2003.

[76] I. Daubechies, *Ten Lectures on Wavelets.* SIAM, 1992.

[77] S. G. Mallat, *A Wavelet Tour of Signal Processing.* New York: Academic Press, 1998.

[78] P. Muller and B. Vidakovic, *Bayesian Inference in wavelet-based models*, vol. 141 of *Lecture Notes in Statistics*. New York: Springer-Verlag, 1999.

[79] O. Michailovich and A. Tannenbaum, "Fast approximation of smooth functions from the samples of their partial derivatives," *Signal Processing*, to appear.

[80] J. A. Cadzow, "Blind deconvolution via cumulant extrema," *IEEE Signal Processing Magazine*, vol. 13, pp. 24–42, May 1996.

[81] R. A. Wiggins, "Minimum entropy deconvolution," *Geoexploration*, vol. 16, pp. 21–35, 1978.

[82] D. Donoho, "On minimum entropy deconvolution," *http://www-stat.stanford.edu/ donoho/Reports/Oldies/index.html*, 1981.

[83] A. K. Nandi, D. Mampel, and B. Roscher, "Blind deconvolution of ultrasound signals in non-destructive testing applications," *IEEE Transactions on Signal Processing*, vol. 45, pp. 1382–1390, May 1997.

[84] J. F. Claerbout, "Parsimonious deconvolution," *Standford Exploration Project*, vol. 13, pp. 1–9, 1977.

[85] M. M. Bronstein, A. M. Bronstein, M. Zibulevsky, and Y. Y. Zeevi, "Blind deconvolution of images using optimal sparse representations," *IEEE Transactions on Image Processing*, vol. 14, pp. 726–736, June 2005.

[86] O. Michailovich and A. Tannenbaum, "Blind deconvolution of medical ultrasound images: Parametric inverse filtering approach," *IEEE Transactions on Medical Imaging*, to appear.

[87] S. Webb, A. P. Long, R. J. Ott, M. O. Leach, and M. A. Flower, "Constrained deconvolution of SPECT liver tomograms by direct digital image restoration," *Medical Physics*, vol. 12, pp. 53–58, 1985.

[88] M. Mignotte, J. Meunier, J.-P. Soucy, and C. Janicki, "Comparison of deconvolution techniques using a distribution mixture parameter estimation: Application in single photon emission computed tomography imagery," *Journal on Electronic Imaging*, vol. 11, pp. 11–24, Jan. 2002.

[89] M. A. King, P. W. Doherty, and R. B. Schwinger, "A Wiener filter for nuclear medicine images," *Medical Physics*, vol. 10, pp. 876–880, 1983.

[90] J. C. Yanch, M. A. Flower, and S. Webb, "A comparison of deconvolution and windowed subtraction techniques for scatter compensation in SPECT," *IEEE Transactions on Medical Imaging*, vol. 7, pp. 13–20, 1988.

[91] M. Mignotte and J. Meunier, "Three-dimensional blind deconvolution of SPECT images," *IEEE Transactions on Biomedical Engineering*, vol. 47, pp. 274–280, Jan. 2000.

[92] S. J. Glick, M. A. King, K. Knesaurek, and K. Burbank, "An investigation of the stationarity of the 3-D modulation transfer function in SPECT," *IEEE Transactions on Nuclear Science*, vol. 36, pp. 973–977, 1989.

[93] J. A. Fessler, University of Michigan, "Personal communications," April 2006.

[94] Y. L. You and M. Kaveh, "A regularization approach to joint blur identification and image restoration," *IEEE Transactions on Image Processing*, vol. 5, pp. 416–428, March 1996.

[95] P. Moulin, "Wavelet thresholding techniques for power spectrum estimation," *IEEE Transactions on Signal Processing*, vol. 42, pp. 3126–3136, Nov. 1994.

[96] T. F. Chan and C.-K. Wong, "Total variation blind deconvolution," *IEEE Transactions on Image Processing*, vol. 7, pp. 370–375, March 1998.

[97] P. C. Hansen and D. P. O'Leary, "The use of the L-curve in the regularization of discrete ill-posed problems," *SIAM Journal on Scientific Computing*, vol. 14, pp. 1487–1503, 1993.

[98] D. Kundur and D. Hatzinakos, "A novel blind deconvolution scheme for image restoration using recursive filtering," *IEEE Transactions on Signal Processing*, vol. 46, pp. 375–390, Feb. 1998.

[99] A. P. Dempster, N. M. Laird, and D. B. Rubin, "Maximum likelihood from incomplete data via the EM algorithm," *Journal of the Royal Statistical Society B*, vol. 39, pp. 1–37, 1977.

[100] L. Shepp and Y. Vardi, "Maximum likelihood reconstruction for emission tomography," *IEEE Transactions on Medical Imaging*, vol. MI-1, pp. 113–122, 1982.

[101] T. J. Holmes, "Maximum-likelihood image restoration adapted for noncoherent optical imaging," *Journal of the Optical Society of America A*, vol. 5, pp. 666–673, 1988.

[102] T. J. Holmes, "Blind deconvolution of quantum-limited incoherent imagery: maximum-likelihood approach," *Journal of the Optical Society of America A*, vol. 9, pp. 1052–1061, July 1992.

[103] V. Krishnamurthi, Y.-H. Liu, S. Bhattacharayya, J. N. Turner, and T. J. Holmes, "Blind deconvolution of fluorescence micrographs by maximum-likelihood approach," *Applied Optics*, vol. 34, pp. 6633–6647, Oct. 1995.

[104] D. L. Snyder, *Random Point Processes*. New York: Wiley, 1975.

[105] T. J. Holmes and N. J. O'Connor, "Blind deconvolution of 3-D transmitted light brightfield micrographs," *Journal of Microscopy*, vol. 200, pp. 114–127, 2000.

6

Bayesian Estimation of Blur and Noise in Remote Sensing Imaging

André Jalobeanu

PASEO research group, MIV team (LSIIT UMR 7005 CNRS-ULP), Illkirch, France

e-mail: jalobeanu@lsiit.u-strasbg.fr

Josiane Zerubia, Laure Blanc-Féraud

Ariana research group (INRIA-I3S), Sophia Antipolis, France

e-mail: (Josiane.Zerubia, Laure.Blanc_Feraud)@sophia.inria.fr

Abstract

We propose a Bayesian approach to estimate the parameters of both blur and noise in remote sensing images. The goal is to infer the modulation transfer function (MTF) related to each observation, including atmospheric effects, optical blur, uniform motion, and pixel-level sampling. The MTF is modeled by a real-valued parametric function with a small number of parameters. The noise is assumed to be an additive, white Gaussian process. Both blur and noise processes are supposed to be stationary. To constrain this ill-posed inverse problem, the unknown scene is modeled by a scale-invariant stochastic process governed by a fractal exponent and a global energy term. The main novelty consists of treating all parameters as random variables whose mean is estimated within a fully Bayesian framework. The chosen approach can be summarized as the computation of the mean posterior marginal related to useful parameters only. This requires integrating the joint probability density function (PDF) with respect to all the nuisance parameters, which is achieved through Laplace approximations. In this chapter we present two approaches; the former is straightforward, and the latter leads to a more efficient, simplified, and optimized estimation algorithm. In addition, we investigate methods of uncertainty estimation and model assessment, in order to validate our approach on real images and to propose further improvements.

6.1   Introduction

In Earth observation, sometimes there are no other data about the scene of interest but a single picture, usually corrupted by blur and noise. In such a case, how is it possible to recover a good-quality image of this scene? This difficult, underdetermined problem is known as blind deconvolution. It covers many applications, ranging from astronomy and remote sensing to microscopy. However, we will restrict ourselves to satellite or aerial imaging, and we will mostly focus on natural images (as opposed to images of man-made structures). Moreover, our main interest will be in the degradation model, not in the image itself; we will assume that existing nonblind deblurring techniques could be used efficiently if we are able to provide both blur and noise models accurately enough. Numerous image deblurring algorithms have been developed over the past ten years [1, 2] but they are generally nonblind and therefore require a good knowledge of both blur and noise in order to work properly.

The large number of unknown variables, as well as the multiplicity and instability of solutions that are compatible with the observed image, make this problem particularily difficult to solve. It is twice ill-posed, since both the image and the blur are unknown. Much effort has been put into finding satisfactory solutions in the past 20 years. Despite the apparent complexity of the task, people have managed to come up with interesting results by different means. Most methods require to constrain both image and blur in different ways. However, such constraints are often application-specific and could not be used in remote sensing. We will show in the next subsection that despite the rich literature in this domain, we had to develop a new method to be able to process satellite images effectively.

In the following, the same blur is denoted by either PSF or MTF, which respectively refer to the point spread function in the image space, and the modulation transfer function in the frequency space. The MTF is the modulus of the Fourier transform of the PSF.

6.1.1 Blind Deconvolution: State of the Art

There is no single way to classify the existing blind deconvolution methods. Several grouping schemes could be used depending on the application, the noise statistics, the blur kernel parametrization, and the proposed estimation method. Reviews can be found in [3,4]. Here we will use a classification based on these reviews.

6.1.1.1 Independent Blur Identification and Deblurring

Let us first consider the group of methods that treat the blur estimation and the image restoration independently. There are a few techniques that are designed to identify the blur directly from an image, for instance from known scenes (such as man-made targets, for instance) or through very strong prior knowledge, such as in astronomy where stars are supposed to be points before being blurred. Therefore, the PSF can be determined by deconvolving the observed image by the underlying object. Anyway, no such object is usually visible in satellite images. Weaker priors could be used to solve the problem in more general cases, for instance in x-ray imaging when some knowledge of the object contours is available [5]. Even in remote sensing, one could think of using geometric features such as straight roads or sea shores, though such features are not always encountered and seldom exhibit more than a few orientations, which impedes an accurate parameter estimation in all directions.

In some special cases, where camera motion or an out-of-focus lens are considered, some methods can be used [6–8] that take advantage of the zeros in the observed image spectrum, related to zeros in the transfer function. The problem therefore reduces to locating these zeros, which can be tricky because of the noise. MTFs encountered in remote sensing rarely exhibit enough zeros to allow for a clear identification.

Some authors have used autoregressive models to represent the unknown image, combined with the blur and noise model to form a so-called autoregressive moving average process. In this framework, the model parameters can be simultaneously identified [9]. The estimation could be performed through expectation maximization [10] or generalized cross-validation [11]. Other methods use the image bispectrum [7] to estimate the phase of the transfer function, but they suffer from instabilities.

The drawback of such methods is the lack of realism of the underlying scene model, which is assumed to be linear and stationary, without even capturing the scale invariance properties proper to natural images. When the scene contains sharp edges or spatially varying textures, these models fail to estimate the transfer function correctly. This kind of global approach, including both scene and blur in the model, can succeed only if both scene and blur are modeled accurately.

Recently, new promising approaches have been developed. In the field of neural networks, the technique described in [12] was applied to satellite images to measure a parametric MTF from random images. However, such a method requires a computationally expensive training step prior to the estimation, and it is unclear how well it will perform when the training data set is limited to the observed image. Other approaches take advantage of the availability of at least two images of the same scene at different resolutions (i.e., 1 m and 2.5 m) to measure the ratio between low- and high-resolution MTFs [13]. Our approach relies upon a rather different theory and is worth developing mainly because it only needs a single input image.

6.1.1.2   Joint Blur Estimation and Deblurring

The second group of methods aims at the restoration of the corrupted image and provides the blur function as a by-product. If any of these methods were able to solve our problem, we would stop here. Let us review them quickly to understand why they are not well suited to remote sensing imagery.

Apart from a few theoretical findings, mathematically interesting but of little interest in practice (e.g., the zero-sheet separation [14]) because of unrealistic assumptions, such as special blur kernels or noise-free images, several methods proved to be useful in some special cases.

A first class of algorithms attempt to estimate the blur kernel pixelwise, assuming a small spatial support. The PSF pixels can be considered as random variables within a probabilistic framework. One will typically try to maximize the probability of both image and blur, which can be achieved through minimizing an energy term, usually nonquadratic. The different constraints on this PSF (such as positivity, normalization, and support, for instance) can be embedded in properly designed prior probability densities. They can alternatively be introduced in an artificial way during the optimization [15] but without any proof of convergence. They can also be enforced during the optimization process via projections, penalty terms, reparametrization, or even by doing constrained optimization. The energy term will then be minimized alternately with respect to (w.r.t.) the image and the PSF, thus forming the series of iterative blind deconvolution (IBD) algorithms.

Such techniques are proposed in [16] and [17] where constraints are enforced in both image and frequency spaces. Stability is not guaranteed. In [18] a nonquadratic penalty term is used on both image and blur, which allows for edge-preserving regularization of the solutions. This is obviously better suited to boxcar PSFs than smoother, more regular and realistic ones. A similar approach is described in [19] in microwave imaging. Although smoother blur kernels can be recovered, such a method is sensitive to the prior choice; usually the prior is defined according to the physics of the problem, and must be close enough to the expected solution. Sometimes the regularization is achieved by stopping the optimization algorithm [20, 21] and projections are used instead of priors. In [22] the constraints are strictly enforced through PSF reparametrization.

Other methods different from IBD exist [3], and seem to converge more rapidly. However, the noise is not taken into account explicitly during the inversion process; therefore, the algorithm is stopped before convergence to avoid noise amplification. Stochastic methods have also been proposed, such as in [23] where the authors use simulated annealing to perform the optimization.

More recently, a method based on sparse representations has been proposed [24]. It requires training on nondegraded images of the same type in order to determine a so-called sparsification kernel. Although this kind of iterative technique performs pretty well on noise-free images, it is unclear whether it is applicable at all in real cases where noise is present. In confocal microscopy, some authors propose to use the coefficients of a steerable pyramid to estimate the PSF [25].

Another class of methods assumes a parametric form of the blur kernel, which is a simple and elegant way to fulfill all the required conditions while taking into account the physics [4]. One may argue that this is too strong a constraint, and therefore does not allow for exotic, data-driven PSFs to be inferred. This is a weak argument: if the physics of the system are known, there is little chance to get odd-shaped blur kernels. Moreover, the theory of model selection would allow for the selection of the best parametrization among different solutions. In the worst case, model checking can help diagnose model failures, and design more appropriate blur models.

In any case, the parametric approach is far less complex and underdetermined than the one relying on a pointwise estimate. The smaller number of parameters can be estimated using classical techniques such as maximum likelihood. For instance, in [26] an expectation maximization (EM) algorithm is used to perform such an estimation, using a scheme designed for astronomical images. Examples of such parametric methods can be found in [27] and [28] where the authors respectively use frequency space and L_1 regularization-based prior models of the unknown image. In [29] the blur is parametrized by the coefficients of a 2×2 matrix; a nonstationary autoregressive image model is used and the estimation is performed within a Bayesian framework.

6.1.2 Constraining a Difficult Problem

Among all the previously developed algorithms, few can be applied to satellite or aerial images representing scenes chosen at random. Except for the inference from known targets, no method is actually able to provide an accurate estimate of the PSF from a single image, suitable for the remote sensing industry. Most methods fail because they are designed to work in a different imaging domain (e.g., astronomy or microwave imaging). The constraints that help recover both blur and image in these cases are of little use on remote sensing scenes. The adequacy between model and image is essential. One of the key issues is to design a relevant image model.

A relevant model is not only useful for image restoration, but also helps discriminate among scene, blur, and noise. This is why we choose to employ a power-law spectrum model; it is different enough from most physics-based

MTFs as well as from white noise spectra. Existing approaches that only use models expressed in the image space (the ones exploiting low-order neighborhoods) fail to capture essential characteristics of natural scenes which exhibit details at all scales. The MTF can be merely seen as the attenuation of the spectrum; therefore spectral properties must be taken into account so that the MTF can be inferred. An obviously wrong spectrum model will systematically impede any attempt to recover it. Sometimes, neither image nor blur models are used at all, hoping that the constraints alone will manage to separate out the blur from the unknown scene. Commonly used constraints are insufficient in satellite imaging, hence our choice of a model-based approach.

Another key point is the reduction of the number of unknown variables. As stated previously, PSF or MTF parametrization help constrain the problem very efficiently. Thus, recovering the blur amounts to estimating a reduced number of parameters instead of determining the dozens or even hundreds of pixels of a discrete blur kernel. We choose to use parametric functions for both MTF and scene spectrum, so that we keep a pretty small parameter count. This will also help integrate out the unwanted variables and analyze the remaining ones easily, as it will be shown in the following sections. No doubt there are enough reasons to reduce the dimensionality of such an inference problem. As long as we do not lose useful information by doing so, we choose to start with the shortest description models; model checking shall tell us if we have to increase the dimensionality or change the parametrization.

6.1.3   The Bayesian Viewpoint

Within a Bayesian framework [30], the model parameters are all assumed to be random variables. To solve a particular problem, one generally seeks to compute the posterior PDF of the variables of interest. It is posterior in the sense that it is conditioned upon all the observations. In this work, we focus on the determination of the mean of this posterior PDF.

However, the users of such probabilistic approaches sometimes do not try to remove the uninteresting variables from the PDF, to simplify the calculations. Though in a purely Bayesian approach, one must integrate over all other variables (also called nuisance variables) [29, 31, 32], the ones that are neither the parameters to be estimated, nor the observations. This step, also known as marginalization, can prove rather tricky in some cases. It is ignored most of the time for that very reason. Nonetheless, we will try to comply to the original methodology and integrate out the image model parameters, since we aim at the inference of the blur and noise parameters only. A few approximations will be required to achieve this integration in a computationally efficient way.

6.2   The Forward Model

In this section, we describe a generative model that is assumed to be a sufficient description of how the spectrum of an observed image is formed from an
original, natural scene. Even though its simplicity prevents us from using it to
deblur images efficiently, it provides a precise enough modeling of both degradations and original scene power spectrum to allow for a blur/noise/signal
separation.

6.2.1   Modeling the Natural Scene Using Fractals

Natural scenes, assumed to be defined on a continuous bounded support
included in $\mathbb{R}^2$, are supposed to follow a fractal model [33,34]. This comes
from the fact that the power spectrum of most images of natural landscapes
(incidentally, it is also true with European cities) exhibits scale-invariance
properties [35]. Natural phenomena are statistically scale-invariant, which
means that rescaling objects does not affect the measured statistics. This can
also be called self-similarity: a scene resembles a scaled version of itself, or
even a subset of itself – of course from a statistical viewpoint. The stochastic approach is then a convenient way to model these properties. Up to a
multiplicative factor, the statistic regarded as a function of spatial variables
does not change, whereas the transformed object can exhibit noticeable differences. Commonly used statistics are the expectations of autocorrelation
functions and power spectra, depending on the working space. Let x and y
denote the image space coordinates, and u and v the normalized spatial frequencies and $r = \sqrt{u^2 + v^2}$ the radial frequency. The normalization is done
w.r.t. the Nyquist frequency [36], so that $(u, v) \in [-1/2, 1/2]^2$.

Let S be the scene, such that $S_{xy} \in \mathbb{R}^+$, defined over the bounded domain
$\Omega \subset \mathbb{R}^2$, and $\mathcal{F}$ the Fourier transform; then our model is

$$\mathcal{F}[S]_{uv} = G_{uv}\, w_0\, r^{-q} \tag{6.1}$$

where G is a Gaussian stationary process (Brownian motion) [34] of marginal
variance 1, and w_0 and q are respectively the energy and fractal exponent
of the spectrum model. The mean power spectrum is the variance of the
process and is equal to $w_0^2\, r^{-2q}$, which is an obviously scale-invariant power-
law function.

There is another important property that somehow has to be accounted for,
namely the nonstationarity. Scenes obviously exhibit a high spatial variability,
since they are made of random arrangements of textured patches, sharp edges,
and sometimes almost constant areas. A simple way to take into account
this variability is to define the process G in the image space, as the product
between an uncorrelated Gaussian process of stationary variance 1 denoted
by $\mathcal{E}$, and a variable field denoted by L, a real-valued function defined over

Ω, possibly scale invariant but whose properties do not really matter for the studied problem. Thus, we set

$$G_{uv} = \mathcal{F}[\mathcal{E} \times L]_{uv} \tag{6.2}$$

This has no direct consequences on the mean power spectrum.

This model is well suited to natural images [37] noncorrupted by blur or noise. It is also very convenient for remote sensing and, surprisingly enough, for urban areas. We performed experiments on a set of images representing different scenes (country, industrial areas, city centers, and so on) and found a good fit between this model and the observations. We plotted $\log E_r(|\mathcal{F}[S]|^2)$ as a function of $\log r$, where E_r is the mean energy within a frequency band of width δr around the radial frequency r. In order to avoid the effects of blur and noise, which corrupt any real image obtained with a sensor through an optical system, the image is subsampled by a factor ranging from 4 to 8, depending on the overall optical quality of the system. This also helps average out the noise. The subsampling is achieved in the frequency space by a sharp band-pass function. An example computed from aerial images is provided in Figure 6.1.

Numerous experiments have shown that on real scenes q is between 0.9 and 1.5, which means that S can be modeled by a fractional Brownian motion [33] in two dimensions.

6.2.2   Understanding the Image Formation

In this section, we briefly explain how to form a discrete image from a known scene defined over a continuous support. The whole physics-based process can be mathematically summarized by a simple canonical transform consisting of blurring, sampling, and noise addition [36, 38].

6.2.2.1   Atmospheric, Optical, and Sensor MTF Modeling

We assume a shift-invariant, convolutional blur, which can be best described by a MTF (no transfer function phase recovery will be attempted here) in the frequency space. Its parameters are denoted by α. This enables us to take into account the physics of light scattering, diffraction, pixel integration, and other nonoptical phenomena that affect the overall blur. Since all the different phenomena are modeled by successive convolutions, the overall MTF is the product of all the individual terms:

$$\text{MTF} = \text{MTF}_{\text{atm}} \times \text{MTF}_{\text{opt}} \times \text{MTF}_{\text{mot}} \times \text{MTF}_{\text{sen}} \tag{6.3}$$

which respectively summarize the atmospheric, optical system, motion, and sensor contributions.

The atmosphere contributes through two major phenomena. Turbulence, which heavily distorts the wavefront, has random effects on both phase and

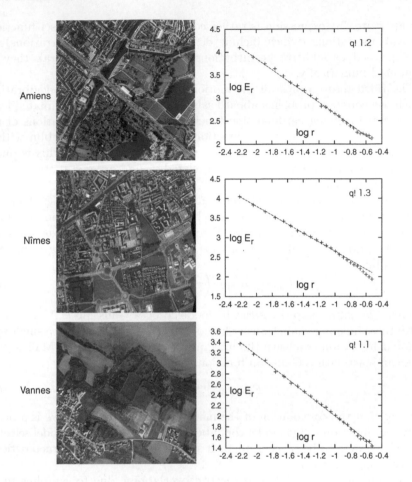

FIGURE 6.1: Left: aerial images, size 512×512, subsampled, no degradation. Right: radial power spectrum (log-log scale) integrated over concentric rings. The dashed line represents the proposed model. Amiens © IGN (French Geographic Institute); Nîmes and Vannes © CNES (French Space Agency).

magnitude of the transfer function. However, for long enough exposures its contribution merely consists of a deterministic space-invariant PSF, equivalent to a MTF expressed as $\exp(-\kappa\, r^{5/3})$ [39]. On the other hand, aerosols contribute to light scattering and attenuation. The effect on the MTF can be modeled by $\exp\left(-\kappa\,(r/r_c)^2\right)$ for r below the cut-off frequency r_c (and a constant above).

The optical system has a limited aperture which causes a diffraction blur, whose MTF is given by the autocorrelation of the pupil function [40]. It is mostly characterized by a cut-off frequency which is an increasing function of the aperture. Its expression is rather complex and involves Bessel functions. Optical aberrations (which include defocus and spherical aberrations) can be expressed as additive perturbations to this MTF. In general, they are computed numerically.

The third term groups all the motion-related blurs: shift and vibrations, which are common problems aboard satellites. It can be approximated by an expression involving cardinal sine functions related to combinations of uniform motions along one or two directions. We assume a motion blur without acceleration and only try to recover a minimum-phase blur. This is due to the chosen Gaussian image model and related second-order statistics that are insensitive to phase shifts.

Finally, the sensor also contributes to the blur, mostly because the irradiance is integrated over pixels, but also because of charge diffusion or transfer problems between adjacent pixels. For a matrix sensor with pixel size p and sampling grid size Δ (allowing for gaps between pixels) the integration MTF is given by

$$(\mathrm{MTF}_{\mathrm{int}})_{uv} = \mathrm{sinc}\left(\pi u \frac{p}{\Delta}\right) \mathrm{sinc}\left(\pi v \frac{p}{\Delta}\right), \tag{6.4}$$

whereas the diffusion process can be modeled by a Gaussian $e^{-\kappa r^2}$.

All these imaging-chain transfer functions can be gathered into a single term denoted by F, and we leave the integration term out: $\mathrm{MTF} = \mathrm{MTF}_{\mathrm{int}} \times F$. One can start with a Gaussian function:

$$F_{uv} = e^{-(\alpha_0 u^2 + \alpha_1 v^2)} \tag{6.5}$$

arguing that the convolution of the many blurs mentioned above is a nearly Gaussian function. This model could be further refined via a model selection step if it proves to be inaccurate, or if some of the physical characteristics of the system are known.

To avoid problems related to spectral overlapping due to sampling in the image space, we use a band-pass boxcar function, Π_c, equal to 1 over the useful frequency range. Then we get:

$$\mathrm{MTF} = \mathrm{MTF}_0 \times F \quad \text{where} \quad \mathrm{MTF}_0 = \Pi_c \times \mathrm{MTF}_{\mathrm{int}} \tag{6.6}$$

6.2.2.2   Sampling and Sensor Noise

We denote the observed image by Y. The observed pixel values are obtained by sampling the deterministic blurred scene on a regular square grid (pixel size p), and adding a stochastic noise denoted by ϵ. The pixel coordinates i and j range from 0 to $N-1$ (square image of $N \times N$ pixels).

$$Y_{ij} = (S * \mathcal{F}^{-1}[\mathrm{MTF}])_{pi,pj} + \epsilon_{ij} \tag{6.7}$$

We make the following assumptions: the random variables ϵ_{ij} are independent, and their distribution is stationary. The noise is due to different independent sources, mostly quantum noise, thermal and readout noise, which yield a combination between Poisson and Gauss processes. A quantization noise also exists, modeled by a uniform distribution; it can be neglected if the quantization step is small enough. If all these perturbations have additive effects, their PDF is then the convolution of several PDFs, therefore it can be efficiently approximated by a Gaussian PDF, defined for d-dimensional variables by:

$$\mathcal{N}_x^d \left(\mu, \sigma^2 \right) = \frac{1}{(2\pi\sigma^2)^{d/2}} e^{-|x-\mu|^2/2\sigma^2}. \tag{6.8}$$

We assume a stationary variance denoted by σ^2, so that the observation model can be expressed in the discrete frequency space with a white, independently distributed Gaussian noise. This brings us to a probabilistic expression of the observed data Y given the scene S, which accounts for the stochastic nature of the noise. The FFT of Y is denoted by $\tilde{Y}$. To ensure the conservation of the number of independent variables, the coordinates in the frequency space (k, l) are such that $k \in \{-N/2 + 1, \cdots N/2\}$ and $l \in \{0, \cdots N/2\}$. We set $u = k/N$ and $v = l/N$.

$$P(Y \mid S, \alpha, \sigma) = \prod_{ij} \mathcal{N}_{Y_{ij}}^1 \left((S * \mathcal{F}^{-1} [\text{MTF}])_{pi,pj}, \sigma^2 \right) \tag{6.9}$$

$$P(\tilde{Y} \mid S, \alpha, \sigma) = \prod_{kl} \mathcal{N}_{\tilde{Y}_{kl}}^2 \left(\mathcal{F} [S]_{uv} \, \text{MTF}_{uv}, \sigma^2 \right) \tag{6.10}$$

If images were accurately described by shift-invariant models, we could simply assume independently distributed Gaussian Fourier coefficients $\tilde{Y}_{kl}$ (for $k > 0$) given the model parameters w_0 and q. From Equations (6.1) and (6.2) we have $\mathcal{F}[S] = \mathcal{E} * \mathcal{F}[L] \, w_0 \, r^{-q}$. The noise $\mathcal{E}$ is no longer white after convolution with $\mathcal{F}[L]$. However, independence is a good approximation since L has a large spatial support. In order to express the likelihood of all parameters we integrate out the Gaussian field G. Finally the forward model can be expressed as follows:

$$P(\tilde{Y} \mid w_0, q, \alpha, \sigma) = \prod_{kl} \mathcal{N}_{\tilde{Y}_{kl}}^2 \left(0, \omega_{kl}^2 \right) \quad \text{where} \quad \omega_{kl}^2 = w_0^2 \, r_{uv}^{-2q} \, \text{MTF}_{uv}^2 + \sigma^2.$$
$$\tag{6.11}$$

6.3 Bayesian Estimation: Invert the Forward Model

The goal is to compute the mode of the posterior PDF of the parameters of interest α and σ, i.e., $P(\alpha \mid \tilde{Y})$ and $P(\sigma \mid \tilde{Y})$. This involves two steps: apply the Bayes rule to invert the forward model, and marginalize or integrate w.r.t. the prior model parameters w_0 and q, which are nuisance variables. This can be summarized as:

$$P(\alpha \,|\, \tilde{Y}) \propto P(\alpha) \int P(\tilde{Y} \,|\, w_0, q, \alpha, \sigma) \, P(w_0, q) \, P(\sigma) \, dw_0 \, dq \, d\sigma. \qquad (6.12)$$

In the following, we will assume flat priors on all parameters, so that the equation above amounts to integrating the likelihood defined by (6.11). Therefore the main problem is to solve states as follows (solving it also provides a solution for σ):

$$\hat{\alpha} = \arg\max_{\alpha} \int P(\tilde{Y} \,|\, w_0, q, \alpha, \sigma) \, dw_0 \, dq \, d\sigma. \qquad (6.13)$$

We can write similar equations for $P(\sigma \,|\, \tilde{Y})$ and $\hat{\sigma}$.

6.3.1   Marginalization and Related Approximations

Computing the integral (6.12), also known as marginalization, is intractable in general. We propose to use a Laplace approximation [41], which involves a quadratic approximation of the -log likelihood or energy U around its optimum in $\theta = (w_0, q, \sigma)$:

$$U(w_0, q, \alpha, \sigma, \tilde{Y}) = -\log P(\tilde{Y} \,|\, w_0, q, \alpha, \sigma) = \sum_{kl} \log\left(2\pi\omega_{kl}^2\right) + \frac{|\tilde{Y}_{kl}|^2}{2\omega_{kl}^2} \qquad (6.14)$$

$$\simeq U(\hat{\theta}, \alpha, \tilde{Y}) + \frac{1}{2}(\theta - \hat{\theta})^t \left[\Sigma_{\theta}^{-1}\right](\theta - \hat{\theta}), \qquad (6.15)$$

where the ω_{kl} are given by (6.11). Therefore we have [41]:

$$\int P(\tilde{Y} \,|\, \theta, \alpha) \, d\theta \simeq \sqrt{|2\pi\Sigma_{\theta}|} \, e^{-U(\hat{\theta}, \alpha, \tilde{Y})}. \qquad (6.16)$$

The problem (6.13) can be reformulated as the minimization of a new energy:

$$\hat{\alpha} = \arg\min_{\alpha} \left[U(\hat{\theta}, \alpha, \tilde{Y}) - \frac{1}{2}\log|\Sigma_{\theta}| \right]. \qquad (6.17)$$

In this equation, the variations of the log-determinant term w.r.t. α can be neglected w.r.t. the variations of the optimal energy $U(\hat{\theta})$. We have checked this assumption experimentally. Although the determinant can be explicitly computed using second derivatives of U, and differentiated w.r.t. α, the drop-off in the computational efficiency was not justified by the enhancement of the estimation results.

Finally, we can write the search for α as two nested optimizations:

$$\hat{\alpha} = \arg\min_{\alpha} U(\hat{\theta}, \alpha, \tilde{Y}), \quad \text{where} \quad \hat{\theta} = \arg\min_{w_0, q, \sigma} U(w_0, q, \alpha, \sigma, \tilde{Y}) \qquad (6.18)$$

which leads to an algorithm with two nested loops, which is fundamentally different from the joint optimization w.r.t. MTF, prior model and noise parameters. As a consequence, the resulting method is more stable than usual joint methods and therefore better copes with such nonlinear cases. As shown in [32], it is necessary to integrate the likelihood w.r.t. the nuisance parameters to get a robust estimate, i.e., to compute the marginalized likelihood instead of the joint likelihood.

6.3.2 A Natural Parameter Estimation Algorithm (BLINDE)

In this section we present an algorithm called BLINDE (BLINd DEconvolution) based on the approach described above, initially described in [42] and successfully applied to simulated remote sensing imagery (SPOT 5 and Pléiades satellites) provided by the French Space Agency (CNES). The general approach is patented[1].

6.3.2.1 Noise Variance Marginalization

The noise variance and the prior model parameters are optimized separately in this approach. We assume that the energy coming from the signal is negligible w.r.t. the contribution of the noise in the highest frequencies, i.e., $r \in [1/2, \sqrt{2}]$. Therefore it is estimated from the power spectrum in these frequencies, after masking the bands $u = 0$ and $v = 0$ as well as the corners such that $r > r''$, in order to avoid artifacts related to bad rows or columns, stripes, interferences, and the finite size of the image (boundary effects). See Figure 6.2 for an illustration. Except for such artifacts, the highest frequencies of $\tilde{Y}$ contain only noise as long as the image is blurred enough, or at least correctly sampled in all directions. Other methods could be used for noise variance estimation [43, 44].

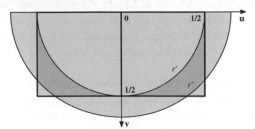

FIGURE 6.2: Noise estimation in the frequency space by measuring the energy in the spectral ring $r' < r < r''$ (assumed to contain only white Gaussian noise).

---

[1]Patents: USA #20040234162 (2002); Europe #02774858.1 (2002); France #0110189 (2001).

6.3.2.2 Prior Model Marginalization

Once the noise standard deviation has been determined, we have to optimize U w.r.t. w_0 and q while α is fixed. This can be achieved through various gradient-based methods (conjugate gradient, for instance). We give here the derivatives of U defined by Equation (6.14) involved in the minimization:

$$\frac{\partial U}{\partial \theta} = \sum_{kl} \frac{1}{\omega_{kl}^2} \left(1 - \frac{|\tilde{Y}_{kl}|^2}{2\omega_{kl}^2} \right) \frac{\partial \omega_{kl}^2}{\partial \theta} \tag{6.19}$$

where $\quad \dfrac{\partial \omega_{kl}^2}{\partial w_0} = r_{uv}^{-2q}\, \mathrm{MTF}_{uv}^2$ and $\dfrac{\partial \omega_{kl}^2}{\partial q} = -2\log(r_{uv})\, w_0^2\, r_{uv}^{-2q}\, \mathrm{MTF}_{uv}^2 \; (6.20)$

To accelerate this optimization step (it is critical for the overall computational efficiency, since this has to be done every time the MTF parameters are updated), we used a linear regression technique which has the advantage to provide a closed-form solution. Recall the log plots from Figure 6.1. For a nondegraded image, the log radial power spectrum denoted by y is a linear function of the log radial frequency denoted by x, such that $y = w_0 - qx$. The trick consists of "deconvolving" the mean radial power spectrum using the current estimate of the MTF and noise variance:

$$y_i = \log E_{r_i}[T_\sigma(\tilde{Y})\,\mathrm{MTF}^{-1}] \qquad x_i = \log r_i \tag{6.21}$$

where T_σ is a soft thresholding function and E denotes the average over the angle φ such that (r, φ) is the polar representation of (u, v). It is applied to both real and imaginary parts of the Fourier coefficients $\tilde{Y}$ to avoid noise amplification when dividing by the MTF. Frequency bands yielding too small energies are not taken into account in the regression. The variances related to each y_i are set proportional to y_i, assuming these variables have a χ^2 distribution since they are obtained by summing up a large number of Gaussian coefficients.

This method is not guaranteed to provide accurate estimates; therefore, it can be used as a quick initialization of an iterative method. This way, only a few steps of conjugate gradient (less than five) are needed.

6.3.2.3 The Algorithm

Let us focus on the highest-level optimization procedure, seeking the value of α that minimizes the energy $U(\hat{\theta}(\alpha))$ where θ is computed as explained above. If we use a gradient-based method, the first derivatives are needed. Since $\hat{\theta}$ satisfies $\partial U/\partial \theta = 0$, even though $\hat{\theta}$ depends on α, we have:

$$\frac{\partial U(\hat{\theta}(\alpha), \alpha, \tilde{Y})}{\partial \alpha} = \left.\frac{\partial U(\theta, \alpha, \tilde{Y})}{\partial \alpha}\right|_{\hat{\theta}} = \sum_{kl} \frac{1}{\omega_{kl}^2} \left(1 - \frac{|\tilde{Y}_{kl}|^2}{2\omega_{kl}^2} \right) \left.\frac{\partial \omega_{kl}^2}{\partial \alpha}\right|_{\hat{\theta}} \tag{6.22}$$

$$\text{where} \quad \frac{\partial \omega_{kl}^2}{\partial \alpha} = w_0^2\, r_{uv}^{-2q}\, (\mathrm{MTF}_0)_{uv}^2\, \frac{\partial F_{kl}^2}{\partial \alpha}. \tag{6.23}$$

For a Gaussian blur kernel F and a bidimensional vector α we have:

$$\frac{\partial \omega_{kl}^2}{\partial \alpha} = -2w_0^2 \, r_{uv}^{-2q} \, \text{MTF}_{uv}^2 \left[u^2 \quad v^2 \right]. \tag{6.24}$$

The optimization is performed using a conjugate gradient (CG) method adapted to non-quadratic functions, which uses line minimizations (involving gradient computation and projection along the search direction) as well as an update iteration for the search direction, once the line minimization has converged. Details can be found in [45]. The convergence properties are well known and beyond the scope of this chapter. It is applied under particular conditions as approximations are made in order to simplify the marginalization-related computations. However, experimental studies have shown a good stability, on simulations from the Gaussian image model with known parameter values, as well as on artificially degraded remote sensing images. No more than 30 iterations are needed when initialized with $\alpha = 0$.

- **Initialization.** The variance of the noise is estimated from $\tilde{Y}$, using only very high frequencies $r > 1/2$.

- **(1) Inner optimization loop:**

  - Initialization: linear regression using the data from Equation (6.21).

  - Iterative optimization: conjugate gradient using the derivatives of Eqns. (6.19) and (6.20) to get $\hat{\theta}$.

- **Evaluate** $U(\hat{\theta})$ and its derivatives w.r.t. α using Eqns. (6.22) and (6.23).

- **Update** α. Update the CG direction. A line minimization step is required and the inner loop has to be run for each value of α involved in this step.

- Go to step **(1)** until convergence is reached.

This method is fundamentally different from alternate minimization schemes usually employed, since we have to optimize w.r.t. θ each time α is changed.

6.3.3 Why Use a Simplified Model?

The nonlinearity of the previous approach and the high number of data points (causing the method to slow down when increasing the image size) motivate the choice of a new, simplified model. This model enables us to reduce the data so that the complexity of the iterative estimation procedure does not depend on the image size any longer (except for an initial fast Fourier transform [FFT]). It should also reduce the dependence between the observed variables.

We reformulate the observation model through a binning process. The spectrum coefficients $\tilde{Y}_{kl}$ are transformed into M new variables z_n. Let each

z_n be the mean squared magnitude of $\tilde{Y}$ over the n-th frequency bin, defined by a square area of size $B \times B$ around the frequential coordinates (k_n^0, l_n^0).

$$z_n = \frac{1}{B^2} \sum_{k=k_n^0-B}^{k_n^0+B-1} \sum_{l=l_n^0-B}^{l_n^0+B-1} \left| \tilde{Y}_{kl} \right|^2 \tag{6.25}$$

This way, if the bins are large enough the dependencies can be substantially reduced and we can assume independent observations $\{z_n\}$. This also reduces the dimensionality of the problem and the computational complexity. The bin size shall be as large as possible to minimize the number of data points, without compromising the assumed stationarity of the mean power spectrum. We will not discuss the optimal choice of the bin size in this chapter; refer to the model assessment section for details about how to validate the proposed model in general. The binning and the selected spatial frequency domain are illustrated by Figure 6.3.

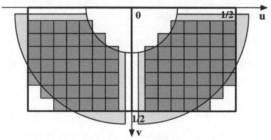

FIGURE 6.3: Dark squares: spatial frequency binning. Shaded area: radial frequency clipping and removal of frequencies close to $u = 0$ and $v = 0$, to ensure the validity of the model while avoiding various artifacts (bad rows or columns, image periodization, structured noise).

Sums of independent Gaussian variables have a χ^2 distribution. Despite the weak correlation between Fourier coefficients (due to the convolution by $\mathcal{F}[s]$) we use an independence assumption, so that the variance of each z_n can be approximated by $2z_n$. Denoting the mean of z_n by μ_n, we have:

$$P(z \mid w_0, q, \alpha, \sigma) = \prod_n \mathcal{N}_{z_n}^1 \left(\mu_n, \varepsilon_n^2 \right) \quad \text{where} \quad \begin{cases} \mu_n = w_0^2 \, r_n^{-2q} \, \mathrm{MTF}_n^2 + \sigma^2 \\ \varepsilon_n^2 \simeq 2z_n \end{cases} . \tag{6.26}$$

Here, r_n and MTF_n denote the radial frequency and MTF averaged over the respective bin; they can be approximated by the values taken at the central spatial frequency of the bin, i.e., at $(k_0^n/N, l_0^n/N)$.

6.3.4 A Simplified, Optimized Algorithm

The joint-log likelihood is now denoted by V and writes:

$$V(w_0, q, \alpha, \sigma, z) = -\log P(z \,|\, w_0, q, \alpha, \sigma) \simeq \sum_n \frac{(\mu_n - z_n)^2}{4z_n} + \text{const.} \quad (6.27)$$

The constant term is related to the normalization and depends only on z if the variance of z only depends on the observed data (we will show in the result section how to reestimate this variance when needed). This yields a much simpler expression of the energy term than we had before. If we set $\phi = (w_0^2, \sigma^2)$ we can see that the optimization problem is linear w.r.t. ϕ. Therefore there is a closed-form solution for the optimal w_0^2 and σ^2 for each value of q.

$$\frac{\partial V}{\partial \phi} = \sum_n \left(\frac{\mu_n - z_n}{2z_n} \right) \frac{\partial \mu_n}{\partial \phi} \quad (6.28)$$

$$\text{where} \quad \frac{\partial \mu_n}{\partial w_0^2} = r_n^{-2q} \text{MTF}_n^2 \quad \text{and} \quad \frac{\partial \mu_n}{\partial \sigma^2} = 1 \quad (6.29)$$

We have the linear system $[A(q)]\phi = c(q)$ where:

$$[A(q)] = \begin{bmatrix} \sum r_n^{-4q} \text{MTF}_n^4 / z_n & \sum r_n^{-2q} \text{MTF}_n^2 / z_n \\ \sum r_n^{-2q} \text{MTF}_n^2 / z_n & \sum 1/z_n \end{bmatrix} \quad c(q) = \begin{bmatrix} \sum r_n^{-2q} \text{MTF}_n^2 & M \end{bmatrix}$$
$$(6.30)$$

which yields $\hat{\phi} = [A(q)]^{-1} c(q)$.

A 1-D line search can be performed to find the optimal value of q. Since $\hat{\phi}$ satisfies $\partial V / \partial \phi = 0$, even though $\hat{\phi}$ depends on q, the derivative of $V(\hat{\phi})$ w.r.t. q writes:

$$\frac{\partial V(\hat{\phi}(q), q, \alpha, z)}{\partial q} = \left. \frac{\partial V(\phi, q, \alpha, z)}{\partial q} \right|_{\hat{\phi}}$$
$$= -\sum_n \left(\frac{\mu_n - z_n}{z_n} \right) \log(r_n) \, w_0^2 \, r_n^{-2q} \, \text{MTF}_n^2 \Big|_{\hat{\phi}}. \quad (6.31)$$

Thus, there is no need for an approximate method (such as the one based on linear regression of Section 6.3.2.2) or any multidimensional gradient descent technique. Moreover, the noise variance is marginalized as well, thus avoiding the initialization step in which it is estimated from the highest frequencies of the power spectrum.

Now we can focus on the estimation of the MTF parameters using this second approach. We keep exactly the same marginalization framework as in the previous approach (see Section 6.3.1), with a different notation. Equation (6.18) remains valid (with V, z instead of $U, \tilde{Y}$). The derivatives w.r.t. α are given by:

$$\frac{\partial V(\hat{\theta}(\alpha), \alpha, z)}{\partial \alpha} = \left.\frac{\partial V(\theta, \alpha, z)}{\partial \alpha}\right|_{\hat{\theta}} = \sum_{kl} \left(\frac{\mu_n - z_n}{2z_n}\right) \left.\frac{\partial \mu_n}{\partial \alpha}\right|_{\hat{\theta}} \qquad (6.32)$$

$$\text{where} \quad \frac{\partial \mu_n}{\partial \alpha} = w_0^2 \, r_n^{-2q} \, \text{MTF}_{0\,n}^2 \, \frac{\partial F_n^2}{\partial \alpha}. \qquad (6.33)$$

For a Gaussian blur, we have:

$$\frac{\partial \mu_n}{\partial \alpha} = -2w_0^2 \, r_n^{-2q} \, \text{MTF}_n^2 \, \begin{bmatrix} u_n^2 & v_n^2 \end{bmatrix}. \qquad (6.34)$$

The optimization algorithm relies again on a conjugate gradient method adapted to nonquadratic functions [45]. Regarding convergence, the remarks we made for the BLINDE algorithm still hold.

- **(1) Inner optimization loop:** Line optimization using the derivative (6.31) where w_0^2 and σ^2 are replaced by their closed-form optimum using (6.30).

- **Evaluate** $V(\hat{\theta})$ and its derivatives w.r.t. α using Eqns. (6.32) and (6.33).

- **Update** α. Update the CG direction. A line minimization step is required and the inner loop has to be run for each value of α involved in this step.

- Go to step **(1)** until convergence is reached.

Other optimization methods could be used. Recently, we have developed a quasi-Newton scheme based on the second derivatives of the marginalized V w.r.t. α. It seems to perform better than the CG, since it does not require any accurate line minimization at each step in order to converge quickly. However, it might be more sensitive to the initialization because it involves an inversion of the Hessian matrix at each update step. Nonetheless, the problem can be stabilized by using a linear approximation of μ when computing this matrix.

6.4   Possible Improvements and Further Development

6.4.1   Computing Uncertainties

The Bayesian formalism can be applied to perform inference instead of point estimation. It is possible to compute a Gaussian approximation to the posterior PDF (6.12), characterized not only by a mean, but also by a covariance matrix. The covariance provides a measure of the uncertainty related to the estimated parameters. If the PDF is proportional to $e^{-V(\alpha)}$, a quadratic approximation is made around the optimum $\hat{\alpha}$ of V in which the second derivatives are the elements of the inverse covariance matrix. Marginalization w.r.t. prior model parameters amounts to keeping the entries of the covariance matrix related to the remaining variables, i.e., α.

However, we encountered difficulties in obtaining error estimates that were consistent with the ground truth. In our experiments (simulations of degraded images using predefined MTF parameters), 4 out of 5 true parameter values were out of the 99% confidence ellipses, meaning that the actual error was bigger than the predicted one. Which brings us to the conclusion that no matter how well uncertainties are computed for a particular model, they are correct only as long as the model provides an accurate description of the underlying scene. Even with an imprecise or inadequate model, computed uncertainties can be made arbitrarily small if a large enough number of data points are measured, thus mistakenly concluding that the inference is accurate. Only an appropriate model-checking procedure could tell us if our modeling is flawed.

6.4.2 Model Assessment and Checking

We can distinguish between two main sources of estimation errors:

- *The lack of robustness of the input data z of Equation (6.25).* The sum of squares statistic is actually inadequate for any non-Gaussian distribution, since it yields estimates that are quite sensitive to outliers. One can observe discrepancies between the assumed Gaussian model for Fourier coefficients and the actual data, occurring if there are unusual spikes at particular frequencies in the spectrum, or holes in the spectrum, due to artificial structures or a lack of features in the scene. These discrepancies could be better detected using wavelet transforms rather than the Fourier transform, since they can be strongly related to the nonstationarity property of the scene. Instead of using the sum of squares, one could use an order-p moment (with $p = 1$, for instance) as a robust estimator of the mean power spectrum.

- *The discrepancy between the assumed $1/f$ model and the actual spectral decay of the scene.* In practice, natural scenes of limited size are not perfectly fractal. One can expect unusually fast decays of the power spectrum in the highest frequencies, or the contrary. This can be seen as a curvature in the log-log plot, which inevitably leads to MTF parameter over- or underestimation, depending on the sign of the curvature. Such behaviors are particularly difficult to detect. This can lead to indetermination: degenerate solutions that fit the data can be found, such that various prior model curvatures and MTF parameter values lead to the exact same predicted mean μ of Equation (6.26).

Model assessment can be performed in various ways; the simplest way to check the statistical validity of a model such as the one used in this work is through residual analysis. There are a variety of tests that can be run on the residuals (differences $z_n - \mu_n$) once the estimation has been done [31]. For instance, the sum of squared, normalized residuals is supposed to have a χ^2 distribution with $M - n$ degrees of freedom, where n is the number of parameters, if the error model is Gaussian. Outliers, or random modeling errors, can typically be detected by such tests. However, epistemic errors (coming from an inappropriate model choice) could not be detected this way. Therefore simple statistical tests are not sufficient and one must resort to model comparison in order to select the best model, or the best combination of models.

Automatic model selection (or model averaging) can be performed in the Bayesian framework [31, 41], by adding a new variable acting as a model label and infer it (assuming we have a choice among a discrete set of models) or by adding new parameters to the model itself until some optimality condition is reached (relating, for instance, to the best compromise between model sparsity and prediction accuracy). In all cases, marginalization will have to be done in order to integrate out these extra variables. To summarize this, determining the posterior PDF (6.12) is still the goal; there are simply a few more variables to be integrated out. The higher computational complexity would certainly be justified by the increased robustness of the resulting algorithm.

6.4.3   Robustness-Related Improvements

Two enhancements can be proposed in a recursive way, in order to improve the overall robustness of the algorithm. First, the error estimates ε can be updated according to the estimated spectrum model, not to the observed values z. Second, outliers can be identified and rejected, according to the statistical significance of the residual related to each data point (for instance, one could reject all points having a residual magnitude greater than 2ϵ, which corresponds to less than 5% statistical significance). Finally, the estimation algorithm has to be run again. This is what was done in the experiments shown in the next section. Ideally, the whole process shall be repeated until convergence. This method should probably handle stochastic modeling errors efficiently; however, it would be of little use in case of a systematic model mismatch (e.g., power spectrum curvature on a log-log scale).

6.5 Results

6.5.1 First Method: BLINDE

The MTF parameter α of an isotropic Gaussian blur (Equation (6.5) with $\alpha_0 = \alpha_1$) was computed from simulations performed on both an aerial image of the city of Amiens provided by IGN (French Geographic Institute), and an aerial image of the city of Nîmes provided by CNES (French Space Agency), with respective subsampling factors 8 and 3. Subsampling through a band-pass boxcar function in the frequency space was necessary in order to avoid artifacts coming from the MTF of the instrument (telescope, motion...). Blurred and noisy satellite image simulations are 512×512 and about $2.5\,m$ in resolution due to the subsampling factor. Figure 6.4 displays results obtained with the BLINDE algorithm, showing a good accuracy (better than 10%) on the MTF value at $r = 1/4$. The overall accuracy seems to increase with the blur size. The method was also directly applied to an area extracted from the real image of Amiens at $30\,cm$ resolution (see Figure 6.5). We found a realistic blur size close to the optimal sampling rate, but could not check it against any ground truth.

6.5.2 Second method

Experimental settings The MTF parameters α_x and α_y of a Gaussian blur of Equation (6.5) were estimated from simulations performed on subsampled Ikonos images (credit Space Imaging). We choose subsampling factors from 2 to 4 to avoid artifacts coming from the MTF of the instrument (telescope, motion...) or the lossy compression scheme. We assumed that after a frequency-space subsampling through a boxcar band-pass function these images exhibit a fractal behavior, an assumption that can be checked by analyzing the power spectrum as shown on Figure 6.10. We choose to use simulations (obtained by applying blur and noise to such subsampled images) because of the lack of availability of real images whose blur is perfectly calibrated.

Figures 6.6 through 6.9 display the four chosen scenes. For each one of the scenes, we show three examples of corrupted observations (small region of interest) for three blur parameter settings (denoted by a, b, c in the summary Tables 6.1 and 6.2) corresponding to $\sigma^2 = 3$ and $\alpha_x, \alpha_y \in \{10, 30\}$. For each setting, a small region of the degraded image is shown (left subfigure). The transfer functions along the horizontal and vertical axes are plotted (respectively for $v = 0$ and $u = 0$, see center and right subfigures) for both true and estimated parameters. Thus the consequence of any parameter misestimation on the overall transfer function can be quickly visualized (systems designers are usually interested in MTF values at u or v equal to $1/4$ or $1/2$, which can be read from these graphs).

Tables 6.1 and 6.2 summarize all our experiments made on the four different scenes, from simulated blurred and noisy images at two different sizes (and two respective binnings 32×32 and 16×16), for two values of the noise variance $\sigma^2 \in \{3, 9\}$ and four combinations of MTF parameters (both isotropic and anisotropic) $(\alpha_x, \alpha_y) \in \{10, 30\}^2$ indexed by (a,b,c...h). The value of the MTF at three spatial frequencies (u and/or v equal to $1/4$) are shown for both true and estimated parameter values. The difference between true and estimated values at these three points is proposed as a simple absolute error measure.

To analyze the residuals, we define the normalized chi-squared residual as $k = (\chi^2 - M)/\sqrt{2M}$ where M is a good approximation of the number of degrees of freedom ($M = 435$ in the experiments and corresponds to a spectrum subdivision into 32×16 bins among which we keep the ones having $0.1 < r < 0.5$, $u \neq 0$ and $v \neq 0$). The rejected outlier rates and k give an idea of the estimated model quality. However, there is no way to assess the appropriateness of the fractal model directly from the blurred images. This is why we also display the power spectra of the used images without degradation (and with an ideal box sampling in the frequency domain) on Figure 6.10 using the same binning z as in the second estimation method. This figure clearly shows that the fractal model is not perfect, since there are many outliers and also some significant systematic deviations. We notice that this is not directly related to the statistics of the underlying scene; unfortunately the images were obtained by a real optical system and sensor, and were processed in a way unknown to us.

Result discussion and comments   Matangi (Figure 6.6) is mostly made of natural features (coastlines and clouds), so the estimation is expected to be accurate since the image content is well described by a fractal model. Helliniko (Figure 6.7) displays a complex series of man-made features, and it surprisingly provides the best results among the four scenes. Whether artificial constructions are located or not on underlying natural structures is far beyond the scope of this study; however, it seems that these constructions obey the proposed power-law decay of the power spectrum, which enables the entire approach to clearly identify the blur parameters. Also, the spectral content is fairly rich as shown on Figure 6.10 which undoubtedly helps estimate the parameters more accurately.

Gooseneck (Figure 6.8) and Stonehenge (Figure 6.9), on the other hand, have a relatively poor spectrum and even some unwanted texture (which is obviously not self-similar), which could explain why our method does not perform so well on these images. We can notice that there is a systematic overestimation of the MTF. Despite the apparent good match between data and model for the Gooseneck image on Figure 6.10, the log-log spectrum of the 512×512 image is probably bent in such a way that the MTF is consistently overestimated. It appears that on these two scenes, there are more high frequencies than predicted by the model (or equivalently, less lower frequencies). This can be due to multiple reasons, including piecewise scale

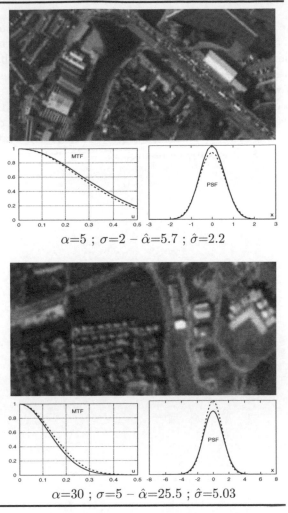

FIGURE 6.4: Images extracted from 512×512 simulations of blurred and noisy satellite images at 2.5m resolution (top: Amiens © IGN, bottom: Nîmes © CNES). The isotropic MTF (left) and PSF (right) are plotted along the horizontal axis ($v = 0$). Dashed lines: estimated blur.

invariance intrinsic to the scene, and preprocessing artifacts we were not aware of (even a slight sharpening would be sufficient to explain this behavior).

As shown on the summary Tables 6.1 and 6.2, normalized residuals are high enough to suggest that the spectrum model is not sufficient to model all kinds of scenes. These residuals are highest for man-made features (Helliniko buildings and Stonehenge fields). Outlier rejection helps reduce the residuals in some cases (as in Helliniko) but fails to solve the model mismatch problem in other cases (as in Stonehenge), especially when the spectral content is poor.

In general, the number of outliers tends to increase with the image size, so we conclude that a subsampling factor 2 is insufficient to reduce the artifacts on the source images, except for Gooseneck where the best results are obtained from the highest-resolution image. On real data, however, one should expect better results when using a larger number of pixels.

In order to check the performance of the proposed method in a more accurate way, we recommend the use of real images (instead of simulations) whenever possible, with a good determination of their degradation model (by traditional means such as target imaging, for instance).

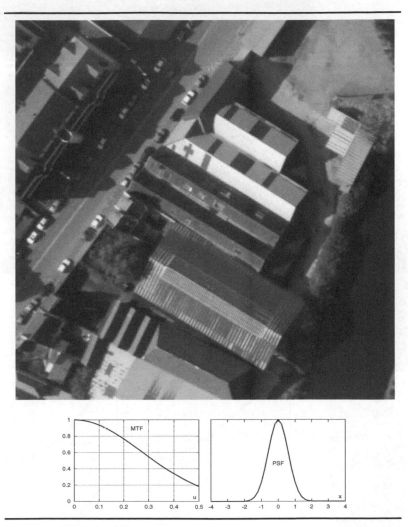

FIGURE 6.5: Region extracted from a real, blurred, and noisy aerial image of Amiens © IGN, 30 *cm* resolution. The estimated isotropic MTF (left) and PSF (right) are plotted in solid lines along the horizontal axis (respectively $v = 0$ and $y = 0$). (We found $\hat{\alpha} \simeq 5$.)

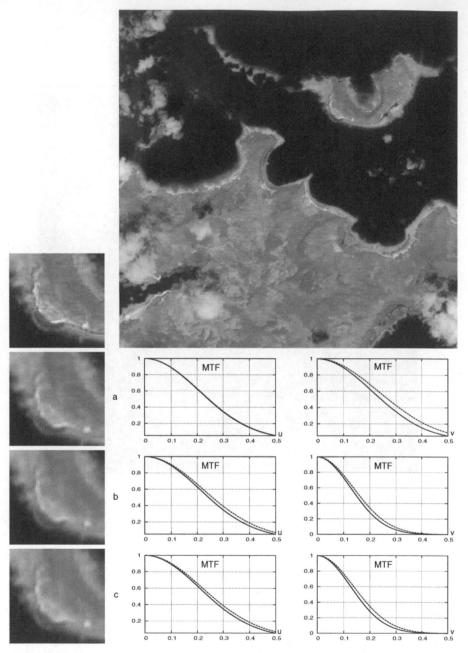

FIGURE 6.6:   Subsampled image of Matangi © Space Imaging, 512×512, 4 m final resolution, and small region of interest (ROI). For each experimental setting a, b, and c (refer to Table 6.1 for the related parameter values), we show from left to right the degraded ROI and MTF plots along horizontal and vertical axes (respectively for $u = 0$ and $v = 0$). Dashed lines represent estimated MTFs.

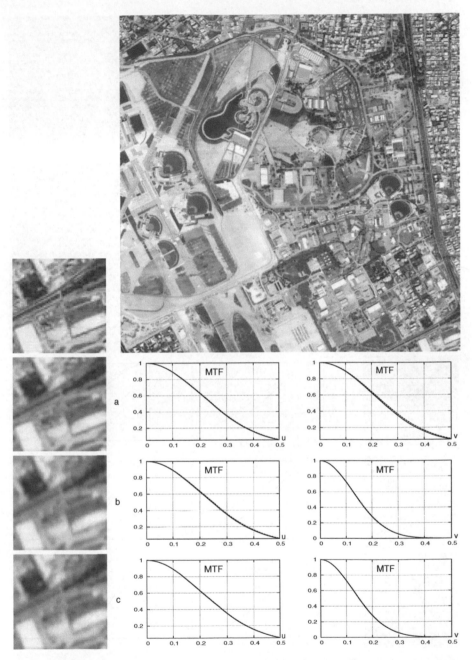

FIGURE 6.7: Subsampled image of Helliniko © Space Imaging, 512×512, 4 m final resolution, and small region of interest (ROI). For each experimental setting a, b, and c (refer to Table 6.1 for the related parameter values), we show from left to right the degraded ROI and MTF plots along horizontal and vertical axes (respectively for $u = 0$ and $v = 0$). Dashed lines represent estimated MTFs.

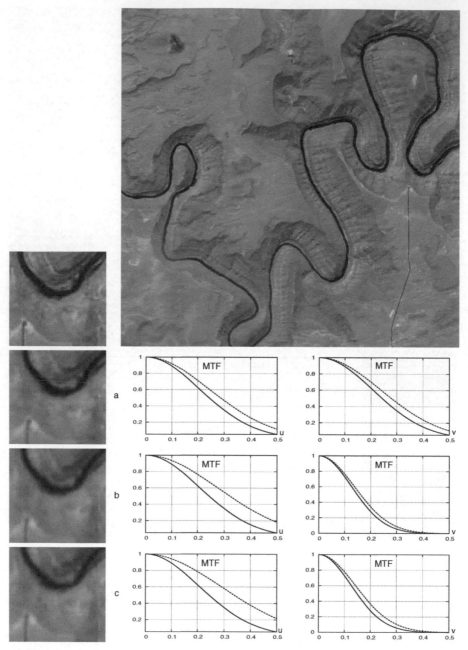

FIGURE 6.8: Subsampled image of Gooseneck ©Space Imaging, 512×512, 4 *m* final resolution, and small region of interest (ROI). For each experimental setting a, b, and c (refer to Table 6.1 for the related parameter values), we show from left to right the degraded ROI and MTF plots along horizontal and vertical axes (respectively for $u = 0$ and $v = 0$). Dashed lines represent estimated MTFs.

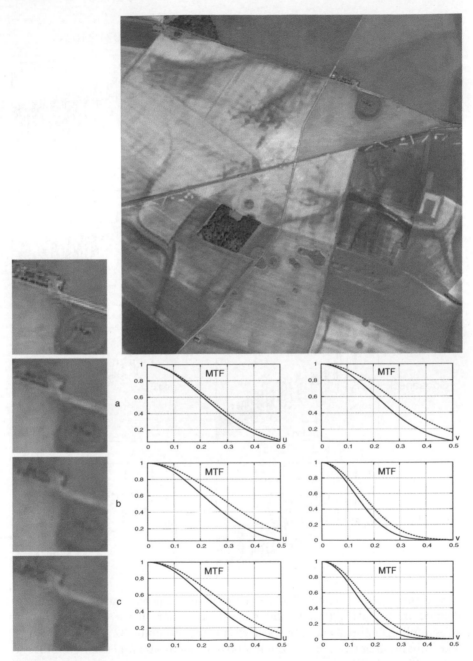

FIGURE 6.9: Subsampled image of Stonehenge ⓒ Space Imaging, 512×512, 4 *m* final resolution, and small region of interest (ROI). For each experimental setting a, b, and c (refer to Table 6.1 for the related parameter values), we show from left to right the degraded ROI and MTF plots along horizontal and vertical axes (respectively for $u = 0$ and $v = 0$). Dashed lines represent estimated MTFs.

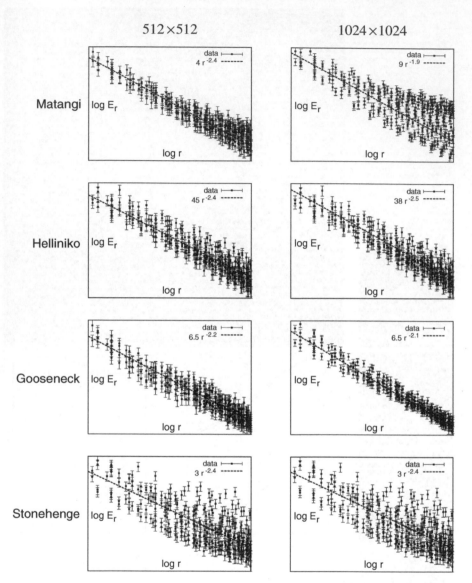

FIGURE 6.10: Log-log plots showing the mean power spectra (z variables defined by Equation (6.25)) and related error bars for $r \in [0.15, 0.5]$, for subsampling factors 2 and 4 (respective spectrum binning 32×32 and 16×16). The subsampling was done through a boxcar band-pass function. No blur, no noise. This helps check the match between real scenes and the proposed fractal image model, shown as a straight dashed line. One should expect a similar behavior at the two different scales for a perfectly fractal image, in addition to the good match.

TABLE 6.1: Summary of the experimental results for images of Matangi and Helliniko, subsampled by factors 2 and 4 using a boxcar band-pass function. For each setting (labeled from a to h), we show the three true parameter values and MTF values at three points in the frequency space, followed by the six corresponding estimated values. Then we show the normalized χ^2 residual (defined in subsection 6.5.2) and the rejected outlier rates, denoted by k and rej. The last three columns display the MTF errors (estimated–true) at the three chosen points in the frequency space.

#	True σ^2	α_x	α_y	MTF $(\frac{1}{4},0)$	MTF $(0,\frac{1}{4})$	MTF $(\frac{1}{4},\frac{1}{4})$	Est σ^2	α_x	α_y	MTF $(\frac{1}{4},0)$	MTF $(0,\frac{1}{4})$	MTF $(\frac{1}{4},\frac{1}{4})$	k	rej	Err MTF $(\frac{1}{4},0)$	MTF $(0,\frac{1}{4})$	MTF $(\frac{1}{4},\frac{1}{4})$
a	3.00	10.0	10.0	0.48	0.48	0.23	3.10	11.7	7.4	0.43	0.57	0.25	38.9	24%	-0.05	+0.08	+0.01
b	3.00	10.0	30.0	0.48	0.14	0.07	3.07	8.9	22.4	0.52	0.22	0.11	11.2	12%	+0.03	+0.08	+0.05
c	3.00	30.0	10.0	0.14	0.48	0.07	3.10	33.4	8.4	0.11	0.53	0.06	16.1	15%	-0.03	+0.05	-0.01
d	3.00	30.0	30.0	0.14	0.14	0.02	3.08	33.0	26.7	0.11	0.17	0.02	6.4	10%	-0.02	+0.03	+0.00
e	9.00	10.0	10.0	0.48	0.48	0.23	9.11	12.2	7.6	0.42	0.56	0.23	19.2	15%	-0.06	+0.08	+0.00
f	9.00	10.0	30.0	0.48	0.14	0.07	9.09	8.1	21.2	0.54	0.24	0.13	6.7	10%	+0.06	+0.10	+0.06
g	9.00	30.0	10.0	0.14	0.48	0.07	9.07	34.1	8.9	0.11	0.51	0.05	8.0	10%	-0.03	+0.03	-0.01
h	9.00	30.0	30.0	0.14	0.14	0.02	9.06	29.3	22.5	0.14	0.22	0.03	5.7	9%	+0.01	+0.08	+0.01

Matangi 1024×1024 (bin size 32×32)

#	True σ^2	α_x	α_y	MTF $(\frac{1}{4},0)$	MTF $(0,\frac{1}{4})$	MTF $(\frac{1}{4},\frac{1}{4})$	Est σ^2	α_x	α_y	MTF $(\frac{1}{4},0)$	MTF $(0,\frac{1}{4})$	MTF $(\frac{1}{4},\frac{1}{4})$	k	rej	Err MTF $(\frac{1}{4},0)$	MTF $(0,\frac{1}{4})$	MTF $(\frac{1}{4},\frac{1}{4})$
a	3.00	10.0	10.0	0.48	0.48	0.23	3.04	9.2	6.9	0.51	0.58	0.30	13.2	15%	+0.03	+0.10	+0.06
b	3.00	10.0	30.0	0.48	0.14	0.07	3.07	9.5	27.2	0.50	0.16	0.08	6.3	9%	+0.01	+0.03	+0.01
c	3.00	30.0	10.0	0.14	0.48	0.07	3.06	30.3	7.5	0.14	0.56	0.08	8.1	10%	-0.00	+0.08	+0.01
d	3.00	30.0	30.0	0.14	0.14	0.02	3.08	32.1	28.7	0.12	0.15	0.02	6.2	8%	-0.02	+0.01	-0.00
e	9.00	10.0	10.0	0.48	0.48	0.23	9.08	10.5	8.3	0.47	0.54	0.25	9.0	11%	-0.02	+0.06	+0.02
f	9.00	10.0	30.0	0.48	0.14	0.07	9.02	7.5	23.1	0.57	0.21	0.12	3.6	7%	+0.08	+0.08	+0.05
g	9.00	30.0	10.0	0.14	0.48	0.07	8.99	32.2	9.0	0.12	0.51	0.06	4.5	9%	-0.02	+0.03	-0.00
h	9.00	30.0	30.0	0.14	0.14	0.02	9.05	36.0	32.2	0.09	0.12	0.01	3.7	7%	-0.04	-0.02	-0.01

Matangi 512×512 (bin size 16×16)

#	True σ^2	α_x	α_y	MTF $(\frac{1}{4},0)$	MTF $(0,\frac{1}{4})$	MTF $(\frac{1}{4},\frac{1}{4})$	Est σ^2	α_x	α_y	MTF $(\frac{1}{4},0)$	MTF $(0,\frac{1}{4})$	MTF $(\frac{1}{4},\frac{1}{4})$	k	rej	Err MTF $(\frac{1}{4},0)$	MTF $(0,\frac{1}{4})$	MTF $(\frac{1}{4},\frac{1}{4})$
a	3.00	10.0	10.0	0.48	0.48	0.23	3.05	9.2	8.6	0.51	0.52	0.27	470.2	58%	+0.03	+0.04	+0.03
b	3.00	10.0	30.0	0.48	0.14	0.07	3.08	9.6	29.5	0.49	0.14	0.07	239.7	30%	+0.01	+0.00	+0.00
c	3.00	30.0	10.0	0.14	0.48	0.07	3.07	29.8	8.5	0.14	0.53	0.07	298.6	39%	+0.00	+0.05	+0.01
d	3.00	30.0	30.0	0.14	0.14	0.02	3.07	29.1	28.4	0.15	0.15	0.02	166.7	23%	+0.01	+0.01	+0.00
e	9.00	10.0	10.0	0.48	0.48	0.23	8.95	8.9	8.4	0.52	0.53	0.27	355.6	48%	+0.03	+0.05	+0.04
f	9.00	10.0	30.0	0.48	0.14	0.07	9.10	9.8	30.0	0.49	0.14	0.07	186.2	25%	+0.00	-0.00	+0.00
g	9.00	30.0	10.0	0.14	0.48	0.07	9.08	29.3	8.0	0.14	0.54	0.08	221.1	32%	+0.01	+0.06	+0.01
h	9.00	30.0	30.0	0.14	0.14	0.02	9.10	31.2	30.9	0.13	0.13	0.02	122.6	19%	-0.01	-0.01	-0.00

Helliniko 1024×1024 (bin size 32×32)

#	True σ^2	α_x	α_y	MTF $(\frac{1}{4},0)$	MTF $(0,\frac{1}{4})$	MTF $(\frac{1}{4},\frac{1}{4})$	Est σ^2	α_x	α_y	MTF $(\frac{1}{4},0)$	MTF $(0,\frac{1}{4})$	MTF $(\frac{1}{4},\frac{1}{4})$	k	rej	Err MTF $(\frac{1}{4},0)$	MTF $(0,\frac{1}{4})$	MTF $(\frac{1}{4},\frac{1}{4})$
a	3.00	10.0	10.0	0.48	0.48	0.23	3.07	10.0	9.8	0.48	0.49	0.23	132.6	43%	-0.00	+0.00	+0.00
b	3.00	10.0	30.0	0.48	0.14	0.07	3.07	10.2	30.0	0.48	0.14	0.07	62.2	24%	-0.01	-0.00	-0.00
c	3.00	30.0	10.0	0.14	0.48	0.07	3.08	29.4	8.5	0.14	0.53	0.08	83.4	28%	+0.00	+0.05	+0.01
d	3.00	30.0	30.0	0.14	0.14	0.02	3.07	30.4	29.5	0.13	0.14	0.02	44.3	20%	-0.00	+0.00	+0.00
e	9.00	10.0	10.0	0.48	0.48	0.23	8.98	9.4	9.2	0.50	0.51	0.25	96.6	34%	+0.02	+0.02	+0.02
f	9.00	10.0	30.0	0.48	0.14	0.07	8.99	9.5	29.2	0.50	0.14	0.07	49.1	21%	+0.02	+0.01	+0.01
g	9.00	30.0	10.0	0.14	0.48	0.07	9.03	30.8	9.8	0.13	0.49	0.06	62.2	25%	-0.01	+0.01	-0.00
h	9.00	30.0	30.0	0.14	0.14	0.02	9.10	37.0	37.6	0.09	0.09	0.01	34.0	16%	-0.05	-0.05	-0.01

Helliniko 512×512 (bin size 16×16)

TABLE 6.2:   Summary of the experimental results for images of Gooseneck and Stonehenge, subsampled by factors 2 and 4 using a boxcar band-pass function. For each setting (labeled from a to h), we show the three true parameter values and MTF values at three points in the frequency space, followed by the six corresponding estimated values. Then we show the normalized χ^2 residual (defined in subsection 6.5.2) and the rejected outlier rates, denoted by k and rej. The last three columns display the MTF errors (estimated–true) at the three chosen points in the frequency space.

#	σ^2	α_x	α_y	True MTF $(\frac{1}{4},0)$	MTF $(0,\frac{1}{4})$	MTF $(\frac{1}{4},\frac{1}{4})$	σ^2	α_x	α_y	Est MTF $(\frac{1}{4},0)$	MTF $(0,\frac{1}{4})$	MTF $(\frac{1}{4},\frac{1}{4})$	k	rej	Err MTF $(\frac{1}{4},0)$	MTF $(0,\frac{1}{4})$	MTF $(\frac{1}{4},\frac{1}{4})$
a	3.00	10.0	10.0	0.48	0.48	0.23	3.08	10.1	9.8	0.48	0.49	0.23	91.6	29%	-0.00	+0.01	+0.00
b	3.00	10.0	30.0	0.48	0.14	0.07	3.08	10.9	30.9	0.46	0.13	0.06	52.4	15%	-0.03	-0.01	-0.01
c	3.00	30.0	10.0	0.14	0.48	0.07	3.07	28.8	9.2	0.15	0.51	0.08	72.6	21%	+0.01	+0.02	+0.01
d	3.00	30.0	30.0	0.14	0.14	0.02	3.08	32.9	33.4	0.12	0.11	0.01	44.0	14%	-0.02	-0.03	-0.01
e	9.00	10.0	10.0	0.48	0.48	0.23	9.06	9.9	9.5	0.48	0.50	0.24	67.0	22%	+0.00	+0.01	+0.00
f	9.00	10.0	30.0	0.48	0.14	0.07	9.07	10.4	30.5	0.47	0.13	0.06	37.2	14%	-0.01	-0.00	-0.00
g	9.00	30.0	10.0	0.14	0.48	0.07	9.06	30.3	10.4	0.14	0.47	0.06	50.1	19%	-0.00	-0.01	-0.00
h	9.00	30.0	30.0	0.14	0.14	0.02	9.07	35.1	36.1	0.10	0.09	0.01	27.5	11%	-0.04	-0.04	-0.01

Gooseneck 1024×1024 (bin size 32×32)

#	σ^2	α_x	α_y	True MTF $(\frac{1}{4},0)$	MTF $(0,\frac{1}{4})$	MTF $(\frac{1}{4},\frac{1}{4})$	σ^2	α_x	α_y	Est MTF $(\frac{1}{4},0)$	MTF $(0,\frac{1}{4})$	MTF $(\frac{1}{4},\frac{1}{4})$	k	rej	Err MTF $(\frac{1}{4},0)$	MTF $(0,\frac{1}{4})$	MTF $(\frac{1}{4},\frac{1}{4})$
a	3.00	10.0	10.0	0.48	0.48	0.23	3.00	5.9	6.0	0.62	0.62	0.39	45.0	26%	+0.14	+0.14	+0.15
b	3.00	10.0	30.0	0.48	0.14	0.07	3.07	4.7	24.9	0.67	0.19	0.13	22.8	15%	+0.19	+0.05	+0.06
c	3.00	30.0	10.0	0.14	0.48	0.07	3.05	25.0	5.5	0.19	0.64	0.12	28.0	17%	+0.05	+0.16	+0.05
d	3.00	30.0	30.0	0.14	0.14	0.02	3.07	15.3	15.5	0.35	0.34	0.12	12.4	10%	+0.21	+0.20	+0.10
e	9.00	10.0	10.0	0.48	0.48	0.23	8.90	4.7	4.9	0.67	0.66	0.45	29.2	20%	+0.19	+0.18	+0.21
f	9.00	10.0	30.0	0.48	0.14	0.07	9.07	5.5	24.9	0.64	0.19	0.12	13.1	12%	+0.16	+0.05	+0.05
g	9.00	30.0	10.0	0.14	0.48	0.07	8.97	24.1	4.7	0.20	0.67	0.13	14.1	12%	+0.06	+0.19	+0.07
h	9.00	30.0	30.0	0.14	0.14	0.02	9.00	15.9	14.8	0.33	0.36	0.12	5.9	9%	+0.19	+0.22	+0.10

Gooseneck 512×512 (bin size 16×16)

#	σ^2	α_x	α_y	True MTF $(\frac{1}{4},0)$	MTF $(0,\frac{1}{4})$	MTF $(\frac{1}{4},\frac{1}{4})$	σ^2	α_x	α_y	Est MTF $(\frac{1}{4},0)$	MTF $(0,\frac{1}{4})$	MTF $(\frac{1}{4},\frac{1}{4})$	k	rej	Err MTF $(\frac{1}{4},0)$	MTF $(0,\frac{1}{4})$	MTF $(\frac{1}{4},\frac{1}{4})$
a	3.00	10.0	10.0	0.48	0.48	0.23	3.08	11.5	8.7	0.44	0.52	0.23	638.7	47%	-0.04	+0.04	-0.00
b	3.00	10.0	30.0	0.48	0.14	0.07	3.06	8.4	23.9	0.53	0.20	0.11	335.0	29%	+0.05	+0.06	+0.04
c	3.00	30.0	10.0	0.14	0.48	0.07	3.09	36.9	12.2	0.09	0.42	0.04	389.1	32%	-0.05	-0.06	-0.03
d	3.00	30.0	30.0	0.14	0.14	0.02	3.08	35.8	29.6	0.10	0.14	0.01	203.5	18%	-0.04	+0.00	-0.01
e	9.00	10.0	10.0	0.48	0.48	0.23	9.10	13.5	10.0	0.39	0.48	0.19	397.4	37%	-0.09	-0.00	-0.05
f	9.00	10.0	30.0	0.48	0.14	0.07	9.05	10.3	25.8	0.47	0.18	0.08	208.5	22%	-0.01	+0.04	+0.02
g	9.00	30.0	10.0	0.14	0.48	0.07	9.06	36.3	11.0	0.09	0.45	0.04	229.5	23%	-0.04	-0.03	-0.02
h	9.00	30.0	30.0	0.14	0.14	0.02	9.09	28.0	24.2	0.16	0.20	0.03	123.8	14%	+0.02	+0.06	+0.01

Stonehenge 1024×1024 (bin size 32×32)

#	σ^2	α_x	α_y	True MTF $(\frac{1}{4},0)$	MTF $(0,\frac{1}{4})$	MTF $(\frac{1}{4},\frac{1}{4})$	σ^2	α_x	α_y	Est MTF $(\frac{1}{4},0)$	MTF $(0,\frac{1}{4})$	MTF $(\frac{1}{4},\frac{1}{4})$	k	rej	Err MTF $(\frac{1}{4},0)$	MTF $(0,\frac{1}{4})$	MTF $(\frac{1}{4},\frac{1}{4})$
a	3.00	10.0	10.0	0.48	0.48	0.23	3.05	7.2	4.4	0.57	0.69	0.39	157.0	33%	+0.09	+0.20	+0.16
b	3.00	10.0	30.0	0.48	0.14	0.07	3.07	3.3	17.9	0.73	0.29	0.22	93.2	24%	+0.25	+0.16	+0.15
c	3.00	30.0	10.0	0.14	0.48	0.07	3.07	33.3	8.8	0.11	0.52	0.06	83.5	24%	-0.03	+0.04	-0.01
d	3.00	30.0	30.0	0.14	0.14	0.02	3.08	22.1	16.6	0.23	0.32	0.07	48.8	16%	+0.09	+0.18	+0.05
e	9.00	10.0	10.0	0.48	0.48	0.23	9.07	9.5	6.1	0.50	0.61	0.31	95.5	27%	+0.02	+0.13	+0.07
f	9.00	10.0	30.0	0.48	0.14	0.07	9.01	1.0	13.6	0.85	0.38	0.32	55.6	21%	+0.36	+0.25	+0.26
g	9.00	30.0	10.0	0.14	0.48	0.07	9.06	33.7	9.2	0.11	0.51	0.06	51.1	19%	-0.03	+0.02	-0.01
h	9.00	30.0	30.0	0.14	0.14	0.02	9.12	30.6	23.0	0.13	0.21	0.03	26.7	12%	-0.00	+0.08	+0.01

Stonehenge 512×512 (bin size 16×16)

6.6 Conclusions

We proposed a new framework for the estimation of the noise and transfer function parameters from any remote sensing image. This involves a fractal parametric model of the underlying natural scene and a parametric MTF that takes into account the optics and the sensor. The stability of the estimation method is due to the marginalization of the image model parameters.

We first proposed an initial algorithm that was successfully applied to both simulated and real data, and patented. Recently, we developed an optimized technique based on a simplified model such that the algorithmic complexity is mainly due to one FFT for large images, not to the estimation itself.

Good-quality estimates were obtained in an entirely unsupervised way. As shown experimentally, random images (from urban to natural areas) could be used, without specific features such as targets or coast lines, as long as their power spectrum content is sufficient. The developed framework is flexible enough to allow different kinds of instruments to be modeled, as long as optics and sensor blur can be accurately described by parametric functions.

In the future, field experiments shall be carried out in order to validate this approach, through the use of real blurred and noisy remote sensing images, as well as precise means of determining the true values of the blur parameters. A performance comparison should also be performed to help determine the net benefit of the proposed technique against some recent ones that have been applied to the exact same type of data, but based on different formalisms and assumptions.

Acknowledgments

The authors would like to thank the French Geographic Institute (IGN), the French Space Agency (CNES), and Space Imaging for providing the data.

References

[1] J-L. Starck, E. Pantin, and F. Murtagh. Deconvolution in astronomy: a review. In *Publications of the Astronomical Society of the Pacific*, volume 114, 2002.

[2] M.R. Banham and A.K. Katsaggelos. Digital image restoration. *IEEE*

Signal Processing Magazine, 14(2), 1997.

[3] D. Kundur and D. Hatzinakos. Blind image deconvolution. *IEEE Signal Processing Magazine*, 13(3), May 1996.

[4] S. Chardon, B. Vozel, and K. Chehdi. A comparative study between parametric blur estimation methods. *Proceedings of ICASSP, Phoenix, AZ, USA*, 1999.

[5] B. Chalmond. PSF estimation for image deblurring. *Graphical Models and Image Processing*, 53(4), July 1991.

[6] M. Cannon. Blind deconvolution of spatially invariant image blurs with phase. *IEEE Transactions on Acoustic, Speech, and Signal Processing*, 24(1), Feb. 1976.

[7] M.M. Chang, A.M. Tekalp, and A.T. Erdem. Blur identification using the bispectrum. *IEEE Transactions on Signal Processing*, 39(10), Oct. 1991.

[8] Q. Li and Y. Yoshida. Parameter-estimation and restoration for motion blurred images. *IEICE Transactions*, E80-A(8), 1997.

[9] S. Stryhanyn-Chardon. *Contribution au problème de la restauration myope des images numériques: analyse et synthèse*. PhD thesis, University of Rennes 1, France, Dec. 1997.

[10] R.L. Lagendijk, J. Biemond, and D.E. Boekee. Identification and restoration of noisy blurred images using the expectation-maximization algorithm. *IEEE Transactions on Acoustic, Speech, and Signal Processing*, 38(7), July 1990.

[11] S.J. Reeves and R.M. Mersereau. Blur identification by the method of generalized cross-validation. *IEEE Transactions on Image Processing*, 1(3), July 1992.

[12] J.M. Delvit, D. Leger, S. Roques, and C. Valorge. Modulation transfer function and noise measurement using neural networks. In *IEEE Proceedings of NNSP*, Toulouse, France, Sept. 2003.

[13] C. Latry, V. Despringre, and C. Valorge. Automatic MTF measurement through a least square method. In *Proceedings of SPIE: Sensors, Systems, and Next-Generation Satellites VIII*, volume 5570, Nov. 2004.

[14] R.G. Lane and R.H.T. Bates. Automatic multidimensional deconvolution. *Journal of the Optical Society of America A* , 4(1), 1987.

[15] R. Gerchberg and W. Saxton. A practical algorithm for the determination of phase from image and diffraction plane pictures. *Optik*, 35, 1972.

[16] S.M. Jefferies and J.C. Christou. Restoration of astronomical images by iterative blind deconvolution. *The Astrophysical Journal*, 415, Oct. 1993.

[17] G.R. Ayers and J.C. Dainty. Iterative blind deconvolution method and its applications. *Optics Letters*, 13(7), July 1988.

[18] Y.-L You and M. Kaveh. A regularization approach to blind restoration of images degraded by shift-variant blurs. In *IEEE Proceedings of ICIP*, Washington, DC, 1995.

[19] A. Mohammad-Djafari, N. Qaddoumi, and R. Zoughi. A blind deconvolution approach for resolution enhancement of near-field microwave images. In *SPIE Conf. on Mathematical Modeling, Bayesian Estimation and Inverse Problems*, volume 3816, Denver, CO, Jul. 1999.

[20] F. Tsumuraya, N. Miura, and N. Baba. Iterative blind deconvolution method using Lucy's algorithm. *Astronomy and Astrophysics*, 282, 1994.

[21] D.S.C. Biggs and M. Andrews. Iterative blind deconvolution of extended objects. In *IEEE Proceedings of ICIP*, volume II, Santa Barbara, CA, Oct. 1997.

[22] E. Thiebaut and J-M. Conan. Strict a priori constraints for maximum likelihood blind deconvolution. *Journal of the Optical Society of America A*, 12(3), March 1995.

[23] B.C. McCallum. Blind deconvolution by simulated annealing. *Optics Communications*, 75(2), Feb. 1990.

[24] A.M. Bronstein, M.M. Bronstein, M. Zibulevsky, and Y.Y. Zeevi. Blind deconvolution of images using optimal sparse representations. *IEEE Transactions on Image Processing*, 14(6), 2005.

[25] F. Rooms, W. Phillips, and S. Lidke. Simultaneous degradation estimation and restoration of confocal images and performance evaluation by colocalization analysis. In *Journal of Microscopy*, volume 218, Apr. 2005.

[26] T.J. Schulz and S.C. Cain. Simultaneous phase retrieval and deblurring for the Hubble space telescope. In R.J. Hanisch and R.L. White, editors, *The restoration of HST images and Spectra II*. STScI, 1994.

[27] A.S. Carasso. Direct blind deconvolution. *SIAM Journal of Applied Math*, 61(6), 2001.

[28] L. Bar, N. Sochen, and N. Kiryati. Semi-blind image restoration via Mumford-Shah regularization. *IEEE Transactions on Image Processing*, 15(2), 2006.

[29] T.E. Bishop and J.R. Hopgood. Blind image restoration using a block-stationary signal model. In *IEEE Proceedings of ICASSP*, Toulouse, France, 2006.

[30] J. Bernardo and A. Smith. *Bayesian Theory*. John Wiley and Sons, Chichester, UK, 1994.

[31] A. Gelman, J.B. Carlin, H.S. Stern, and D.B. Rubin. *Bayesian Data Analysis*. Chapman & Hall, 1995.

[32] J.O. Berger, B. Liseo, and R. Wolpert. Integrated likelihood methods for eliminating nuisance parameters. *Statistical Science*, 14(1), 1999.

[33] B. Mandelbrot and J. Van Ness. Fractional Brownian motion, fractional noises and applications. *SIAM review*, 10(4), 1968.

[34] G. Wornell. *Signal processing with fractals: a wavelet-based approach*. Signal processing series. Prentice Hall, 1995.

[35] J. Huang and D. Mumford. Statistics for natural images and models. *IEEE Proceedings of CVPR*, 1999.

[36] A.K. Jain. *Fundamentals of digital image processing*. Prentice Hall, 1989.

[37] D. Ruderman. Origins of scaling in natural images. *Vision Research*, 37(23), 1997.

[38] B. Rougé. Théorie de la chaîne image et restauration d'image optique à bruit final fixé. *Mémoire d'Habilitation à Diriger des Recherches, France*, 1997.

[39] D. Fried. Optical resolution through a randomly inhomogeneous medium for very long and very short exposures. *Journal of the Optical Society of America*, 56(10), 1966.

[40] J.W. Goodman. *Introduction to Fourier Optics*. McGraw-Hill, 1968.

[41] D.J.C. MacKay. *Information Theory, Inference, and Learning Algorithms*. Cambridge University Press, 2003.

[42] A. Jalobeanu, L. Blanc-Féraud, and J. Zerubia. Estimation of blur and noise parameters in remote sensing. In *IEEE Proceedings of ICASSP*, Orlando, FLA, USA, May 2002.

[43] J.M. Delvit, D. Leger, S. Roques, and C. Valorge. Blind estimation of noise variance. In *Proceedings of OPTIX 2001, Marseille, France*, Nov. 2001.

[44] B. Vozel, K. Chehdi, L. Klaine, V.V. Lukin, and S.K. Abramov. Noise identification and estimation of its statistical parameters by using unsupervised variational classification. In *IEEE Proceedings of ICASSP*, Toulouse, France, May 2006.

[45] W.H. Press, S.A. Teukolsky, W.T. Vetterling, and B.P. Flannery. *Numerical Recipes in C: The Art of Scientific Computing*. Cambridge University Press, 2nd edition, 1993.

7

Deconvolution and Blind Deconvolution in Astronomy

Eric Pantin, Jean-Luc Starck

Service d'Astrophysique CEA-Saclay, 91191 Gif-sur-Yvette, France
e-mail: (epantin, jstarck)@cea.fr

Fionn Murtagh

Dept. Computer Science, Royal Holloway, University of London, Egham, UK
e-mail: fmurtagh@acm.org

Abstract

This chapter reviews different astronomical deconvolution methods. The all-pervasive presence of noise is what makes deconvolution particularly difficult. The diversity of resulting algorithms reflects different ways of estimating the true signal under various idealizations of its properties. Different ways of approaching signal recovery are based on different instrumental noise models, whether the astronomical objects are point-like or extended, and indeed on the computational resources available to the analyst. We present a number of recent results in this survey of signal restoration, including in the areas of super-resolution and dithering. In particular we show that most recent published work has consisted of incorporating some form of multiresolution in the deconvolution process. Finally we show how the deconvolution techniques can be extended to the case of blind deconvolution.

7.1 Introduction

Deconvolution is a key area in signal and image processing. It can include deblurring of an observed signal to remove atmospheric effects. More generally, it means correcting for instrumental effects or observing conditions.

Research in image deconvolution has recently seen considerable work, partly triggered by the Hubble space telescope (HST) optical aberration problem at the beginning of its mission that motivated astronomers to improve current algorithms or develop new and more effective ones. Since then, deconvolution of astronomical images has proven in some cases to be crucial for extracting scientific content. For instance, infrared astronomical satellite (IRAS) images can be effectively reconstructed thanks to a new pyramidal maximum entropy algorithm [1]. Io volcanism can be studied with a lower resolution of 0.15 arcsec, or 570 km on Io [2]. Deconvolved mid-infrared images at 20 μm revealed the innermost structure of the active galactic nucleus (AGN) in NGC1068 surrounding the giant blackhole, some of which enshrouded emission regions unobservable at lower wavelength because of higher dust extinction [3]: see Figure 7.1. Research on gravitational lenses is easier and more effective when applying deconvolution methods [4].

Deconvolution will be even more crucial in the future in order to fully take advantage of increasing numbers of high-quality ground-based telescopes, for which images are strongly limited in resolution by the seeing.

The HST provided a leading example of the need for deconvolution, in the period before the detector system was refurbished. Two proceedings [5, 6] provide useful overviews of this work, and a later reference is [7]. While an atmospheric seeing point spread function (PSF) may be relatively tightly distributed around the mode, this was not the case for the spherically aberrated HST PSF. Whenever the PSF "wings" are extended and irregular, deconvolution offers a straightforward way to mitigate the effects of this and to upgrade the core region of a point source. One usage of deconvolution of continuing importance is in information fusion from different detectors. For example, Faure et al. [8] deconvolve HST images when correlating with ground-based observations. In Radomski et al. [9], Keck data are deconvolved, for study with HST data. VLT (Very Large Telescope) data are deconvolved in [10], with other ESO (european southern observatory) and HST data used as well.

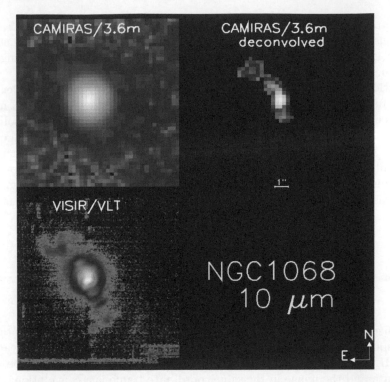

FIGURE 7.1: The active galaxy nucleus of NGC1068 observed at 20 μm. Upper left: the raw image is highly blurred by telescope diffraction (1.4 arcsec resolution on a 3.6-m telescope) and seeing. Right: the restored image using the multiscale entropy method reveals the inner structure in the vicinity of the nucleus. Lower left: the same object observed later on a larger telescope (VLT, 8 m-class telescope) whose sharper resolution confirms the structures found previously by image deconvolution. (**See color insert following page 140.**)

In planetary work, Coustenis et al. [11] discuss CFHT (Canada France Hawaii telescope) data as well as HST and other observations.

What emerges very clearly from this small sample – which is in no way atypical – is that a major use of deconvolution is to help in cross-correlating image and signal information.

An observed signal is never in pristine condition, and improving it involves inverting the spoiling conditions, i.e., finding a solution to an inverse equation. The problem is usually an ill-conditioned one, which leads to the need for constraints on what is to be an acceptable solution. Constraints related to the type of signal we are dealing with play an important role in the development of effective and efficient algorithms. The use of constraints to provide for a stable and unique solution is termed regularization.

Our review opens in Section 7.2 with a formalization of the problem. Section 7.3 considers the issue of regularization. In Section 7.4, the CLEAN method, which is central to radio astronomy, is described. Bayesian modeling and inference in deconvolution are reviewed in Section 7.5.

Section 7.6 further considers regularization, surveying more complex and powerful regularization methods. Section 7.7 introduces wavelet-based methods as used in deconvolution. These methods are based on multiple resolution or scale. In Sections 7.8 and 7.9, important issues related to resolution of the output result image are discussed. Section 7.8 is based on the fact that it is normally not worthwhile to target an output result with better resolution than some limit, for instance a pixel size. Section 7.9 investigates when, where, and how missing information can be inferred to provide a super-resolution output image.

7.2   The Deconvolution Problem

Consider an image characterized by its intensity distribution (the "data") I, corresponding to the observation of a "real image" O through an optical system. If the imaging system is linear and shift-invariant, the relation between the data and the image in the same coordinate frame is a convolution:

$$
I(x, y) = \int\limits_{x_1=-\infty}^{+\infty} \int\limits_{y_1=-\infty}^{+\infty} P(x - x_1, y - y_1)O(x_1, y_1)dx_1dy_1 + N(x, y)
$$

$$
= (P * O)(x, y) + N(x, y) \tag{7.1}
$$

P is the point spread function, PSF, of the imaging system, and N is additive noise.

In Fourier space we have:

$$
\hat{I}(u, v) = \hat{O}(u, v)\hat{P}(u, v) + \hat{N}(u, v). \tag{7.2}
$$

We want to determine $O(x, y)$ knowing I and P. This inverse problem has led to a large amount of work, the main difficulties being the existence of: (i) a cut-off frequency of the point spread function and (ii) the additive noise. See for example [12–15].

A solution can be obtained by computing the Fourier transform of the deconvolved object $\hat{O}$ by a simple division between the image $\hat{I}$ and the PSF $\hat{P}$:

$$
\hat{\hat{O}}(u, v) = \frac{\hat{I}(u, v)}{\hat{P}(u, v)} = \hat{O}(u, v) + \frac{\hat{N}(u, v)}{\hat{P}(u, v)}. \tag{7.3}
$$

This method, sometimes called the *Fourier-quotient method*, is very fast. We only need to do a Fourier transform and an inverse Fourier transform. For frequencies close to the frequency cut-off, the noise term becomes important, and the noise is amplified. Therefore in the presence of noise, this method cannot be used.

Equation 7.1 is usually in practice an ill-posed problem. This means that there is no unique and stable solution.

Other topics related to deconvolution are:

- Super-resolution: object spatial frequency information outside the spatial bandwidth of the image formation system is recovered.

- Blind deconvolution: the PSF P is unknown.

- Myopic deconvolution: the PSF P is partially known.

- Image reconstruction: an image is formed from a series of projections (computed tomography, positron emission tomography or PET, and so on).

We will discuss the first three points in this chapter.

In the deconvolution problem, the PSF is assumed to be known. In practice, we have to construct a PSF from the data, or from an optical model of the imaging telescope. In astronomy, the data may contain stars, or one can point towards a reference star in order to reconstruct a PSF. The drawback is the "degradation" of this PSF because of unavoidable noise or spurious instrument signatures in the data. So, when reconstructing a PSF from experimental data, one has to reduce very carefully the images used (background removal, for instance) or otherwise any spurious feature in the PSF would be repeated around each object in the deconvolved image. Another problem arises when the PSF is highly variable with time, as is the case for adaptive optics images. This means usually that the PSF estimated when observing a reference star, after or before the observation of the scientific target, has small differences from the perfect one. In this particular case, one has to turn towards myopic deconvolution methods [16] in which the PSF is also estimated in the iterative algorithm using a first guess deduced from observations of reference stars.

Another approach consists of constructing a synthetic PSF. Several studies [17–20] have suggested a radially symmetric approximation to the PSF:

$$P(r) \propto \left(1 + \frac{r^2}{R^2}\right)^{-\beta} \tag{7.4}$$

The parameters β and R are obtained by fitting the model with stars contained in the data.

7.3 Linear Regularized Methods

7.3.1 Least Squares Solution

It is easy to verify that the minimization of $\| I(x,y) - P(x,y) * O(x,y) \|^2$ leads to the solution:

$$\hat{\tilde{O}}(u,v) = \frac{\hat{P}^*(u,v)\hat{I}(u,v)}{|\hat{P}(u,v)|^2} \tag{7.5}$$

which is defined only if $\hat{P}(u,v)$ is different from zero. Here $\hat{P}^*$ is the complex conjugate of $\hat{P}$. The problem is generally ill-posed and we need to introduce *regularization* in order to find a unique and stable solution.

7.3.2 Tikhonov Regularization

Tikhonov regularization [21] consists of minimizing the term:

$$J_T(O) = \| I(x,y) - (P * O)(x,y) \|^2 + \lambda \| H * O \|^2 \tag{7.6}$$

where H corresponds to a high-pass filter, and $\| \, . \, \|^2$ is norm squared. This criterion contains two terms. The first, $\| I(x,y) - (P * O)(x,y) \|^2$, expresses fidelity to the data $I(x,y)$, and the second, $\lambda \| H * O \|^2$, expresses smoothness of the restored image. λ is the regularization parameter and represents the trade-off between fidelity to the data and the smoothness of the restored image.

The solution is obtained directly in Fourier space:

$$\hat{\tilde{O}}(u,v) = \frac{\hat{P}^*(u,v)\hat{I}(u,v)}{|\hat{P}(u,v)|^2 + \lambda |\hat{H}(u,v)|^2}. \tag{7.7}$$

Finding the optimal value λ necessitates use of numerical techniques such as cross-validation [22, 23]. This method works well, but computationally it is relatively lengthy and produces smoothed images. This second point can be a real problem when we seek compact structures such as is the case in astronomical imaging. This method can be generalized such that

$$\hat{\tilde{O}}(u,v) = \hat{W}(u,v)\frac{\hat{I}(u,v)}{\hat{P}(u,v)}, \tag{7.8}$$

W being a window function satisfying some conditions [14]. These window functions can be, for instance, the usual Gaussian, Blackmann, Hamming, or Hanning functions. The window function can also be derived directly from the PSF [24].

Linear regularized methods have the advantage of being very attractive from a computation point of view. Furthermore, the noise in the solution can easily be derived from the noise in the data and the window function. For example, if the noise in the data is Gaussian with a standard deviation σ_d, the noise in the solution is $\sigma_s^2 = \sigma_d^2 \sum W_k^2$. But this noise estimation does not take into account errors relative to inaccurate knowledge of the PSF, which limits its interest in practice.

Linear regularized methods present also a number of severe drawbacks:

- Creation of Gibbs oscillations in the neighborhood of the discontinuities contained in the data. The visual quality is therefore degraded.

- No *a priori* information can be used. For example, negative values can exist in the solution, while in most cases we know that the solution must be positive.

- Since the window function is a low-pass filter, the resolution is degraded. There is trade-off between the resolution we want to achieve and the noise level in the solution. Other methods such as wavelet-based methods do not have such a constraint.

7.4 CLEAN

The CLEAN method [25] is a mainstream one in radio astronomy. This approach assumes the object is only composed of point sources. It tries to decompose the image (called the dirty map) into a set of δ-functions. This is done iteratively by finding the point with the largest absolute brightness and subtracting the PSF (dirty beam) scaled with the product of the loop gain and the intensity at that point. The resulting residual map is then used to repeat the process. The process is stopped when some prespecified limit is reached. The convolution of the δ-functions with an ideal PSF (clean beam) plus the residual equals the restored image (clean map). This solution is only possible if the image does not contain large-scale structures.

In the work of [26] and [27], the restoration of an object composed of peaks, called *sparse spike trains*, has been treated in a rigorous way.

7.5 Bayesian Methodology

A Bayesian formulation may well be just another way of describing the classical minimization problem with constraints, as seen in Section 7.3. However, it is important for the following reasons: it incorporates procedures for the estimation of parameters, and prior information on the values of such parameters can also be incorporated easily.

7.5.1 Definition

The Bayesian approach consists of constructing the conditional probability density relationship:

$$p(O \mid I) = \frac{p(I \mid O)p(O)}{p(I)}. \tag{7.9}$$

The Bayes solution is found by maximizing the right part of the equation. Now since we are maximizing $p(O \mid I)$, the image I is always the same for the maximization, and so $p(I)$ is constant here. The maximum likelihood solution (ML) maximizes only the density $p(I \mid O)$ over O:

$$\mathrm{ML}(O) = \max_{O} p(I \mid O). \tag{7.10}$$

The maximum *a posteriori* solution (MAP) maximizes over O the product $p(I \mid O)p(O)$ of the ML and a prior:

$$\mathrm{MAP}(O) = \max_{O} p(I \mid O)p(O). \tag{7.11}$$

$p(I)$ is considered as a constant value which has no effect in the maximization process, and is ignored. The ML solution is equivalent to the MAP solution assuming a uniform probability density for $p(O)$.

7.5.2 Maximum Likelihood with Gaussian Noise

The probability $p(I \mid O)$ is

$$p(I \mid O) = \frac{1}{\sqrt{2\pi}\sigma_I} \exp - \frac{(I - P * O)^2}{2\sigma_I^2} \tag{7.12}$$

where σ_I^2 is image variance and, assuming that $p(O)$ is a constant, maximizing $p(O \mid I)$ is equivalent to minimizing

$$J(O) = \frac{\| I - P * O \|^2}{2\sigma_I^2} \tag{7.13}$$

where the denominator is unimportant for the optimum value of J. Using the steepest descent minimization method, a typical iteration is

$$O^{n+1} = O^n + \gamma P^* * (I - P * O^n) \tag{7.14}$$

where $P^*(x, y) = P(-x, -y)$. P^* is the transpose of the PSF, and O^n is the current estimate of the desired "real image." This method is usually called the

Landweber method [28], but sometimes also the *successive approximations* or Jacobi method [14].

The solution can also be found directly using the Fourier fast transform (FFT) by

$$\hat{O}(u,v) = \frac{\hat{P}^*(u,v)\hat{I}(u,v)}{\hat{P}^*(u,v)\hat{P}(u,v)} \qquad (7.15)$$

which amounts to the same as Equation 7.3. Such a straightforward approach, unfortunately, amplifies noise.

7.5.3 Gaussian Bayes Model

If the object and the noise are assumed to follow Gaussian distributions with zero mean and variance respectively equal to σ_O^2 and σ_N^2, then a Bayes solution leads to the Wiener filter:

$$\hat{O}(u,v) = \frac{\hat{P}^*(u,v)\hat{I}(u,v)}{\mid \hat{P}(u,v) \mid^2 + \frac{\sigma_N^2(u,v)}{\sigma_O^2(u,v)}}. \qquad (7.16)$$

Wiener filtering has serious drawbacks (artifact creation such as ringing effects) and needs spectral noise estimation. Its advantage is that it is very fast.

7.5.4 Maximum Likelihood with Poisson Noise

The probability $p(I \mid O)$ is

$$p(I \mid O) = \prod_{x,y} \frac{((P*O)(x,y))^{I(x,y)} \exp\{-(P*O)(x,y)\}}{I(x,y)!}. \qquad (7.17)$$

The maximum can be computed by taking the derivative of the logarithm:

$$\frac{\partial \ln p(I \mid O)(x,y)}{\partial O(x,y)} = 0 \qquad (7.18)$$

which leads to the result (assuming the PSF is normalized to unity)

$$\left[\frac{I}{P*O} * P^*\right](x,y) = 1. \qquad (7.19)$$

Multiplying both sides by $O(x,y)$

$$O(x,y) = \left[\frac{I}{(P*O)} * P^*\right](x,y)O(x,y), \qquad (7.20)$$

and using Picard iteration [29] leads to

$$O^{n+1}(x,y) = \left[\frac{I}{(P * O^n)} * P^* \right] (x,y) O^n(x,y), \qquad (7.21)$$

which is the Richardson–Lucy algorithm [30–32], also sometimes called the *expectation maximization*, or EM method [33]. This method is commonly used in astronomy. Flux is preserved and the solution is always positive. The positivity of the solution can be obtained too with Van Cittert's and the one-step gradient methods by thresholding negative values in O^n at each iteration.

7.5.5 Maximum *a Posteriori* with Poisson Noise

We formulate the object PDF (probability density function) as

$$p(O) = \prod_{x,y} \frac{M(x,y)^{O(x,y)} \exp\{-M(x,y)\}}{O(x,y)!}. \qquad (7.22)$$

The MAP solution is

$$O(x,y) = M(x,y) \exp \left\{ \left[\frac{I(x,y)}{(P * O)(x,y)} - 1 \right] * P^*(x,y) \right\}, \qquad (7.23)$$

and choosing $M = O^n$ and using Picard iteration leads to

$$O^{n+1}(x,y) = O^n(x,y) \exp \left\{ \left[\frac{I(x,y)}{(P * O^n)(x,y)} - 1 \right] * P^*(x,y) \right\}. \qquad (7.24)$$

7.5.6 Maximum Entropy Method

In the absence of any information on the solution O except its positivity, a possible course of action is to derive the probability of O from its entropy, which is defined from information theory. Then if we know the entropy H of the solution, we derive its probability as

$$p(O) = \exp(-\alpha H(O)). \qquad (7.25)$$

The most commonly used entropy functions are:

- Burg [34]: $H_b(O) = -\sum_x \sum_y \ln(O(x,y))$
- Frieden [35]: $H_f(O) = -\sum_x \sum_y O(x,y) \ln(O(x,y))$
- Gull and Skilling [36]:

$$H_g(O) = \sum_x \sum_y O(x,y) - M(x,y) - O(x,y) \ln(O(x,y)/M(x,y)).$$

The last definition of the entropy has the advantage of having a zero maximum when O equals the model M, usually taken as a flat image.

7.5.7 Other Regularization Models

Molina et al. [15] present an excellent review of taking the spatial context of image restoration into account. Some appropriate prior is used for this. One such regularization constraint is:

$$\|CI\|^2 = \sum_x \sum_y I(x,y) - \frac{1}{4}(I(x,y+1) + I(x,y-1) + I(x+1,y) + I(x-1,y))$$

(7.26)

which is equivalent to defining the prior

$$p(O) \propto \exp\left\{-\frac{\alpha}{2}\|CI\|^2\right\}.$$

(7.27)

α is the inverse of the prior variance and it controls the smoothness of the final solution.

Given the form of Equation (7.26), such regularization can be viewed as setting a constraint on the Laplacian of the restoration. In statistics this model is a simultaneous autoregressive model (SAR) [37].

Alternative prior models can be defined, related to the SAR model of Equation (7.26):

$$p(O) \propto \exp\left\{-\frac{\alpha}{2}\sum_x \sum_y \left(I(x,y) - I(x,y+1)\right)^2 + \left(I(x,y) - I(x+1,y)\right)^2\right\}$$

(7.28)

where constraints are set on first derivatives. This is a conditional autoregressive, or CAR, model, discussed further below.

Blanc-Féraud and Barlaud [38], and Charbonnier et al. [39] consider the following prior:

$$p(O) \propto \exp\left\{-\alpha \sum_x \sum_y \phi(\| \nabla I \|)(x,y)\right\}$$

$$\propto \exp\left\{-\alpha \sum_x \sum_y \phi(I(x,y) - I(x,y+1))^2 + \phi(I(x,y) - I(x+1,y))^2\right\}.$$

(7.29)

The function ϕ, called *potential function*, is an edge-preserving function.

Generally, ϕ functions are chosen with a quadratic part which ensures good smoothing of small gradients [40], and a linear behavior which cancels the penalization of large gradients [41]:

1. $\lim_{t \to 0} \frac{\phi'(t)}{2t} = M < +\infty$ to smooth faint gradients.

2. $\lim_{t \to \infty} \frac{\phi'(t)}{2t} = 0$ to preserve strong gradients.

3. $\frac{\phi'(t)}{2t}$ is continuous and strictly decreasing.

Such functions are often called L_2-L_1 functions. Examples of ϕ functions:

1. $\phi_q(x) = x^2$: quadratic function.

2. $\phi_{TV}(x) = \mid x \mid$: Total Variation.

3. $\phi_2(x) = 2\sqrt{1 + x^2} - 2$: Hyper-Surface [42].

4. $\phi_3(x) = x^2/(1 + x^2)$ [43].

5. $\phi_4(x) = 1 - e^{-x^2}$ [44].

6. $\phi_5(x) = \log(1 + x^2)$ [45].

The ARTUR method [39], which has been used for helioseismic inversion [46], uses the function $\phi(t) = \log(1 + t^2)$. Anisotropic diffusion [44, 47] uses similar functions, but in this case the solution is computed using *partial differential equations*.

The function $\phi(t) = t$ leads to the *total variation* method [48, 49], the constraints are on first derivatives, and the model is a special case of a conditional autoregressive or CAR model. Molina et al. [15] discuss the applicability of CAR models to image restoration involving galaxies. They argue that such models are particularly appropriate for the modeling of luminosity exponential and $r^{1/4}$ laws.

The priors reviewed above can be extended to more complex models. In Molina et al. [50, 51], a compound Gauss Markov random field (CGMRF) model is used, one of the main properties of which is to target the preservation and improvement of line processes.

Another prior again was used in Molina and Cortijo [52] for the case of planetary images.

7.6   Iterative Regularized Methods

7.6.1   Constraints

We assume now that there exists a general operator, $\mathcal{P}_C(.)$, which enforces a set of constraints on a given object O, such that if O satisfies all the constraints, we have: $O = \mathcal{P}_C(O)$. Commonly used constraints are positivity, spatial support constraint, band-limitation in Fourier space. These constraints can be incorporated easily in the basic iterative scheme.

7.6.2 Jansson–Van Cittert Method

Van Cittert [53] restoration is relatively easy to write. We start with $n = 0$ and $O^{(0)} = I$ and we iterate:

$$O^{n+1} = O^n + \alpha(I - P * O^n) \tag{7.30}$$

where α is a convergence parameter generally taken as 1. When n tends to infinity, we have $O = O + I - P * O$, so $I = P * O$. In Fourier space, the convolution product becomes a product

$$\hat{O}^{n+1} = \hat{O}^n + \alpha(\hat{I} - \hat{P}\hat{O}^n). \tag{7.31}$$

In this equation, the object distribution is modified by adding a term proportional to the residual. The algorithm converges quickly, often after only five or six iterations. But the algorithm generally diverges in the presence of noise. Jansson [54] modified this technique in order to give it more robustness by considering constraints on the solution. If we wish that $A \leq O_k \leq B$, the iteration becomes

$$O^{n+1}(x, y) = O^n(x, y) + r(x, y) \left[I - (P * O^n) \right](x, y) \tag{7.32}$$

with:

$$r(x, y) = C \left[1 - 2(B - A)^{-1} \mid O^n(x, y) - 2^{-1}(A + B) \mid \right]$$

and with C constant.

More generally the constrained Van Cittert method is written as:

$$O^{n+1} = \mathcal{P}_C \left[O^n + \alpha(I - P * O^n) \right]. \tag{7.33}$$

7.6.3 Other Iterative Methods

Other iterative methods can be constrained in the same way:

- Landweber:

$$O^{n+1} = \mathcal{P}_C \left[O^n + \gamma P^*(I - P * O^n) \right] \tag{7.34}$$

- Richardson–Lucy method:

$$O^{n+1}(x, y) = \mathcal{P}_C \left[O^n(x, y) \left[\frac{I}{(P * O^n)} * P^* \right] (x, y) \right] \tag{7.35}$$

- Tikhonov: the Tikhonov solution can be obtained iteratively by computing the gradient of Equation (7.6):

$$\nabla(J_T(O)) = \left[P^* * P + \mu H^* * H \right] * O - P^* * I \tag{7.36}$$

and applying the following iteration:

$$O^{n+1}(x,y) = O^n(x,y) - \gamma\nabla(J_T(O))(x,y). \tag{7.37}$$

The constrained Tikhonov solution is therefore obtained by:

$$O^{n+1}(x,y) = \mathcal{P}_C\left[O^n(x,y) - \gamma\nabla(J_T(O))(x,y)\right]. \tag{7.38}$$

The number of iterations plays an important role in these iterative methods. Indeed, the number of iterations can be discussed in relation to regularization. When the number of iterations increases, the iterates first approach the unknown object, and then potentially go away from it [14]. The Landweber and Richardson–Lucy methods converge to useless solutions due to noise amplification. The main problem is that the space of solutions of these methods is almost infinite. Regularized methods force the solution obtained at convergence to be constrained to images having the desirable properties of the original (unknown) image, i.e. being nonnoisy.

7.7 Wavelet-Based Deconvolution

7.7.1 Introduction

Deconvolution and Fourier Domain

The Fourier domain diagonalizes the convolution operator, and we can identify and reduce the noise which is amplified during the inversion. When the signal can be modeled as stationary and Gaussian, the Wiener filter is optimal. But when the signal presents spatially localized features such as singularities or edges, these features cannot be well represented with Fourier basis functions, which extend over the entire spatial domain. Other basis functions, such as wavelets, are better suited to represent a large class of signals.

Towards Multiresolution

The concept of multiresolution was first introduced for deconvolution by Wakker and Schwarz [55] when they proposed the Multiresolution CLEAN algorithm for interferometric image deconvolution. During the last ten years, many developments have taken place in order to improve the existing methods (CLEAN, Landweber, Lucy, MEM, and so on), and these results have led to the use of different levels of resolution.

The Lucy algorithm was modified [56] in order to take into account *a priori* information about stars in the field where both position and brightness are known. This is done by using a two-channel restoration algorithm, one channel representing the contribution relative to the stars, and the second to the background. A smoothness constraint is added on the background channel. This method was then refined firstly (and called *CPLUCY*) for considering subpixel positions [57], and a second time [58] (and called *GIRA*) for modifying the smoothness constraint.

A similar approach has been followed by Magain et al. [59], but more in the spirit of the CLEAN algorithm. Again, the data are modeled as a set of point sources on top of a spatially varying background, leading to a two-channel algorithm.

The MEM method has also been modified by several authors [1, 60–63]. First, Weir proposed the *Multi-channel MEM* method, in which an object is modeled as the sum of objects at different levels of resolution. The method was then improved by Bontekoe et al. [1] with the *Pyramid MEM*. In particular, many regularization parameters were fixed by the introduction of the dyadic pyramid. The link between *Pyramid MEM* and wavelets was underlined in [61,63], and it was shown that all the regularization parameters can be derived from the noise modeling. Wavelets were also used in [62] in order to create a segmentation of the image, each region being then restored with a different smoothness constraint, depending on the resolution level where the region was found. This last method has, however, the drawback of requiring user interaction for deriving the segmentation threshold in the wavelet space.

The *Pixon* method [64, 65] is relatively different to the previously described methods. This time, an object is modeled as the sum of pseudo-images smoothed locally by a function with position-dependent scale, called the pixon shape function. The set of pseudo-images defines a dictionary, and the image is supposed to contain only features included in this dictionary. But the main problem lies in the fact that features which cannot be detected directly in the data, nor in the data after a few Lucy iterations, will not be modeled with the pixon functions, and will be strongly regularized as background. The result is that the faintest objects are over-regularized while strong objects are well restored. This is striking in the example shown in Figure 7.2.

The *total variation* method has a close relation with the Haar transform [66,67], and more generally, it has been shown that potential functions, used in Markov Random Field and PDE methods, can be applied on the wavelet coefficients as well. This leads to multiscale regularization, and the original method becomes a specific case where only one decomposition level is used in the wavelet transform.

Wavelets offer a mathematical framework for the multiresolution processing. Furthermore, they furnish an ideal way to include noise modeling in the deconvolution methods. Since the noise is the main problem in deconvolution, wavelets are very well adapted to the regularization task.

7.7.2 Regularization from the Multiresolution Support

7.7.2.1 Noise Suppression Based on the Wavelet Transform

We have noted how, in using an iterative deconvolution algorithm such as Van Cittert or Richardson–Lucy, we define $R^{(n)}(x, y)$, the residual at iteration n:

$$R^n(x, y) = I(x, y) - (P * O^n)(x, y). \tag{7.39}$$

By using the *à trous* wavelet transform algorithm, R^n can be defined as the sum of its J wavelet scales and the last smooth array:

$$R^n(x, y) = c_J(x, y) + \sum_{j=1}^{J} w_{j,x,y} \tag{7.40}$$

where the first term on the right is the last smoothed array, and w denotes a wavelet scale.

The wavelet coefficients provide a mechanism to extract only the significant structures from the residuals at each iteration. Normally, a large part of these residuals is statistically nonsignificant. The significant residual [68,69] is then:

$$\bar{R}^n(x, y) = c_{J,x,y} + \sum_{j=1}^{J} M(j, x, y) w_{j,x,y} \tag{7.41}$$

where $M(j, x, y)$ is the multiresolution support, and is defined by:

$$M(j, x, y) = \begin{cases} 1 & \text{if } w_{j,x,y} \text{ is significant} \\ 0 & \text{if } w_{j,x,y} \text{ is nonsignificant} \end{cases} \tag{7.42}$$

This describes in a logical or Boolean way if the data contain information at a given scale j and at a given position (x, y).

An alternative approach was outlined in [70] and [71]: the support was initialized to zero, and built up at each iteration of the restoration algorithm. Thus in Equation (7.41) above, $M(j, x, y)$ was additionally indexed by n, the iteration number. In this case, the support was specified in terms of significant pixels at each scale, j; and in addition pixels could become significant as the iterations proceeded, but could not be made nonsignificant.

7.7.2.2 Regularization of Van Cittert's Algorithm

Van Cittert's iteration [53] is:

$$O^{n+1}(x, y) = O^n(x, y) + \alpha R^n(x, y) \tag{7.43}$$

with $R^n(x, y) = I^n(x, y) - (P * O^n)(x, y)$. Regularization using significant structures leads to:

$$O^{n+1}(x, y) = O^n(x, y) + \alpha \bar{R}^n(x, y). \tag{7.44}$$

The basic idea of this regularization method consists of detecting, at each scale, structures of a given size in the residual $R^n(x, y)$ and putting them in the restored image $O^n(x, y)$. The process finishes when no more structures are detected. Then, we have separated the image $I(x, y)$ into two images $\tilde{O}(x, y)$ and $R(x, y)$. $\tilde{O}$ is the restored image, which ought not to contain any noise, and $R(x, y)$ is the final residual which ought not to contain any structure. R is our estimate of the noise $N(x, y)$.

7.7.2.3 Regularization of the One-Step Gradient Method

The one-step gradient iteration is:

$$O^{n+1}(x, y) = O^n(x, y) + (P^* * R^n)(x, y) \tag{7.45}$$

with $R^n(x, y) = I(x, y) - (P * O^n)(x, y)$. Regularization by significant structures leads to:

$$O^{n+1}(x, y) = O^n(x, y) + \left(P^* * \bar{R}^n\right)(x, y). \tag{7.46}$$

7.7.2.4 Regularization of the Richardson–Lucy Algorithm

From Equation (7.1), we have $I^n(x, y) = (P * O^n)(x, y)$. Then $R^n(x, y) = I(x, y) - I^n(x, y)$, and hence $I(x, y) = I^n(x, y) + R^n(x, y)$.
The Richardson–Lucy equation is:

$$O^{n+1}(x, y) = O^n(x, y) \left[\frac{I^n + R^n}{I^n} * P^*\right](x, y)$$

and regularization leads to:

$$O^{n+1}(x, y) = O^n(x, y) \left[\frac{I^n + \bar{R}^n}{I^n} * P^*\right](x, y).$$

7.7.2.5 Convergence

The standard deviation of the residual decreases until no more significant structures are found. Convergence can be estimated from the residual. The algorithm stops when a user-specified threshold is reached:

$$(\sigma_{R^{n-1}} - \sigma_{R^n})/(\sigma_{R^n}) < \epsilon. \tag{7.47}$$

7.7.2.6 Examples

A simulated Hubble Space Telescope Wide Field Camera image of a distant cluster of galaxies is shown in Figure 7.2, upper left. The image used was one of a number described in [72, 73]. The simulated data are shown in Figure 7.2, upper right. Four deconvolution methods were tested: Richardson–Lucy, Pixon, wavelet–vaguelette, Wavelet–Lucy. Deconvolved images are presented respectively in Figure 7.2 middle left, middle right, bottom left, and

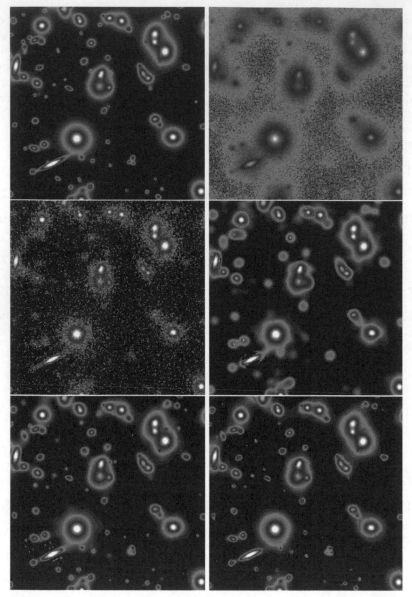

FIGURE 7.2: Simulated Hubble Space Telescope Wide Field Camera image of a distant cluster of galaxies. Six quadrants. Upper left: original, unaberrated and noise-free. Upper right: input, aberrated, noise added. Middle left: restoration, Richardson–Lucy. Middle right: restoration, Pixon method. Lower left, restoration wavelet–vaguelette. Lower right, restoration wavelet–Lucy. (**See color insert following page 140.**)

bottom right. The Richardson–Lucy method amplifies the noise, which implies that the faintest objects disappear in the deconvolved image. The Pixon method introduces regularization, and the noise is under control, while objects where "pixons" have been detected are relatively well protected from the regularization effect. Since the "pixon" features are detected from noisy partially deconvolved data, the faintest objects are not in the pixon map and are strongly regularized. The wavelet–vaguelette method is very fast and produces relatively high-quality results when compared to Pixon or Richardson–Lucy, but the Wavelet–Lucy method seems clearly the best of the four methods. There are fewer spurious objects than in the wavelet–vaguelette method, it is stable for any kind of PSF, and any kind of noise modeling can be considered.

7.7.3 Multiresolution CLEAN

The CLEAN solution is only available if the image does not contain large-scale structures. Wakker and Schwarz [55] introduced the concept of Multiresolution Clean (MRC) in order to alleviate the difficulties occurring in CLEAN for extended sources. The MRC approach consists of building two intermediate images, the first one (called the smooth map) by smoothing the data to a lower resolution with a Gaussian function, and the second one (called the difference map) by subtracting the smoothed image from the original data. Both these images are then processed separately. By using a standard CLEAN algorithm on them, the smoothed clean map and difference clean map are obtained. The recombination of these two maps gives the clean map at the full resolution. This algorithm may be viewed as an artificial recipe, but it has been shown [74–76] that it is linked to multiresolution analysis. Wavelet analysis leads to a generalization of MRC from a set of scales. The Wavelet Clean Method (WCLEAN) consists of the following steps:

- Apply the wavelet transform to the image: we get W_I.

- Apply the wavelet transform to the PSF: we get W_P.

- Apply the wavelet transform to the clean beam: we get W_C.

- For each scale j of the wavelet transform, apply the CLEAN algorithm using the wavelet scale j of both W_I and W_P.

- Apply an iterative reconstruction algorithm using W_C.

More details can be found in [74, 76].

7.7.4 The Wavelet Constraint

We have seen previously that many regularized deconvolution methods (MEM, Tikhonov, total variation, and so on) can be expressed by two terms (i.e., $\| I - P * O \|^2 + \lambda \mathcal{C}(O)$), the first representing the fidelity to the data and the second (i.e., $\mathcal{C}(O)$) the smoothness constraint on the solution. The parameter λ fixes the trade-off between the fit to the data and the smoothness. Using a wavelet-based penalizing term $\mathcal{C}_w$, we want to minimize

$$J(O) = \parallel I - P * O \parallel^2 + \lambda \mathcal{C}_w(O). \tag{7.48}$$

If ϕ is a potential function which was applied on the gradients (see Section 7.5.7), it can also be applied to the wavelet coefficients and the constraint on the solution is now expressed in the wavelet domain by [77]:

$$J(O) = \parallel I - P * O \parallel^2 + \lambda \sum_{j,k,l} \phi(\parallel (WO)_{j,k,l} \parallel_p). \tag{7.49}$$

When $\phi(x) = x$ and $p = 1$, it corresponds to the l_1 norm of the wavelet coefficients. In this framework, the multiscale entropy deconvolution method (see below) is only one special case of the wavelet constraint deconvolution method.

7.7.4.1 Multiscale Entropy

In [63, 78, 79], the benchmark properties for a good "physical" definition of entropy were discussed. The multiscale entropy, which fulfills these properties, consists of considering that the entropy of a signal is the sum of the information at each scale of its wavelet transform [78], and the information of a wavelet coefficient is related to the probability of it being due to noise.

For Gaussian noise, the multiscale entropy penalization function is:

$$h_n(w_{j,k}) = \frac{1}{\sigma_j^2} \int\limits_0^{|w_{j,k}|} u \ \text{erfc}\left(\frac{|w_{j,k}| - u}{\sqrt{2}\sigma_j}\right) du \tag{7.50}$$

where erfc is the complementary error function. A complete description of this method is given in [80]. Figure 7.3 shows the multiscale entropy penalization function. The dashed line corresponds to a l_1 penalization (i.e., $\phi(w) = \mid w \mid$), the dotted line to a l_2 penalization $\phi(w) = \frac{w^2}{2}$, and the continuous line to the multiscale entropy function. We can immediately see that the multiscale entropy function presents quadratic behavior for small values, and is closer to the l_1 penalization function for large values. Penalization functions with a l_2-l_1 behavior are generally a good choice for image restoration.

The Beta Pictoris image [61] was obtained by integrating 5 hours on-source using a mid-infrared camera, TIMMI, placed on the 3.6 ESO telescope (La Silla, Chile). The raw image has a peak signal-to-noise ratio of 80. It is strongly blurred by a combination of seeing, diffraction (0.7 arcsec on a 3 m-class telescope), and additive Gaussian noise. The initial disk shape in the original image has been lost after the convolution with the PSF (see Figure 7.3). Thus we need to deconvolve such an image to get the best information on this object, i.e., the exact profile and thickness of the disk, and subsequently to compare the results to models of thermal dust emission.

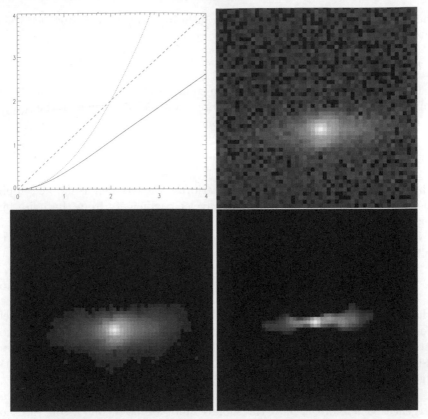

FIGURE 7.3: Upper left: penalization functions: dashed, l_1 norm (i.e., $\phi(w) = \mid w \mid$); dotted, l_2 norm $\phi(w) = \frac{w^2}{2}$; continuous, multiscale entropy function. Upper right: Beta Pictoris raw data. Lower left: filtered image using multiscale entropy. Lower right: deconvolved one. (**See color insert following page 140.**)

After filtering (see Figure 7.3, lower left), the disk appears clearly. For detection of faint structures (the disk here), one can calculate that the application of such a filtering method to this image provides a gain of observing time of a factor of around 60. The deconvolved image (Figure 7.3, lower right) shows that the disk is relatively flat at 10 μm and asymmetrical.

7.7.4.2 Total Variation and Undecimated Haar Transform

A link between the total variation (TV) and the undecimated Haar wavelet soft thresholding has been studied in [66, 67], arguing that in the 1D case the TV and the undecimated single-resolution Haar are equivalent. When going to 2D, this relation does not hold any more, but the two approaches share some similarities. Whereas the TV introduces translation- and rotation-invariance, the undecimated 2D Haar presents translation- and scale-invariance (being multi-scale).

7.7.4.3   Minimization Algorithm

Recent works [81–83] show that the solution of Equation 7.49 can be obtained in a very efficient way, by applying a wavelet denoising on the solution at each step of the Landweber iteration:

$$O^{(n+1)} = \mathbf{WDen}_\lambda \left(O^{(n)} + P^* * \left(I - P * O^{(n)} \right) \right) \qquad (7.51)$$

where **WDen** is the operator which performs wavelet denoising, i.e., applies the wavelet transform, corrects the wavelet coefficients from the noise, and applies the inverse wavelet transform.

If $\phi(x) = x$ and $p = 1$ (i.e., l_1 norm), the solution is obtained by the following iteration:

$$O^{(n+1)} = \mathbf{soft}_\lambda (O^{(n)} + P^* * (I - P * O^{(n)})) \qquad (7.52)$$

where **soft** is the soft thresholding. (A hard threshold retains wavelet coefficients whose absolute value is above the threshold, whereas a soft threshold attenuates the wavelet coefficient, using the absolute value minus the threshold so long as this is > 0.) If the Haar wavelet transform is chosen, this algorithm is a fast method for minimizing the total variation.

The penalty function needs to be continuous in order to guarantee the convergence. Therefore, a hard threshold cannot be used, but a soft threshold as well as many other shrinkage techniques can be used. If the penalty function is strictly convex (as in soft thresholding), then it converges to a global minimum [81].

7.7.4.4   Constraints in the Object or Image Domains

Let us define the *object domain* $\mathcal{O}$ as the space to which the solution belongs, and the *image domain* $\mathcal{I}$ as the space to which the data belong (i.e., if $X \in \mathcal{O}$ then $P * X \in \mathcal{I}$). In Section 7.7.2, it was shown that the multiresolution support constraint leads to a powerful regularization method. In this case, the constraint was applied in the image domain. Here, we have considered constraints on the solution. Hence, two different wavelet-based strategies can be chosen in order to regularize the deconvolution problem.

The constraint in the image domain through the multiresolution support leads a very robust way to control the noise. Indeed, whatever the nature of the noise, we can always derive robust detection levels in the wavelet space and determine scales and positions of the important coefficients. A drawback of the image constraints is that there is no guarantee that the solution is free of artifacts such as ringing around point sources. A second drawback is that image constraints can be used only if the point spread function is relatively compact, i.e., does not smear the information over the whole image. If it does do so, the concept of localization of information does not make sense any more.

The property of introducing a robust noise modeling is lost when applying the constraint in the object domain. For example, in the case of Poisson noise, there is no way (except using time-consuming Monte Carlo techniques) to estimate the level of the noise in the solution and to adjust properly the thresholds. The second problem with this approach is that we try to solve two problems (noise amplification and artifact control in the solution) with one parameter (i.e., λ). The choice of this parameter is crucial, while such a parameter does not exist when using the multiresolution support.

Constraints can be added in both the object and image domains in order to better control the noise by using the multiresolution support. This gives us a guarantee that the solution is free of artifacts when using the wavelet constraint on the solution [61, 63, 84]. This leads to the following equation to be minimized:

$$J(O) = \| M.\mathcal{W}_1 (I - P * O) \|^2 + \lambda \sum_{j,k,l} \phi(\| (\mathcal{W}_2 O)_{j,k,l} \|_p) \qquad (7.53)$$

where M is the multiresolution support derived from I and $\mathcal{W}_1$. $\mathcal{W}_1$ and $\mathcal{W}_2$ are the wavelet transforms used in the object and image domains. We may want to use two different wavelet decompositions: $\mathcal{W}_1$ for detecting the significant coefficients and $\mathcal{W}_2$ for removing the artifacts in the solution. Since the noise is controlled by the multiscale transforms, the regularization parameter λ does not have the same importance as in standard deconvolution methods. A much lower value is enough to remove the artifacts relative to the use of the wavelets. The positivity constraint can be applied at each iteration. The iterative scheme is now:

$$O^{(n+1)} = \mathbf{WDen}_\lambda(\left(O^{(n)} + P^* * \bar{R}^n \right) \qquad (7.54)$$

where $\bar{R}^n$ is the significant residual, i.e., $\bar{R}^n = \mathcal{W}_1^{-1} M[\mathcal{W}_1(I - P * O^{(n)})]$ (see Equation (7.41)).

7.7.4.5 The Combined Deconvolution Method

One may want to benefit from the advantages of both the wavelet and the curvelet transforms for detecting the significant features contained in the data. More generally, we assume we use K transforms $T_1, \ldots, T_K$, and we derive K multiresolution supports $M_1, \ldots, M_K$ from the input image I using noise modeling. Following determination of a set of multiresolution supports, we can solve the following optimization problem [84]:

$$\min_{\tilde{O}} \mathcal{C}(\tilde{O}), \quad \text{subject to} \quad M_k T_k [P * \tilde{O}] = M_k T_k I \quad \text{for all } k, \qquad (7.55)$$

where $\mathcal{C}$ is the smoothness constraint.

The constraint imposes fidelity on the data, or more exactly, on the significant coefficients of the data, obtained by the different transforms. Nonsignificant (i.e., noisy) coefficients are not taken into account, preventing any noise amplification in the final algorithm.

A solution for this problem could be obtained by relaxing the constraint to become an approximate one:

$$\min_{\tilde{O}} \quad \sum_k \|M_k T_k I - M_k T_k [P * \tilde{O}]\|_2 + \lambda \mathcal{C}(\tilde{O}). \tag{7.56}$$

The solution is computed by using the projected Landweber method [14]:

$$\tilde{O}^{n+1} = \tilde{O}^n + \alpha \left(P^* * \bar{R}^n - \lambda \frac{\partial \mathcal{C}}{\partial O}(\tilde{O}^n) \right) \tag{7.57}$$

where $\bar{R}^n$ is the significant residual which is obtained using the following algorithm:

- Set $I_0^n = I^n = P * \tilde{O}^n$.

- For $k = 1, \ldots, K$ do $I_k^n = I_{k-1}^n + \mathcal{R}_k \left[M_k (\mathcal{T}_k I - \mathcal{T}_k I_{k-1}^n) \right]$.

- The significant residual $\bar{R}^n$ is obtained by: $\bar{R}^n = I_K^n - I^n$.

This can be interpreted as a generalization of the multiresolution support constraint to the case where several transforms are used. The order in which the transforms are applied has no effect on the solution. We extract in the residual the information at scales and pixel indices where significant coefficients have been detected.

α is a convergence parameter, chosen either by a line search minimizing the overall penalty function or as a fixed step size of moderate value that guarantees convergence.

If the $\mathcal{C}$ is a wavelet-based penalization function, then the minimization can again be done using the previous wavelet denoising approach:

$$\tilde{O}^{n+1} = \mathbf{WDen}\left(\tilde{O}^n + (P^* * \bar{R}^n) \right). \tag{7.58}$$

The positivity is introduced in the following way:

$$O^{n+1} = \mathcal{P}_c \left[\mathbf{WDen}\left(\tilde{O}^n + (P^* * \bar{R}^n) \right) \right]. \tag{7.59}$$

7.8   Deconvolution and Resolution

In many cases, there is no sense in trying to deconvolve an image at the resolution of the pixel (especially when the PSF is very large). The idea to limit the resolution is relatively old, because it is already this concept which is used in the CLEAN algorithm [25]. Indeed the clean beam fixes the resolution in the final solution. This principle was also developed by Lannes and Roques [85] in a different form. This concept was reinvented, first by Gull and Skilling [36] who called the clean beam the *Intrinsic Correlation Function* (ICF), and more recently by Magain et al. [59] and Pijpers [24].

The ICF is usually a Gaussian, but in some cases it may be useful to take another function. For example, if we want to compare two images I_1 and I_2 which are obtained with two wavelengths or with two different instruments, their PSFs P_1 and P_2 will certainly be different. The classic approach would be to deconvolve I_1 with P_2 and I_2 with P_1, so we are sure that both are at the same resolution. But unfortunately we lose some resolution in doing this. Deconvolving both images is generally not possible because we can never be sure that both solutions O_1 and O_2 will have the same resolution.

A solution would be to deconvolve only the image which has the worse resolution (say I_1), and to limit the deconvolution to the second image resolution (I_2). Then, we just have to take P_2 for the ICF. The deconvolution problem is to find $\tilde{O}$ (hidden solution) such that:

$$I_1 = P_1 * P_2 * \tilde{O} \qquad (7.60)$$

and our real solution O_1 at the same resolution as I_2 is obtained by convolving $\tilde{O}$ with P_2. O_1 and I_2 can then be compared.

Introducing an ICF G in the deconvolution equation leads to just considering a new PSF P' which is the convolution of P and G. The deconvolution is carried out using P', and the solution must be reconvolved with G at the end. In this way, the solution has a constrained resolution, but aliasing may occur during the iterative process, and it is not sure that the artifacts will disappear after the reconvolution with G. Magain et al. [59] proposed an innovative alternative to this problem, by assuming that the PSF can be considered as the convolution product of two terms, the ICF G and an unknown S, $P = G * S$. Using S instead of P in the deconvolution process, and a sufficiently large full width half maximum (FWHM) value for G, implies that the Shannon sampling theorem [86] is never violated. But the problem is now to calculate S, knowing P and G, which is again a deconvolution problem. Unfortunately, this delicate point was not discussed in the original paper. Propagation of the error on the S estimation in the final solution has also until now not been investigated, even if this issue seems to be quite important.

7.9 Myopic and Blind Deconvolution

In the field of astronomy the PSF is, in many cases, variable with time or within the observed field. For instance, when observing in the visible or near-

infrared range from the ground, the atmospheric turbulence produces images
with seeing-limited spatial resolution, that are much lower than the theoretical diffraction limit. This resolution ranges typically for the visible radiations
between 0.4 and 1 arcsec in the best sites for astronomical observations. Because of the stochastic nature of the seeing (due to random changes of the
optical index of the different layers of the atmosphere), several timescales are
involved. First, the PSF is highly unstable over a timescale of several tens of
milliseconds in the visible, but its time-averaged value over typically a few sec-
onds can remain stable over a much longer time (several minutes). Depending
on the site, the average seeing can vary significantly with typical timescales of
a few tens of minutes. Long exposures, seeing averaged, would usually provide
an estimate of the PSF for long exposures on scientific targets.

For a given observation, one can define a parameter, called the Fried parameter, r_0. Its value defines the equivalent diameter of a telescope that would
produce a diffraction-limited image with the same resolution as the seeing-
limited one. This parameter varies as a function of the wavelength such that
$r_0 \propto (\lambda)^{5/6}$cm; a typical value of this parameter is 20 cm at 0.5 μm. The image resolution is then of the order of r_0/λ in the case of seeing-limited data,
instead of a value around D/λ (D being the telescope diameter) for diffraction-
limited observations. If visible/near-infrared data are always seeing-limited,
one can note that mid-infrared data around 10 μm are mainly diffraction limited on a 4 m-class telescope. However, for one 8 m-class telescope such as the
Very Large Telescope in Chile and given the seeing conditions at this site, one
is just at the transition regime between purely diffraction-limited and seeing-
limited. Hence mid-infrared data on 8 m-class telescopes require now the use
of myopic deconvolution methods.

The principle of adaptive optics (AO) observations was developed in order
to try to get rid of the seeing limitation and recover diffraction-limited images. However, this is true to a certain extent. First, only a small field of
view (typically, a few arcsec in the visible range), around the AO reference
(a sufficiently bright object; typically a bright star with magnitude V lower
than 12), currently benefits from the best image correction which degrades as
$\exp(-\Delta\theta)^{5/3}$ ($\Delta\theta$ is the distance to the center field). This field extent is defined by the isoplanatic angle, in which the wave-front distorsion is corrected
accurately. In order to reach larger fields of view, multiconjugated adaptive
optics (MCAO) systems are under development in parallel with laser guide
stars which will allow one to synthetically produce AO references at any place
in the sky. The Strehl ratio, which is defined as the ratio of the central intensity of the currently observed image to the central intensity of a perfect
diffraction-limited image, can achieve typical values around 30–50% for K
band observations, 2.2 μm). As one observes an object far from the AO reference, this correction assessment parameter degrades, reaching a value half its
value at the center field (the location of the AO reference), at a distance of 40
arcsec from it. Since AO performances are intimately linked to the seeing (or
r_0) value, the AO correction will also strongly vary with time. In summary,

AO observations deal with quite significantly varying PSFs: varying in time because of seeing time variations, and varying in space because of the isoplanatic patch. As a consequence, AO astronomical observations do require also myopic deconvolution methods. Much effort in astronomical blind deconvolution have been thus naturally devoted to this type of data.

Astronomical short exposures, or speckle images, have usually unknown PSF because of strong and random changes of the image phase. Although it is essentially the phase of the object which is affected, while its amplitude is barely modified, some specific methods of blind deconvolution in which the PSF is assumed unknown have to be used. The challenge that faces any method is that it should be able to incorporate as much as possible by way of constraints and *a priori* knowledge on the solution in order to avoid being trapped in trivial solutions such as $\{O,P\}=\{I,\delta\}$, where δ is the Dirac function.

7.9.1 Myopic Deconvolution

Myopic deconvolution assumes that an estimate of the PSF, which is not too "far" from the true one, is known. A measure of this PSF is usually performed by observing an object supposed to be point-like, in the same observational conditions (same telescope/instrument configuration, and weather conditions as close as possible). This is achieved by observing frequently, before and after any scientific target, some unresolved star, at an airmass close to that of the science target, and if possible, not too far in the sky from it. Astronomical data usually deal then with multiframe data: multiple records of the blurred object, multiple records of estimates of the PSF P_i. Within this framework some new methods have emerged in the last decade and are presented in the following.

Iterative deconvolution algorithm in C (IDAC) is a myopic, Bayesian deconvolution method derived from a blind deconvolution method ([87]; see Section 7.9.2 below for the details), and applied mainly to adaptive optics data. It is a multiframe deconvolution method in which *loose* constraints are put on the solution.

The MISTRAL method was developed by Mugnier et al. [88] in order to deconvolve AO partially corrected data. An estimate of the PSF P_{est} can be derived from observations of an unresolved star shortly before or after observing the object of interest. Using the usual Bayesian framework and the MAP approach, one ends up with the following functional to minimize, to find O and P simultaneously:

$$J(O, P) = J_{res} + J_P(P) + J_O(O) \tag{7.61}$$

where J_{res} is the "fidelity to the data" term such that:

$$J_{res}(O, P) = \frac{1}{2\sigma^2} \parallel (I - P * O) \parallel^2 \text{ in the case of Gaussian noise} \quad (7.62)$$

and

$$J_{res}(O, P) = \sum_{pixels} (P * O - I \ln[P * O]) \text{ in the case of Poisson noise.} \quad (7.63)$$

$J_P(P)$ is the penalty term expressing that solution P should not be "too different" from the estimate of the PSF P_{est}:

$$J_p(P) = \sum_{u,v} \frac{|\hat{P}(u, v) - \hat{P}_{est}(u, v)|^2}{S_p(u, v)} \quad (7.64)$$

where $S_p(u, v) = E[|\hat{P}(u, v) - \hat{P}_{est}(u, v)|]$ is the spatial power spectral density. P_{est} is computed as $E[P_i]$, P_i being the different estimates of the PSF.

In addition, a reparametrization $(O = \psi^2)$ ensures the positivity of the solution.

Some very nice results obtained with this method applied to images of the satellite of Jupiter, Io, are shown in Figure 7.4, allowing us to study Io volcanism with ground-based AO data.

The deconvolution from wave-front sensing was originally proposed in 1985 by Fontanella [89]. The idea behind it is that in AO observations wave-front data are simultaneously recorded in the wave-front sensor (WFS); these data contain information about the PSF but in an unusual form (often projected onto a base of Zernike polynomials). Since we deal with short exposure images (less than 10 ms typically), the atmospheric turbulence is assumed to be frozen, so that the PSF at a time t can be fully characterized by the turbulent phase $\phi(t)$ in the pupil plane:

$$PSF(t) = \mathcal{F}^{-1}(Pe^{j\phi(t)}) \quad (7.65)$$

where P is the pupil function.

The usual techniques consist of firstly estimating the wave fronts from the WFS, and then obtaining the deconvolved image by maximum *a posteriori* (MAP) estimation [90]. Since wave-front estimates are inevitably noisy, Mugnier et al. [91] proposed a robust joint estimator within a Bayesian framework to reconstruct the true wave-front data and the restored object simultaneously. This estimator uses all statistical information that one has either on the noise, or on the stochastic processes that control the turbulent phase (Kolmogorov statistics). A functional containing constraints on $\phi(t)$ (complying with Kolmogorov statistics) and on the object O is minimized. This method is more efficient than speckle interferometry because it is not limited by speckle noise at high photon flux and the signal-to-noise ratio is better for extended objects. Figure 7.5 shows how spectacular the result can be on experimental data of the binary star Capella.

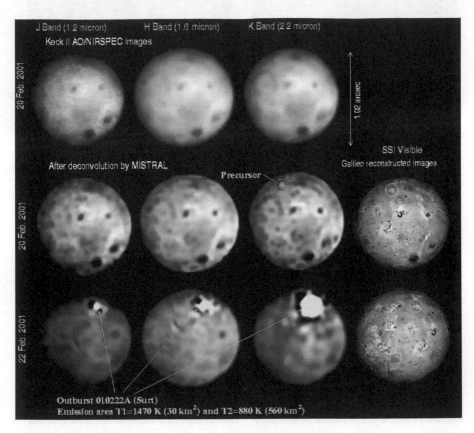

FIGURE 7.4: Jupiter-facing of Io observed with the Keck AO system in J, H, and K bands. The second row shows the former images deconvolved with the MISTRAL method, and are compared to Galileo probe images in which a precursor to a volcanic burst is detected. The last row shows deconvolved images 2 days later in which a major outburst, seen also on Galileo images but fainter (Galileo observes in the visible range and is less sensitive to "thermal" events), is also detected from Earth. (**See color insert following page 140.**)

7.9.2 Blind Deconvolution

In 1994, Tsumuraya et al. [92] proposed a simple method for blind deconvolution based on Lucy's algorithm. The idea is to alternatively perform a Lucy iteration on the object O and then on the PSF P. However, although attractive because of its simplicity, this process (i) can be highly unstable, and (ii) puts no constraint on the PSF, making it difficult to prevent it tending towards the trivial solution $\{I, \delta\}$.

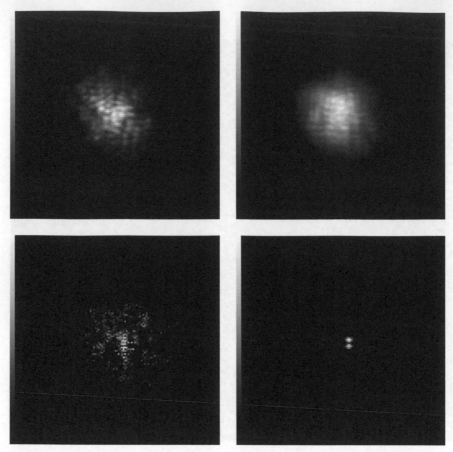

FIGURE 7.5: Experimental short exposure of the binary system Capella taken at the 4.2-m William Herschel telescope, La Palma, Canarias islands (upper left) and corresponding long-exposure, 10 times longer, showing the loss of resolution due to atmospheric turbulence blurring (upper right). Restored image using a MAP estimate of the wavefront followed by a quadratic restoration with positivity constraint (lower left). Image restored by the myopic deconvolution method developed by Mugnier et al. [91] (lower right).

Jefferies and Christou [93] have proposed an iterative blind deconvolution method of multiframe data based on the minimization of a penalty functional putting physical and "reasonable" *loose* constraints on the solution (O, P_i). Assuming that one deals with $i = 1, \ldots$ frames, this method minimizes the functional:

$$J(O, P) = E_{im} + E_{conv} + E_{bl} + E_{Fm} \tag{7.66}$$

where:

1. $E_{im} = \sum_{n \in \gamma_O} O(n)^2 + \sum_{m \in \gamma_P} P(m)^2$ is the image domain error which penalizes negative values (γ_O and γ_P sets) in the solution O or in the PSF P.

2. $E_{conv} = \sum_i \sum_{u,v} |\hat{I}(u,v) - \hat{P}(u,v)\hat{O}(u,v)|^2 M(u,v)$ is called the convolution error and expresses the fidelity of the reconvolved image to the data in Fourier space, M being a mask excluding spatial frequencies higher than the physical cut-off frequency (by diffraction (D/λ) or seeing (r_0/λ)) of the instrumental set. Depending on the conventions on the Fourier transform, one has to normalize this term by the number of pixels in the image.

3. E_{bl} is the band-limit error defined by $E_{bl} = \sum_i \sum_{u,v} |\hat{P}_i(u,v)|^2 M'_i(u,v)$ where M' is a spatial frequencies mask greater than 1.39 of the cut-off frequency. The same normalization rule applies as above.

4. E_{Fm} is the Fourier modulus error, such that $E_{Fm} = \sum_i \sum_{u,v} |\hat{O}(u,v) - \hat{O}_{est}(u,v)|^2 \Phi(u,v)$ where $O_{est}(u,v)$ is a crude estimate of the object's Fourier modulus given by:

$$|\hat{O}_{est}| = \sqrt{\frac{\langle|\hat{I}|^2\rangle - \langle|\hat{N}|^2\rangle}{\langle|\hat{P}|^2\rangle}}. \tag{7.67}$$

Thiebaut and Conan [94] compared the application of loose (similar to some extent to [93]) and of strict *a priori* constraints on the solution when maximizing the likelihood. They found that much better solutions can be achieved when applying the second option. Strict *a priori* constraints are applied in this case by a reparametrization ensuring the sought physical properties of the solution, e.g.:

1. Positivity of the object O making the change of variable:

$$O = \psi_O^2 \tag{7.68}$$

2. Positivity and normalization of the PSF P by setting:

$$P = \frac{\psi_h^2}{\sum_{pixels} \psi_h^2}. \tag{7.69}$$

The gradients of the functional are then recomputed, and depend now on the new unknowns ψ_P (representing P) and ψ_O (representing O). This method has been shown to give very good results in the case of speckle data on the Capella binary system [94].

In summary, several methods are currently available to carry out myopic or blind deconvolution. For best results we recommend the Bayesian formalism because it offers a nice way to incorporate any type of constraint on the solution. These methods offer much better results than any other algebraic method, but at some computational cost (the minimization of the functional requires intensive computing and efficient minimization schemes).

7.10   Conclusions and Chapter Summary

As in many fields, simple methods can be availed of – for example, the solution provided by Equation (7.16) – but at the expense of quality in the solution and a full understanding of one's data. Often a simple solution can be fully justified. However, if our data or our problems are important enough, then appropriate problem-solving approaches have to be adopted. The panoply of methods presented in this review provide options for high-quality image and signal restoration.

We have noted how the wavelet transform offers a powerful mathematical and algorithmic framework for multiple resolution analysis. Furthermore noise modeling is very advantageously carried out in wavelet space. Finally, and of crucial importance in this chapter, noise is the main problem in deconvolution.

Progress has been significant in a wide range of areas related to deconvolution. One thinks of Bayesian methods, the use of entropy, and issues relating to super-resolution, for example.

We will conclude with a short look at how multiscale methods used in deconvolution are evolving and maturing.

We have seen that the recent improvement in deconvolution methods has led to use of a multiscale approach. Finally, wavelet-based constraints can be added in both domains [63]. This allows us to separate the deconvolution problem into two separate problems: noise control from one side, and solution smoothness control on the other side. The advantage is that noise control is better carried out in the image domain, while smoothness control can only be carried out in the object domain.

The reason for the success of wavelets is due to the fact that wavelet bases represent well a large class of signals, especially astronomical data where most of the objects are more or less isotropic. When the data contain anisotropic features (solar, planetary images, and so on), other multiscale methods, such as the ridgelet or the curvelet transform [95–98], are good candidates for replacing the wavelet transform. The ultimate step is the combination of the different multiscale decompositions.

Very nice results have been obtained based on myopic or blind deconvolution. However, there is currently no algorithm based on multiscale methods in the field of myopic or blind deconvolution for the regularization. New methods taking advantage of these tools, as was already done for standard image deconvolution, should appear soon.

Acknowledgments

We are grateful to Eric Thiebaut and Laurent Mugnier for use of their code, and to the referees of this chapter.

References

[1] T. Bontekoe, E. Koper, and D. Kester, "Pyramid maximum entropy images of IRAS survey data," *Astronomy and Astrophysics*, vol. 284, pp. 1037–1053, 1994.

[2] F. Marchis, R. Prangé, and J. Christou, "Adaptive optics mapping of Io's volcanism in the thermal IR (3.8 μm)," *Icarus*, vol. 148, pp. 384–396, Dec. 2000.

[3] D. Alloin, E. Pantin, P. O. Lagage, and G. L. Granato, "0.6 resolution images at 11 and 20 mu m of the active galactic nucleus in NGC 1068," *Astronomy and Astrophysics*, vol. 363, pp. 926–932, Nov. 2000.

[4] F. Courbin, C. Lidman, and P. Magain, "Detection of the lensing galaxy in HE 1104-1805," *Astronomy and Astrophysics*, vol. 330, pp. 57–62, Feb. 1998.

[5] R. L. White and R. J. Allen, eds., *The restoration of HST images and spectra*, 1991.

[6] R. J. Hanisch and R. L. White, eds., *The restoration of HST images and spectra - II*, Space Telescope Science Institute, Baltimore, 1994.

[7] H. Adorf, R. Hook, and L. Lucy, "HST image restoration developments at the ST-ECF," *International Journal of Imaging Systems and Technology*, vol. 6, pp. 339–349, 1995.

[8] C. Faure, F. Courbin, J. P. Kneib, D. Alloin, M. Bolzonella, and I. Burud, "The lensing system towards the doubly imaged quasar SBS 1520+530," *Astronomy and Astrophysics*, vol. 386, pp. 69–76, Apr. 2002.

[9] J. T. Radomski, R. K. Piña, C. Packham, C. M. Telesco, and C. N. Tadhunter, "High-resolution Mid-infrared Morphology of Cygnus A," *Astrophysical Journal*, vol. 566, pp. 675–681, Feb. 2002.

[10] I. Burud, F. Courbin, P. Magain, C. Lidman, D. Hutsemékers, J.-P. Kneib, J. Hjorth, J. Brewer, E. Pompei, L. Germany, J. Pritchard, A. O. Jaunsen, G. Letawe, and G. Meylan, "An optical time-delay for the lensed BAL quasar HE 2149-2745," *Astronomy and Astrophysics*, vol. 383, pp. 71–81, Jan. 2002.

[11] A. Coustenis, E. Gendron, O. Lai, J. Véran, J. Woillez, M. Combes, L. Vapillon, T. Fusco, L. Mugnier, and P. Rannou, "Images of Titan at 1.3 and 1.6 μ m with Adaptive Optics at the CFHT," *Icarus*, vol. 154, pp. 501–515, 2001.

[12] T. J. Cornwell, "Image Restoration," in *NATO ASIC Proc. 274: Diffraction-Limited Imaging with Very Large Telescopes*, pp. 273–292, 1989.

[13] A. Katsaggelos, *Digital Image Processing*. Springer-Verlag, 1993.

[14] M. Bertero and P. Boccacci, *Introduction to Inverse Problems in Imaging*. Institute of Physics, 1998.

[15] R. Molina, J. Núñez, F. Cortijo, and J. Mateos, "Image restoration in astronomy: a Bayesian review," *IEEE Signal Processing Magazine*, vol. 18, pp. 11–29, 2001.

[16] J. C. Christou, D. Bonnacini, N. Ageorges, and F. Marchis, "Myopic deconvolution of adaptive optics images," *The Messenger*, vol. 97, pp. 14–22, Sept. 1999.

[17] R. Buonanno, G. Buscema, C. Corsi, I. Ferraro, and G. Iannicola, "Automated photographic photometry of stars in globular clusters," *Astronomy and Astrophysics*, vol. 126, pp. 278–282, Oct. 1983.

[18] A. Moffat, "A theoretical investigation of focal stellar images in the photographic emulsion and application to photographic photometry," *Astronomy and Astrophysics*, vol. 3, p. 455, Dec. 1969.

[19] S. Djorgovski, "Modelling of seeing effects in extragalactic astronomy and cosmology," *Journal of Astrophysics and Astronomy*, vol. 4, pp. 271–288, Dec. 1983.

[20] R. Molina, B. Ripley, A. Molina, F. Moreno, and J. Ortiz, "Bayesian deconvolution with prior knowledge of object location – applications to ground-based planetary images," *Astrophysical Journal*, vol. 104, pp. 1662–1668, Oct. 1992.

[21] A. Tikhonov, A. Goncharski, V. Stepanov, and I. Kochikov, "Ill-posed image processing problems," *Soviet Physics – Doklady*, vol. 32, pp. 456–458, 1987.

[22] G. Golub, M. Heath, and G. Wahba, "Generalized cross-validation as a method for choosing a good ridge parameter," *Technometrics*, vol. 21, pp. 215–223, 1979.

[23] N. Galatsanos and A. Katsaggelos, "Methods for choosing the regularization parameter and estimating the noise variance in image restoration and their relation," *IEEE Transactions on Image Processing*, vol. 1, pp. 322–336, 1992.

[24] F. P. Pijpers, "Unbiased image reconstruction as an inverse problem," *Monthly Notices of the Royal Astronomical Society*, vol. 307, pp. 659–668, Aug. 1999.

[25] J. Högbom, "Aperture synthesis with a non-regular distribution of interferometer baselines," *Astronomy and Astrophysics Supplement Series*, vol. 15, pp. 417–426, 1974.

[26] F. Champagnat, Y. Goussard, and J. Idier, "Unsupervised deconvolution of sparse spike trains using stochastic approximation," *IEEE Transactions on Image Processing*, vol. 44, pp. 2988–2997, 1996.

[27] K. Kaaresen, "Deconvolution of sparse spike trains by iterated window maximization," *IEEE Transactions on Image Processing*, vol. 45, pp. 1173–1183, 1997.

[28] L. Landweber, "An iteration formula for Fredholm integral equations of the first kind," *American Journal of Mathematics*, vol. 73, pp. 615–624, 1951.

[29] E. Issacson and H. Keller, *Analysis of Numerical Methods*. Wiley, 1966.

[30] W. Richardson, "Bayesian-based iterative method of image restoration," *Journal of the Optical Society of America*, vol. 62, pp. 55–59, 1972.

[31] L. Lucy, "An iteration technique for the rectification of observed distributions," *Astronomical Journal*, vol. 79, pp. 745–754, 1974.

[32] L. Shepp and Y. Vardi, "Maximum likelihood reconstruction for emission tomography," *IEEE Transactions on Medical Imaging*, vol. MI-2, pp. 113–122, 1982.

[33] A. Dempster, N. Laird, and D. Rubin, "Maximum likelihood from incomplete data via the EM algorithm," *Journal of the Royal Statistical Society, Series B*, vol. 39, pp. 1–22, 1977.

[34] J. Burg, "Multichannel maximum entropy spectral analysis." Annual Meeting International Society Exploratory Geophysics, Reprinted in Modern Spectral Analysis, D.G. Childers, ed., IEEE Press, 34–41, 1978.

[35] B. Frieden, *Image Enhancement and Restoration*. Springer-Verlag, 1978.

[36] S. Gull and J. Skilling, *MEMSYS5 Quantified Maximum Entropy User's Manual*. Royston, England, 1991.

[37] B. Ripley, *Spatial Statistics*. Wiley, 1981.

[38] L. Blanc-Féraud and M. Barlaud, "Edge preserving restoration of astro-physical images," *Vistas in Astronomy*, vol. 40, pp. 531–538, 1996.

[39] P. Charbonnier, L. Blanc-Féraud, G. Aubert, and M. Barlaud, "Deterministic edge-preserving regularization in computed imaging," *IEEE Transactions on Image Processing*, vol. 6, pp. 298–311, 1997.

[40] P. J. Green, "Bayesian reconstruction from emission tomography data using a modified EM algorithm," *IEEE Transactions on Medical Imaging*, vol. 9, no. 1, pp. 84–93, 1990.

[41] C. A. Bouman and K. Sauer, "A generalized Gaussian image model for edge-preserving MAP estimation," *IEEE Transactions on Image Processing*, vol. 2, no. 3, pp. 296–310, 1993.

[42] P. Charbonnier, L. Blanc-Féraud, G. Aubert, and M. Barlaud, "Deterministic edge-preserving regularization in computed imaging," *IEEE Transactions on Image Processing*, vol. 6, no. 2, pp. 298–311, 1997.

[43] S. Geman and D. McClure, "Bayesian image analysis: an application to single photon emission tomography," in *Proc. Statist. Comput. Sect.*, (Washington DC), American Statistical Association, 1985.

[44] P. Perona and J. Malik, "Scale-space and edge detection using anisotropic diffusion," *IEEE Transactions on Pattern Analysis and Machine Intelligence*, vol. 12, pp. 629–639, 1990.

[45] T. Hebert and R. Leahy, "A generalized EM algorithm for 3-d bayesian reconstruction from poisson data using Gibbs priors," *IEEE Transactions on Medical Imaging*, vol. 8, no. 2, pp. 194–202, 1989.

[46] T. Corbard, L. Blanc-Féraud, G. Berthomieu, and J. Provost, "Nonlinear regularization for helioseismic inversions. Application for the study of the solar tachocline," *Astronomy and Astrophysics*, vol. 344, pp. 696–708, Apr. 1999.

[47] L. Alvarez, P.-L. Lions, and J.-M. Morel, "Image selective smoothing and edge detection by nonlinear diffusion," *SIAM Journal on Numerical Analysis*, vol. 29, pp. 845–866, 1992.

[48] L. Rudin, S. Osher, and E. Fatemi, "Nonlinear total variation noise removal algorithm," *Physica D*, vol. 60, pp. 259–268, 1992.

[49] R. Acar and C. Vogel, "Analysis of bounded variation penalty methods for ill-posed problem," *Physica D*, vol. 10, pp. 1217–1229, 1994.

[50] R. Molina, A. Katsaggelos, J. Mateos, and J. Abad, "Compound Gauss-Markov random fields for astronomical image restoration," *Vistas in Astronomy*, vol. 40, pp. 539–546, 1996.

[51] R. Molina, A. Katsaggelos, J. Mateos, A. Hermoso, and A. Segall, "Restoration of severely blurred high range images using stochastic and

deterministic relaxation algorithms in compound Gauss Markov random fields," *Pattern Recognition*, vol. 33, pp. 555–571, 2000.

[52] R. Molina and F. Cortijo, "On the Bayesian deconvolution of planets," in *Proc. International Conference on Pattern Recognition, ICPR'92*, vol. 3, pp. 147–150, 1992.

[53] P. V. Cittert, "Zum Einfluß der Spaltbreite auf die Intensitätsverteilung in Spektrallinien II," *Zeitschrift für Physik*, vol. 69, pp. 298–308, 1931.

[54] P. Jansson, R. Hunt, and E. Peyler, "Resolution enhancement of spectra," *Journal of the Optical Society of America*, vol. 60, pp. 596–599, 1970.

[55] B. Wakker and U. Schwarz, "The multi-resolution Clean and its application to the short-spacing problem in interferometry," *Annual Reviews of Astronomy and Astrophysics*, vol. 200, p. 312, 1988.

[56] L. Lucy, "Image restoration of high photometric quality," in *The Restoration of HST Images and Spectra II* (R. J. Hanisch and R. L. White, eds.), p. 79, Space Telescope Science Institute, 1994.

[57] R. Hook, "An overview of some image restoration and combination methods," *ST-ECF Newsletter No. 26*, pp. 3–5, 1999.

[58] N. Pirzkal, R. Hook, and L. Lucy, "GIRA – two channel photometric restoration," in *Astronomical Data Analysis Software and Systems IX* (N. Manset, C. Veillet, and D. Crabtree, eds.), p. 655, Astronomical Society of the Pacific, 2000.

[59] P. Magain, F. Courbin, and S. Sohy, "Deconvolution with correct sampling," *Astrophysical Journal*, vol. 494, p. 472, 1998.

[60] N. Weir, "A multi-channel method of maximum entropy image restoration," in *Astronomical Data Analysis Software and System 1* (D. Worral, C. Biemesderfer, and J. Barnes, eds.), pp. 186–190, Astronomical Society of the Pacific, 1992.

[61] E. Pantin and J.-L. Starck, "Deconvolution of astronomical images using the multiscale maximum entropy method," *Astronomy and Astrophysics, Supplement Series*, vol. 315, pp. 575–585, 1996.

[62] J. Núñez and J. Llacer, "Bayesian image reconstruction with space-variant noise suppression," *Astronomy and Astrophysics, Supplement Series*, vol. 131, pp. 167–180, July 1998.

[63] J.-L. Starck, F. Murtagh, P. Querre, and F. Bonnarel, "Entropy and astronomical data analysis: Perspectives from multiresolution analysis," *Astronomy and Astrophysics*, vol. 368, pp. 730–746, 2001.

[64] D. Dixon, W. Johnson, J. Kurfess, R. Pina, R. Puetter, W. Purcell, T. Tuemer, W. Wheaton, and A. Zych, "Pixon-based deconvolution,"

Astronomy and Astrophysics, Supplement Series, vol. 120, pp. 683–686, Dec. 1996.

[65] R. Puetter and A. Yahil, "The pixon method of image reconstruction," in *ASP Conference Series 172: Astronomical Data Analysis Software and Systems VIII*, p. 307, Astronomical Society of the Pacific, 1999.

[66] A. Cohen, R. DeVore, P. Petrushev, and H. Xu, "Nonlinear approximation and the space $BV(R^2)$," *American Journal of Mathematics*, vol. 121, pp. 587–628, 1999.

[67] G. Steidl, J. Weickert, T. Brox, P. Mrzek, and M. Welk, "On the equivalence of soft wavelet shrinkage, total variation diffusion, total variation regularization, and sides," Tech. Rep. 26, Department of Mathematics, University of Bremen, Germany, 2003.

[68] F. Murtagh and J.-L. Starck, "Multiresolution image analysis using wavelets: Some recent results and some current directions," *ST-ECF Newsletter No. 21*, pp. 19–20, 1994.

[69] J.-L. Starck and F. Murtagh, "Image restoration with noise suppression using the wavelet transform," *Astronomy and Astrophysics*, vol. 288, pp. 343–348, 1994.

[70] F. Murtagh, J.-L. Starck, and A. Bijaoui, "Image restoration with noise suppression using a multiresolution support," *Astronomy and Astrophysics, Supplement Series*, vol. 112, pp. 179–189, 1995.

[71] J.-L. Starck, A. Bijaoui, and F. Murtagh, "Multiresolution support applied to image filtering and deconvolution," *CVGIP: Graphical Models and Image Processing*, vol. 57, pp. 420–431, 1995.

[72] A. Caulet and W. Freudling, "Distant galaxy cluster simulations – HST and ground-based," *ST-ECF Newsletter No. 20*, pp. 5–7, 1993.

[73] W. Freudling and A. Caulet, "Simulated HST observations of distant clusters of galaxies," in *Proceedings of the 5th ESO/ST-ECF Data Analysis Workshop* (P. Grosbøl, ed.), pp. 63–68, European Southern Observatory, 1993.

[74] J.-L. Starck, A. Bijaoui, B. Lopez, and C. Perrier, "Image reconstruction by the wavelet transform applied to aperture synthesis," *Astronomy and Astrophysics*, vol. 283, pp. 349–360, 1994.

[75] J.-L. Starck and A. Bijaoui, "Filtering and deconvolution by the wavelet transform," *Signal Processing*, vol. 35, pp. 195–211, 1994.

[76] J.-L. Starck, F. Murtagh, and A. Bijaoui, *Image Processing and Data Analysis: The Multiscale Approach.* Cambridge University Press, 1998.

[77] A. Jalobeanu, *Models, Bayesian estimation and algorithms for remote sensing data deconvolution.* PhD thesis, Université de Nice Sophia Antipolis, December 2001.

[78] J.-L. Starck, F. Murtagh, and R. Gastaud, "A new entropy measure based on the wavelet transform and noise modeling," *IEEE Transactions on Circuits and Systems II*, vol. 45, pp. 1118–1124, 1998.

[79] J.-L. Starck and F. Murtagh, "Multiscale entropy filtering," *Signal Processing*, vol. 76, pp. 147–165, 1999.

[80] J.-L. Starck and F. Murtagh, *Astronomical Image and Data Analysis.* Springer-Verlag, 2002.

[81] M. Figueiredo and R. Nowak, "An EM algorithm for wavelet-based image restoration," *IEEE Transactions on Image Processing*, vol. 12, no. 8, pp. 906–916, 2003.

[82] I. Daubechies, M. Defrise, and C. D. Mol, "An iterative thresholding algorithm for linear inverse problems with a sparsity constraint," *Communications on Pure and Applied Mathematics*, vol. 57, pp. 1413–1541, 2004.

[83] P. Combettes and V. Vajs, "Signal recovery by forward-backward splitting," *SIAM Journal on Multiscale Modeling and Simulation,* vol. 4, no. 4, pp. 1168-1200, Nov. 2005.

[84] J.-L. Starck, M. Nguyen, and F. Murtagh, "Wavelets and curvelets for image deconvolution: a combined approach," *Signal Processing*, vol. 83, no. 10, pp. 2279–2283, 2003.

[85] A. Lannes and S. Roques, "Resolution and robustness in image processing: a new regularization principle," *Journal of the Optical Society of America*, vol. 4, pp. 189–199, 1987.

[86] C. Shannon, "A mathematical theory for communication," *Bell System Technical Journal*, vol. 27, pp. 379–423, 1948.

[87] J. Christou, "IDAC – Iterative Deconvolution Algorithm in C web page," *http://babcock.ucsd.edu/cfao_ucsd/idac/idac_package/idac_index.html*, 2000.

[88] L. M. Mugnier, T. Fusco, and J. M. Conan, "MISTRAL: a myopic edge-preserving image restoration method, with application to astronomical adaptive-optics-corrected long-exposure images," *Journal of the Optical Society of America A*, vol. 10, pp. 1841–1853, Oct. 2004.

[89] J. C. Fontanella, "Analyse de surfaces d'onde, déconvolution et optique active," *Journal of Modern Optics*, vol. 16, pp. 257–268, 1985.

[90] P. A. Bakut, V. E. Kirakosyants, V. A. Loginov, C. J. Solomon, and J. C. Dainty, "Optimal wavefront reconstruction from a Shark-Hartmann sensor by use of a Bayesian algorithm," *Optics Communications*, vol. 109, pp. 10–15, 1994.

[91] L. M. Mugnier, C. Robert, J. M. Conan, V. Michaud, and S. Salem, "Myopic deconvolution from wave-front sensing," *Journal of the Optical Society of America A* , vol. 18, pp. 862–872, Apr. 2001.

[92] F. Tsumuraya, N. Miura, and N. Baba, "Iterative blind deconvolution method using Lucy's algorithm," *Astronomy and Astrophysics*, vol. 282, pp. 699–708, Feb. 1994.

[93] S. M. Jefferies and J. C. Christou, "Restoration of astronomical images by iterative blind deconvolution," *Astrophysical Journal*, vol. 415, pp. 862–874, Oct. 1993.

[94] E. Thiebaut and J. M. Conan, "Strict a priori constraints for maximum likehood blind deconvolution," *Journal of the Optical Society of America A,* vol. 12, pp. 485–492, Oct. 1996.

[95] E. Candès and D. Donoho, "Ridgelets: the key to high dimensional intermittency?," *Philosophical Transactions of the Royal Society of London A*, vol. 357, pp. 2495–2509, 1999.

[96] E. Candès and D. Donoho, "Curvelets, multiresolution representation, and scaling laws," in *SPIE Conference on Signal and Image Processing: Wavelet Applications in Signal and Image Processing VIII*, 2000.

[97] D. Donoho and M. Duncan, "Digital curvelet transform: strategy, implementation and experiments," in *Proc. Aerosense 2000, Wavelet Applications VII* (H. Szu, M. Vetterli, W. Campbell, and J. Buss, eds.), vol. 4056, pp. 12–29, SPIE, 2000.

[98] J.-L. Starck, E. Candès, and D. Donoho, "The curvelet transform for image denoising," *IEEE Transactions on Image Processing*, vol. 11, no. 6, pp. 131–141, 2002.

8

Multiframe Blind Deconvolution Coupled with Frame Registration and Resolution Enhancement

Filip Šroubek, Jan Flusser

Institute of Information Theory and Automation
Academy of Sciences of the Czech Republic
Pod vodárenskou věží 4, Prague 8, 182 08, Czech Republic
e-mail: (sroubekf, flusser)@utia.cas.cz

Gabriel Cristóbal

Instituto de Óptica
Consejo Superior de Investigaciones Científicas (CSIC)
Serrano 121, 28006 Madrid, Spain
email: gabriel@optica.csic.es

Abstract

The chapter addresses problems of image registration, blind deconvolution, and superresolution of multiple degraded low-resolution frames in one unifying framework. We propose a method that simultaneously estimates shifts and blurs and recovers the original undistorted image, all in high resolution, without any prior knowledge of the blurs and original image. We accomplish this by formulating the problem as a constrained least squares energy minimization with appropriate regularization terms, which guarantees a close-to-perfect solution. Several experiments on synthetic and real data illustrate the robustness and utilization of the proposed technique in real applications. A discussion on limitations of the proposed method concludes the chapter.

8.1   Introduction

Imaging devices have limited achievable resolution due to many theoretical and practical restrictions. An original scene with a continuous intensity function $o[x, y]$ warps at the camera lens because of the scene motion or change of the camera position. In addition, several external effects blur images: atmospheric turbulence, camera lens, relative camera–scene motion, and so on. We will call these effects *volatile blurs* to emphasize their unpredictable and transitory behavior, yet we will assume that we can model them as convolution with an unknown point spread function (PSF) $v[x, y]$. This is a reasonable assumption if the original scene is flat and perpendicular to the optical axis. Finally, the CCD (charge-coupled device) discretizes the images and produces digitized noisy images $g[i, j]$ (frame). We refer to $g[i, j]$ as a *low-resolution* (LR) *image*, since the spatial resolution is too low to capture all the details of the original scene. In conclusion, the acquisition model becomes

$$g[i, j] = D((v * o[W(n_1, n_2)])[x, y]) + n[i, j], \qquad (8.1)$$

where $n[i, j]$ is additive noise and W denotes the geometric deformation (warping). $D(\cdot) = S(g*\cdot)$ is the *decimation operator* that models the function of the CCD sensors. It consists of convolution with the *sensor PSF* $g[i, j]$ followed by the *sampling operator* S, which we define as multiplication by a sum of delta functions placed on an evenly spaced grid. The above model for one single observation $g[i, j]$ is extremely ill-posed. Instead of taking a single image we can take K $(K > 1)$ images of the original scene and this way partially overcome the equivocation of the problem. Hence we write

$$g_k[i, j] = D((v_k * o[W_k(n_1, n_2)])[x, y]) + n_k[i, j], \qquad (8.2)$$

where $k = 1, \ldots, K$, and D remains the same in all the acquisitions. In the perspective of this multiframe model, the original scene $o[x, y]$ is a single input and the acquired LR images $g_k[i, j]$ are multiple outputs. The model is therefore called a single-input multiple-output (SIMO) formation model. The upper part of Figure 8.1 illustrates the multiframe LR acquisition process. To our knowledge, this is the most accurate, state-of-the-art model, as it takes all possible degradations into account.

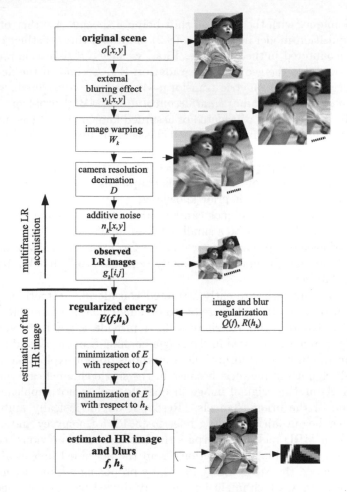

FIGURE 8.1: Low-resolution acquisition (top) and reconstruction flow (bottom).

Superresolution (SR) is the process of combining a sequence of LR images in order to produce a higher-resolution image or sequence. It is unrealistic to assume that the superresolved image can recover the original scene $o[x, y]$ exactly. A reasonable goal of SR is a discrete version of $o[x, y]$ that has a higher spatial resolution than the resolution of the LR images and that is free of the volatile blurs (deconvolved). In the sequel, we will refer to this superresolved image as a *high-resolution* (HR) *image* $f[i, j]$. The standard SR approach consists of subpixel registration, overlaying the LR images on a HR grid, and interpolating the missing values. The subpixel shift between images thus constitutes the essential assumption. We will demonstrate that assuming volatile blurs in the model explicitly leads to a more general and robust technique, with the subpixel shift being a special case thereof.

The acquisition model in Equation (8.2) embraces three distinct cases frequently encountered in the literature. First, we face a registration problem, if we want to resolve the geometric degradation W_k. Second, if the decimation operator D and the geometric transform W_k are not considered, we face a *multichannel* (or multiframe) *blind deconvolution* (MBD) problem. Third, if the volatile blur v_k is not considered or assumed known, and W_k is suppressed up to a subpixel translation, we obtain a classical SR formulation. In practice, it is crucial to consider all three cases at once. We are then confronted with a problem of *blind superresolution* (BSR), which is the subject of this investigation. The approach presented in this chapter is one of the first attempts to solve BSR with only little prior knowledge.

Proper registration techniques can suppress large and complex geometric distortions (usually just up to a small between-image shift). There have been hundreds of methods proposed; see, e.g., [1] for a survey. We will assume in the sequel that the LR images are roughly registered and that W_k reduces to small translations.

The MBD problem has recently attracted considerable attention. The first blind deconvolution attempts were based on single-channel formulations, such as in [2–5]. Kundur and Hatzinakos [6, 7] provide a good overview. The problem is extremely ill-posed in the single-channel framework and cannot be resolved in the fully blind form. These methods do not exploit the potential of the multichannel framework, because in the single-channel case missing information about the original image in one channel is not supplemented by information in the other channels. Research on intrinsically multichannel methods has begun fairly recently; refer to [8–12] for a survey and other references. Such MBD methods overpass the limitations of previous techniques and can recover the blurring functions from the degraded images alone. We further developed the MBD theory in [13] by proposing a blind deconvolution method for images, which might be mutually shifted by unknown vectors.

A countless number of papers address the standard SR problem. A good survey can be found for example in [14,15]. Maximum likelihood (ML), maximum *a posteriori* (MAP), the set theoretic approach using POCS (projection on convex sets), and fast Fourier techniques can all provide a solution to the SR problem. Earlier approaches assumed that subpixel shifts are estimated by other means. More advanced techniques, such as in [16–18], include the shift estimation of the SR process. Other approaches focus on fast implementation [19], space–time SR [20], or SR of compressed video [17]. In general, most of the SR techniques assume *a priori* known blurs. However, few exceptions exist. Authors in [21, 22] proposed BSR that can handle parametric PSFs, i.e., PSFs modeled with one parameter. This restriction is unfortunately very limiting for most real applications. To our knowledge, the first attempts for BSR with an arbitrary PSF appeared in [23, 24]. The interesting idea proposed therein is the conversion of the SR problem from SIMO to multiple-input multiple-output (MIMO) using so-called polyphase components. We will adopt the same idea here as well. Other preliminary results of the BSR problem with focus on fast calculation are given in [25], where the authors propose a modification of the Richardson–Lucy algorithm.

Current multiframe blind deconvolution techniques require no or very little prior information about the blurs; they are sufficiently robust to noise and provide satisfying results in most real applications. However, they can hardly cope with the downsampling operator, which violates the standard convolution model. On the contrary, state-of-the-art SR techniques achieve remarkable results in resolution enhancement in the case of no blur. They accurately estimate the subpixel shift between images but lack any apparatus for calculating the blurs.

We propose a unifying method that simultaneously estimates the volatile blurs and HR image without any prior knowledge of the blurs and the original image. We accomplish this by formulating the problem as a minimization of a regularized energy function, where the regularization is carried out in both the image and blur domains. The image regularization is based on variational integrals and a consequent anisotropic diffusion with good edge-preserving capabilities. A typical example of such regularization is total variation. However, the main contribution of this work lies in the development of the blur regularization term. We show that the blurs can be recovered from the LR images up to small ambiguity. One can consider this as a generalization of the results proposed for blur estimation in the case of MBD problems. This fundamental observation enables us to build a simple regularization term for the blurs even in the case of the SR problem. To tackle the minimization task, we use an alternating minimization approach (see Figure 8.1) consisting of two simple linear equations.

The rest of the chapter is organized as follows. Section 8.2 outlines the degradation model. Section 8.3 reformulates the degradation model using polyphase formalism, which we utilize in the next section and develop a procedure for the volatile blur estimation. These results effortlessly blend in a regularization term of the BSR algorithm as described in Section 8.5. Finally, Section 8.6 illustrates the applicability of the proposed method to real situations.

8.2   Mathematical Model

To simplify the notation, we will assume only images and PSFs with square supports. An extension to rectangular images is straightforward. Let $f[x, y]$ be an arbitrary discrete image of size $F \times F$, then $\mathbf{f}$ denotes an image column vector of size $F^2 \times 1$ and $\mathbf{C}_A\{f\}$ denotes a matrix that performs convolution of f with a kernel of size $A \times A$, i.e., $\mathbf{C}_A\{f\}\mathbf{k}$ is the vector form of $f * k$, where k is of size $A \times A$. The convolution matrix can have a different output size. Adopting the MATLAB® naming convention, we distinguish two cases: "full" convolution $\mathbf{C}_A\{f\}$ of size $(F + A - 1)^2 \times A^2$ and "valid" convolution $\mathbf{C}_A^v\{f\}$ of size $(F - A + 1)^2 \times A^2$. In both cases the convolution matrix is a Toeplitz-block-Toeplitz (TBT) matrix. In the sequel we will not specify dimensions of convolution matrices, if it is obvious from the size of the right argument.

Let us assume we have K different LR frames $\{g_k\}$ (each of size $G \times G$) that represent degraded (blurred and noisy) versions of the original scene. Our goal is to estimate the HR representation of the original scene, which we denoted as the HR image f of size $F \times F$. The LR frames are linked with the HR image through a series of degradations similar to those between $o[x, y]$ and g_k in (8.2). First f is geometrically warped ($\mathbf{W}_k$), then it is convolved with a volatile PSF ($\mathbf{V}_k$), and finally it is decimated ($\mathbf{D}$). The formation of the LR images in vector-matrix notation is then described as

$$\mathbf{g}_k = \mathbf{D}\mathbf{V}_k\mathbf{W}_k\mathbf{f} + \mathbf{n}_k, \tag{8.3}$$

where $\mathbf{n}_k$ is additive noise present in every channel. The decimation matrix $\mathbf{D} = \mathbf{S}\mathbf{U}$ simulates the behavior of digital sensors by performing first convolution with the $U \times U$ sensor PSF ($\mathbf{U}$) and then downsampling ($\mathbf{S}$). The Gaussian function is widely accepted as an appropriate sensor PSF and it is also used here. Its justification is experimentally verified in [26]. A physical interpretation of the sensor blur is that the sensor is of finite size and it integrates impinging light over its surface. The sensitivity of the sensor is the highest in the middle and decreases towards its borders with Gaussian-like decay. Further, we assume that the subsampling factor (or SR factor, depending on the point of view), denoted by ε, is the same in both x and y directions. It is important to underline that ε is a user-defined parameter. In principle, $\mathbf{W}_k$ can be a very complex geometric transform that must be estimated by image registration or motion detection techniques. We have to keep in mind that subpixel accuracy in $\mathbf{g}_k$ is necessary for SR to work. Standard image registration techniques can hardly achieve this and they leave a small misalignment behind. Therefore, we will assume that complex geometric transforms are removed in the preprocessing step and $\mathbf{W}_k$ reduces to a small translation. Hence $\mathbf{V}_k\mathbf{W}_k = \mathbf{H}_k$, where $\mathbf{H}_k$ performs convolution with the shifted version of the volatile PSF v_k, and the acquisition model becomes

$$\mathbf{g}_k = \mathbf{D}\mathbf{H}_k\mathbf{f} + \mathbf{n}_k = \mathbf{S}\mathbf{U}\mathbf{H}_k\mathbf{f} + \mathbf{n}_k. \tag{8.4}$$

The BSR problem then adopts the following form: we know the LR images $\{g_k\}$ and we want to estimate the HR image f for the given $\mathbf{S}$ and the sensor blur $\mathbf{U}$. To avoid boundary effects, we assume that each observation g_k captures only a part of f. Hence $\mathbf{H}_k$ and $\mathbf{U}$ are "valid" convolution matrices $\mathbf{C}_F^v\{h_k\}$ and $\mathbf{C}_{F-H+1}^v\{u\}$, respectively. In general, the PSFs h_k are of different size. However, we postulate that they all fit into an $H \times H$ support.

In the case of $\varepsilon = 1$, the downsampling $\mathbf{S}$ is not present and we face a slightly modified MBD problem that has been solved elsewhere [8, 13]. Here we are interested in the case of $\varepsilon > 1$, when the downsampling occurs. Can we estimate the blurs like in the $\varepsilon = 1$ case? The presence of $\mathbf{S}$ prevents us from using the cited results directly. In the next section we use the polyphase formulation and transfer the problem from SIMO to MIMO. We then show that conclusions obtained for MBD apply here in a slightly modified form as well.

8.3 Polyphase Formulation

Polyphase formulation is an elegant way to rewrite the acquisition model and thus get a better insight into BSR. First we will assume integer SR factors, for which the model is simple and easy to understand. Then we will generalize it for rational SR factors, for which we will take the full advantage of polyphase formalism. It will allow us to formulate the model for rational factors using a combination of integer factors.

Before we proceed, it is necessary to define precisely the sampling matrix $\mathbf{S}$. Let $\mathbf{S}_i^\varepsilon$ denote a 1-D sampling matrix, where ε is the integer subsampling factor and $i = 1, \ldots, \varepsilon$. Each row of the sampling matrix is a unit vector whose nonzero element is at such position that, if the matrix multiplies an arbitrary vector b, the result of the product is every ε-th element of b starting from b_i. If the vector length is M then the size of the sampling matrix is $(M/\varepsilon) \times M$. If M is not divisible by ε, we can pad the vector with an appropriate number of zeros to make it divisible. A 2-D sampling matrix is defined by

$$\mathbf{S}_{i,j}^\varepsilon := \mathbf{S}_i^\varepsilon \otimes \mathbf{S}_j^\varepsilon. \tag{8.5}$$

If the starting index (i, j) will be $(1, 1)$ we will omit the subscript and simply write $\mathbf{S}^\varepsilon$. Note that the transposed matrix $(\mathbf{S}^\varepsilon)^T$ behaves as an upsampling operator that interlaces the original samples with $(\varepsilon - 1)$ zeros. Now, we are ready to define *polyphase components* of an image $f[x, y]$ as

$$\mathbf{f}^{ij} := \mathbf{S}_{i,j}^\varepsilon \mathbf{f}, \tag{8.6}$$

which is equivalent to

$$\mathbf{f}^{ij} := [f[i, j], f[i + \varepsilon, j], f[i + 2\varepsilon, j], \ldots, f[i, j + \varepsilon], f[i + \varepsilon, j + \varepsilon], \ldots]^T.$$

Therefore, each image breaks into ε^2 distinct polyphase components (downsampled versions of the image); see Figure 8.2. We will refer to this decomposition as a *polyphase decomposition*, and write $\mathbf{P}^\varepsilon \mathbf{f}$, where

$$\mathbf{P}^\varepsilon := [(\mathbf{S}_{1,1}^\varepsilon)^T, \ldots, (\mathbf{S}_{\varepsilon,1}^\varepsilon)^T, (\mathbf{S}_{1,2}^\varepsilon)^T, \ldots, (\mathbf{S}_{\varepsilon,\varepsilon}^\varepsilon)^T]^T. \tag{8.7}$$

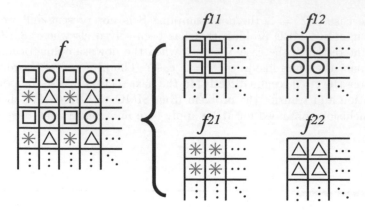

FIGURE 8.2: Polyphase decomposition for $\varepsilon = 2$: original image f decomposes into four downsampled images.

A similar decomposition was proposed in [27]. Note that $\mathbf{P}$ is a permutation matrix and therefore $\mathbf{P}^T\mathbf{P} = \mathbf{P}\mathbf{P}^T = \mathbf{I}$. We first derive polyphase formulation for integer SR factors and then, using a simple trick, we extend it to rational ones.

8.3.1   Integer Downsampling Factor

Let us consider a simple convolution equation

$$\mathbf{g} = \mathbf{H}\mathbf{f}\,, \tag{8.8}$$

and explore the benefits of the polyphase decomposition. Multiplying by $\mathbf{P}^\varepsilon$, we get

$$[\mathbf{P}^\varepsilon\mathbf{g}] = [\mathbf{P}^\varepsilon\mathbf{H}(\mathbf{P}^\varepsilon)^T][\mathbf{P}^\varepsilon\mathbf{f}]\,. \tag{8.9}$$

The permutation matrix $\mathbf{P}^\varepsilon$ decomposes an image into ε^2 polyphase components, and in our case,

$$\mathbf{P}^\varepsilon\mathbf{g} = [(\mathbf{g}^{11})^T,\ldots,(\mathbf{g}^{\varepsilon\varepsilon})^T]^T \text{ and } \mathbf{P}^\varepsilon\mathbf{f} = [(\mathbf{f}^{11})^T,\ldots,(\mathbf{f}^{\varepsilon\varepsilon})^T]^T.$$

For the next discussion it is fundamental to make the observation that $[\mathbf{P}^\varepsilon\mathbf{H}(\mathbf{P}^\varepsilon)^T]$ consists of $\varepsilon^2 \times \varepsilon^2$ blocks. Each block retains the TBT shape of $\mathbf{H}$ but performs the convolution with one polyphase component of h.

We see that (8.9) is just a permutation of rows and columns of (8.8). The advantage of the polyphase formulation resides in the fact that downsampling is equivalent to a section of (8.9) that corresponds to one polyphase component. We conclude this part by reformulating the acquisition model (8.4) using polyphase components and obtain

$$\mathbf{g}_k = \mathbf{S}^\varepsilon\mathbf{U}\mathbf{H}_k\mathbf{f} + \mathbf{n}_k = [\mathbf{S}^\varepsilon\mathbf{U}\mathbf{H}_k(\mathbf{P}^\varepsilon)^T][\mathbf{P}^\varepsilon\mathbf{f}] + \mathbf{n}_k\,, \tag{8.10}$$

for $k = 1, \ldots, K$. Instead of $\mathbf{S}^\varepsilon = \mathbf{S}_{1,1}^\varepsilon$ one can use any $\mathbf{S}_{i,j}^\varepsilon$. However, they are all equivalent from the reconstruction point of view as they correspond to different translations of the HR image f. In the introduction we regarded the acquisition model as SIMO, with one input channel f and K output channels g_k. Under closer examination of the above polyphase formulation, one can see that $[\mathbf{S}^\varepsilon \mathbf{U} \mathbf{H}_k (\mathbf{P}^\varepsilon)^T]$ consists of $1 \times \varepsilon^2$ convolution matrices and that in reality the model is of the MIMO type with ε^2 input channels (polyphase components of f) and K output channels g_k.

8.3.2 Rational Downsampling Factor

Integer SR factors are too limiting. From the practical point of view, we would like to have noninteger SR factors as well. We can extend the above results to factors expressed as a fraction p/q where p and q are positive integers and $p > q$ (p and q are reduced so that they do not have any common factor).

Let $\varepsilon = p/q$ and the sampling frequency of the LR images g_k be q, then the sampling frequency (number of pixels per unit distance) of the HR image f is p. From each LR image g_k we generate q^2 polyphase components. We consider these polyphase components as new output (downsampled LR) images with the sampling frequency 1. Now, to obtain the HR image from the downsampled LR images, we must solve a SR problem with the integer factor $\varepsilon = p$ and not with the rational one as before. In other words, in order to get an integer SR factor we downsample the LR images and thus artificially increase the number of channels. However, the number of unknown PSFs h_k remains the same. We still have K PSFs since every pack of q^2 downsampled LR images contains the same blur. An illustrative diagram of the process in 1-D for $\varepsilon = 3/2$ is given in Figure 8.3.

It is important to understand the discretization of the sensor PSF u in the case of fractional SR factors. Since p is not divisible by q, the product $\mathbf{SU}$ is shift-variant and it depends on a relative shift between the HR and LR pixels. One can readily see that the relative shift repeats every q-th pixels (in both directions x and y) of the LR image and therefore we have q^2 distinct PSF discretizations. To better understand this concept, see the configuration for $\varepsilon = 3/2$ in Figure 8.4.

Similarly to (8.10), we reformulate the acquisition model (8.4) using polyphase components and write

$$[\mathbf{P}^q \mathbf{g}_k] = \left[\begin{bmatrix} \mathbf{S}^p \mathbf{U}_{1,1} \\ \vdots \\ \mathbf{S}^p \mathbf{U}_{q,q} \end{bmatrix} \mathbf{H}_k (\mathbf{P}^p)^T \right] [\mathbf{P}^p \mathbf{f}] + \mathbf{n}_k , \qquad (8.11)$$

where each $\mathbf{U}_{i,j}$ performs the convolution with one of the q^2 discretizations of the sensor PSF u. We see that the rational and integer SR factors lead to similar expressions. Only in the rational case, the resulting MIMO problem has Kq^2 output channels and p^2 input channels.

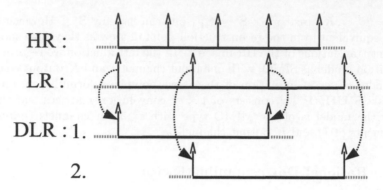

FIGURE 8.3: Rational downsampling $\varepsilon = 3/2$ in 1-D: we have LR signals (middle row) with the sampling frequency 2 and we want to obtain a HR signal (top row) with the sampling frequency 3. We convert this scenario to the one with the integer SR factor by considering every second sample of the LR signal and thus creating from each LR signal two signals (bottom row) of half size. These downsampled LR signals are then used in the SR problem with the integer factor 3.

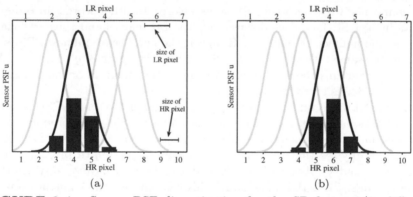

FIGURE 8.4: Sensor PSF discretization for the SR factor $3/2$: different discretizations of the PSF reside in a varying relative shift between LR and HR pixels. If the LR pixel is $1.5\times$ the size of the HR pixel, then two distinct discretizations (a) and (b) show up in 1-D (in 2-D we have four discretizations). The plotted curves depict the sensor PSF in the continuous domain at different locations and the bar plots its discrete version.

8.4 Reconstruction of Volatile Blurs

Estimation of blurs in the MBD case (no downsampling) attracted considerable attention in the past. A wide variety of methods were proposed, such as in [8,9], that provide a satisfactory solution. For these methods to work correctly, a certain channel disparity is necessary. The disparity is defined as weak coprimeness of the channel blurs, which states that the blurs have no common factor except of a scalar constant. In other words, if the channel blurs can be expressed as a convolution of two subkernels then there is no subkernel that is common to all blurs. An exact definition of weakly coprime blurs can be found in [9]. The channel coprimeness is satisfied for many practical cases, since the necessary channel disparity is mostly guaranteed by the nature of the acquisition scheme and random processes therein. We refer the reader to [8] for a relevant discussion. This channel disparity is also necessary for the BSR case.

Let us first recall how to estimate blurs in the MBD case and then we will show how to generalize the results for integer and rational downsampling factors. In the following, we will assume that noise n is not present and wait till the next Section 8.5, where we will address noise appropriately.

8.4.1 The MBD Case

The downsampling matrix $\mathbf{S}$ is not present in (8.4) and only convolution binds the input with the outputs. The acquisition model is of the SIMO type with one input channel f and K output channels g_k. Under the assumption of channel coprimeness, we can see that any two correct blurs h_i and h_j satisfy

$$\|g_i * h_j - g_j * h_i\|^2 = 0 \,. \tag{8.12}$$

Considering all possible pairs of blurs, we can arrange the above relation into one system

$$\mathcal{N}'\mathbf{h} = \mathbf{0} \,, \tag{8.13}$$

where $\mathbf{h} = [\mathbf{h}_1^T, \ldots, \mathbf{h}_K^T]^T$ and $\mathcal{N}'$ is constructed solely by matrices that perform the convolution with g_k. In most real situations the correct blur size (we have assumed square size $H \times H$) is not known in advance and therefore we can generate the above equation for different blur dimensions $\hat{H}_1 \times \hat{H}_2$. The nullity (null space dimension) of $\mathcal{N}'$ is exactly 1 if the blur size is correctly estimated. By applying SVD (singular value decomposition), we recover precisely the blurs except to a scalar factor. One can eliminate this magnitude ambiguity by stipulating that $\sum_{x,y} h_k[x,y] = 1$, which is a common brightness-preserving assumption. If the blur size is underestimated, the above equation has no solution. If the blur size is overestimated, then nullity($\mathcal{N}'$) = $(\hat{H}_1 - H + 1)(\hat{H}_2 - H + 1)$.

8.4.2 The BSR Case

A naive approach, e.g., proposed in [28,29], is to modify (8.13) in the MBD case by applying downsampling and formulating the problem as

$$\min_{\mathbf{h}} \|\mathcal{N}'[\mathbf{I}_K \otimes \mathbf{S}^\varepsilon \mathbf{U}]\mathbf{h}\|^2 \,, \tag{8.14}$$

where $\mathbf{I}_K$ is the $K \times K$ identity matrix. One can easily verify that the condition in (8.12) is not satisfied for the BSR case as the presence of downsampling operators violates the commutative property of convolution. Even more disturbing is the fact that minimizers of (8.14) do not have to correspond to the correct blurs. However, if we use the MIMO polyphase formulation in (8.10) or in (8.11), we will show that the reconstruction of the volatile PSFs h_k is possible even in the BSR case. We will see that for the integer SR factors ε, some ambiguity in the solution of h_k is inevitable, irrespective of the knowledge of the sensor blur u. For the rational ε, a solution is possible if and only if the sensor blur u is known, and surprisingly the solution is without any ambiguity. Note that for correct reconstruction of the HR image, the sensor blur is necessary in any case.

First, we need to rearrange the acquisition model (8.4) and construct from the LR images g_k a convolution matrix $\mathcal{G}$ with a predetermined nullity. Then we take the null space of $\mathcal{G}$ and construct a matrix $\mathcal{N}$, which will contain the correct PSFs h_k in its null space.

Let $E \times E$ be the size of the "nullifying" filters. The meaning of this name will be clear later. Define $\mathcal{G} := [\mathbf{G}_1, \ldots, \mathbf{G}_K]$, where $\mathbf{G}_k := \mathbf{C}_E^v\{g_k\}$ are "valid" convolution matrices. Using (8.10) without noise, we can express $\mathcal{G}$ in terms of f, u, and h_k as

$$\mathcal{G} = \mathbf{S}^\varepsilon \mathbf{F} \mathbf{U} \mathcal{H}, \tag{8.15}$$

where

$$\mathcal{H} := [\mathbf{C}_{\varepsilon E}\{h_1\}(\mathbf{S}^\varepsilon)^T, \ldots, \mathbf{C}_{\varepsilon E}\{h_K\}(\mathbf{S}^\varepsilon)^T], \tag{8.16}$$

$\mathbf{U} := \mathbf{C}_{\varepsilon E + H - 1}\{u\}$ and $\mathbf{F} := \mathbf{C}_{\varepsilon E + H + U - 2}^v\{f\}$.

The convolution matrix $\mathcal{U}$ has more rows than columns and therefore it is of full column rank (see proof in [8] for general convolution matrices). We assume that $\mathbf{S}^\varepsilon \mathbf{F}$ has full column rank as well. This is almost certainly true for real images if $\mathbf{F}$ has at least ε^2-times more rows than columns. Thus $\text{Null}(\mathcal{G}) \equiv \text{Null}(\mathcal{H})$ and the difference between the number of columns and the rows of $\mathcal{H}$ bounds from below the null space dimension, i.e.,

$$\text{nullity}(\mathcal{G}) \geq KE^2 - (\varepsilon E + H - 1)^2. \tag{8.17}$$

Setting $N := KE^2 - (\varepsilon E + H - 1)^2$ and $\mathbf{N} := \text{Null}(\mathcal{G})$, we visualize the null space as

$$\mathbf{N} = \begin{bmatrix} \boldsymbol{\eta}_{1,1} & \cdots & \boldsymbol{\eta}_{1,N} \\ \vdots & \ddots & \vdots \\ \boldsymbol{\eta}_{K,1} & \cdots & \boldsymbol{\eta}_{K,N} \end{bmatrix}, \tag{8.18}$$

where $\boldsymbol{\eta}_{kn}$ is the vector representation of the nullifying filter η_{kn} of size $E \times E$, $k = 1, \ldots, K$ and $n = 1, \ldots, N$. Let $\tilde{\eta}_{kn}$ denote upsampled η_{kn} by factor ε, i.e., $\tilde{\boldsymbol{\eta}}_{kn} := (\mathbf{S}^\varepsilon)^T \boldsymbol{\eta}_{kn}$. Then, we define

$$\mathcal{N} := \begin{bmatrix} \mathbf{C}_H\{\tilde{\eta}_{1,1}\} & \cdots & \mathbf{C}_H\{\tilde{\eta}_{K,1}\} \\ \vdots & \ddots & \vdots \\ \mathbf{C}_H\{\tilde{\eta}_{1,N}\} & \cdots & \mathbf{C}_H\{\tilde{\eta}_{K,N}\} \end{bmatrix} \tag{8.19}$$

and conclude that

$$\mathcal{N}\mathbf{h} = \mathbf{0}\,, \tag{8.20}$$

where $\mathbf{h}^T = [\mathbf{h}_1, \ldots, \mathbf{h}_K]$. We have arrived at an equation that is of the same form as (8.13) in the MBD case. Here we have the solution to the blur estimation problem for the BSR case. However, since it was derived from (8.10), which is of the MIMO type, the ambiguity of the solution is higher. It has been shown in [30] that the solution of the blind 1-D MIMO case is unique apart from a mixing matrix of input signals. The same holds true here as well. Without proofs we provide the following statements. For the correct blur size, nullity($\mathcal{N}$) $= \varepsilon^4$. For the underestimated blur size, (8.20) has no solution. For the overestimated blur size $\hat{H}_1 \times \hat{H}_2$, nullity($\mathcal{N}$) $= \varepsilon^2(\hat{H}_1 - H + \varepsilon)(\hat{H}_2 - H + \varepsilon)$.

The conclusion may seem to be pessimistic. For example, for $\varepsilon = 2$ the nullity is at least 16, and for $\varepsilon = 3$ the nullity is already 81. Nevertheless, Section 8.5 shows that $\mathcal{N}$ plays an important role in the regularized restoration algorithm and its ambiguity is not a serious drawback.

It remains to describe the procedure for the rational downsampling factors $\varepsilon = p/q$. The analysis starts by rearranging the acquisition model in (8.11). Again, let $E \times E$ be the size of the nullifying filters. In the previous section, we have seen that there are q^2 distinct discretizations of the sensor PSF u that depend on the relative shift between HR and LR pixels. Let $u_{i,j}$ ($i, j = 1, \ldots, q$) denote such discretizations. We define "full" convolution matrix $\mathbf{U}_{i,j} := \mathbf{C}_{pE+H-1}\{u_{i,j}\}$ and "valid" convolution matrices $\mathbf{F} := \mathbf{C}_{pE+U+H-2}^v\{f\}$, $\mathbf{G}_k := \mathbf{C}_{qE}^v\{g_k\}$. Then define

$$\mathcal{G} := [\mathbf{G}_1, \ldots, \mathbf{G}_K]\,,$$
$$\mathcal{H}' := [\mathbf{I}_{q^2} \otimes \mathbf{C}_{pE}\{h_1\}, \ldots, \mathbf{I}_{q^2} \otimes \mathbf{C}_{pE}\{h_K\}][\mathbf{I}_{Kq^2} \otimes (\mathbf{S}^\varepsilon)^T]\,.$$

The degradation model for the rational SR factor $\varepsilon = p/q$ becomes

$$\mathbf{S}^q\mathcal{G}[\mathbf{I}_K \otimes \mathbf{P}^q] = \mathbf{S}^p\mathbf{F}[\mathbf{U}_{1,1}, \ldots, \mathbf{U}_{q,q}]\mathcal{H}'\,. \tag{8.21}$$

The integer SR factor is a special case of this equation. By setting $q = 1$ we obtain (8.15).

In analogy with the derivation steps for the integer case, we proceed as follows. Set $\mathbf{N} := \text{Null}(\mathbf{S}^q\mathcal{G})$. The size of $\mathbf{N}$ is $K(qE)^2 \times N$, where we assume $N \geq K(qE)^2 - (pE + H + U - 1)^2 > 0$. We visualize the null space as

$$\mathbf{N} = \begin{bmatrix} \boldsymbol{\eta}_{1,1} & \cdots & \boldsymbol{\eta}_{1,N} \\ \vdots & \ddots & \vdots \\ \boldsymbol{\eta}_{q^2,1} & \cdots & \boldsymbol{\eta}_{q^2,N} \\ \vdots & \ddots & \vdots \\ \boldsymbol{\eta}_{Kq^2,1} & \cdots & \boldsymbol{\eta}_{Kq^2,N} \end{bmatrix}\,, \tag{8.22}$$

where $\boldsymbol{\eta}_{kn}$ is the vector representation of the nullifying filter η_{kn} of size $E \times E$. Let $\tilde{\eta}_{kn}$ denote upsampled η_{kn} by factor p. Then

$$\mathcal{N} := \begin{bmatrix} \mathbf{C}_{U+H-1}\{\tilde{\eta}_{1,1}\} & \cdots & \mathbf{C}_{U+H-1}\{\tilde{\eta}_{Kq^2,1}\} \\ \vdots & \ddots & \vdots \\ \mathbf{C}_{U+H-1}\{\tilde{\eta}_{1,N}\} & \cdots & \mathbf{C}_{U+H-1}\{\tilde{\eta}_{Kq^2,N}\} \end{bmatrix} \times \mathbf{I}_K \otimes \begin{bmatrix} \mathbf{C}_H\{u_{1,1}\} \\ \vdots \\ \mathbf{C}_H\{u_{q,q}\} \end{bmatrix} \quad (8.23)$$

and we conclude that

$$\mathcal{N}\mathbf{h} = \mathbf{0}. \quad (8.24)$$

The presence of shifted versions of u eliminates any ambiguity of the solution and we can prove that for the correctly estimated blur size the nullity of $\mathcal{N}$ is 1.

While this conclusion may appear optimistic, one should realize an important detail that distinguishes $\mathcal{N}$ for the rational factors from $\mathcal{N}$ for the integer factors. The matrix $\mathcal{N}$ in the integer case does not depend on u and therefore the reconstruction of h_k, though ambiguous, can be carried out even without the knowledge of the sensor PSF. On the other hand, $\mathcal{N}$ in the rational case contains q^2 distinct discretizations of the sensor PSF and the reconstruction of h_k can fail if the sensor PSF is incorrectly estimated.

Another interesting consequence of the above derivation is the minimum necessary number of LR images for the blur reconstruction to work. The condition of the $\mathcal{G}$ nullity in (8.17) implies that the minimum number is $K > \varepsilon^2$. For example, for $\varepsilon = 3/2$, three LR images are sufficient; for $\varepsilon = 2$, we need at least five LR images to perform blur reconstruction.

8.5　Blind Superresolution

In order to solve the BSR problem, i.e, determine the HR image f and volatile PSFs h_k, we adopt a classical approach of minimizing a regularized energy function. This way the method will be less vulnerable to noise and better posed. The energy consists of three terms and takes the form

$$E(\mathbf{f}, \mathbf{h}) = \sum_{k=1}^{K} \|\mathbf{D}\mathbf{H}_k\mathbf{f} - \mathbf{g}_k\|^2 + \alpha Q(\mathbf{f}) + \beta R(\mathbf{h}). \quad (8.25)$$

The first term measures the fidelity to the data and emanates from our acquisition model (8.4). The remaining two are regularization terms with positive weighting constants α and β that attract the minimum of E to an admissible set of solutions. The form of E very much resembles the energy proposed in [13] for MBD. Indeed, this should not come as a surprise since MBD and SR are related problems in our formulation.

Regularization $Q(\mathbf{f})$ is a smoothing term of the form

$$Q(\mathbf{f}) = \mathbf{f}^T \mathbf{L} \mathbf{f}, \qquad (8.26)$$

where $\mathbf{L}$ is a high-pass filter. A common strategy is to use convolution with the Laplacian for $\mathbf{L}$, which in the continuous case corresponds to $Q(f) = \int |\nabla f|^2$. Recently, variational integrals $Q(f) = \int \phi(|\nabla f|)$ were proposed, where ϕ is a strictly convex, nondecreasing function that grows at most linearly. Examples of $\phi(s)$ are s (total variation), $\sqrt{1 + s^2} - 1$ (hypersurface minimal function), $\log(\cosh(s))$, or nonconvex functions, such as $\log(1 + s^2)$, $s^2/(1 + s^2)$, and $\arctan(s^2)$ (Mumford–Shah functional). The advantage of the variational approach is that while in smooth areas it has the same isotropic behavior as the Laplacian, it also preserves edges in images. The disadvantage is that it is highly nonlinear and to overcome this difficulty, one must use, e.g., a half-quadratic algorithm [31]. For the purpose of our discussion it suffices to state that after discretization we arrive again at (8.26), where this time $\mathbf{L}$ is a positive semidefinite block tridiagonal matrix constructed of values depending on the gradient of f. The rationale behind the choice of $Q(f)$ is to constrain the local spatial behavior of images; it resembles a Markov Random Field. Some global constraints may be more desirable but are difficult (often impossible) to define, since we develop a general method that should work with any class of images.

The PSF regularization term $R(\mathbf{h})$ directly follows from the conclusions of the previous section. Since the matrix $\mathcal{N}$ in (8.20) (integer factor) or in (8.24) (rational factor) contains the correct PSFs h_k in its null space, we define the regularization term as a least squares fit

$$R(\mathbf{h}) = \|\mathcal{N}\mathbf{h}\|^2 = \mathbf{h}^T \mathcal{N}^T \mathcal{N} \mathbf{h}. \qquad (8.27)$$

The product $\mathcal{N}^T \mathcal{N}$ is a positive semidefinite matrix. More precisely, R is a consistency term that binds the different volatile PSFs to prevent them from moving freely and, unlike the fidelity term (the first term in (8.25)), it is based solely on the observed LR images. A good practice is to include with a small weight a smoothing term $\mathbf{h}^T \mathbf{L} \mathbf{h}$ in $R(\mathbf{h})$. This is especially useful in the case of less noisy data to overcome the higher nullity of integer-factor $\mathcal{N}$.

The complete energy then takes the form

$$E(\mathbf{f}, \mathbf{h}) = \sum_{k=1}^{K} \|\mathbf{D}\mathbf{H}_k \mathbf{f} - \mathbf{g}_k\|^2 + \alpha \mathbf{f}^T \mathbf{L} \mathbf{f} + \beta_1 \|\mathcal{N}\mathbf{h}\|^2 + \beta_2 \mathbf{h}^T \mathbf{L} \mathbf{h}. \qquad (8.28)$$

To find a minimizer of the energy function, we perform alternating minimizations (AM) of E over $\mathbf{f}$ and $\mathbf{h}$. The advantage of this scheme lies in its simplicity. Each term of (8.28) is quadratic and therefore convex (but not necessarily strictly convex) and the derivatives with respect to $\mathbf{f}$ and $\mathbf{h}$ are easy to calculate. This AM approach is a variation on the steepest-descent algorithm. The search space is a concatenation of the blur subspace and the image subspace. The algorithm first descends in the image subspace and after reaching the minimum, i.e., $\nabla_{\mathbf{f}} E = 0$, it advances in the blur subspace in the direction $\nabla_{\mathbf{h}} E$ orthogonal to the previous one, and this scheme repeats. In conclusion, starting with some initial $\mathbf{h}^0$, the two iterative steps are

step 1) $\mathbf{f}^m = \arg\min_{\mathbf{f}} E(\mathbf{f}, \mathbf{h}^m)$

$$\Leftrightarrow (\sum_{k=1}^{K} \mathbf{H}_k^T \mathbf{D}^T \mathbf{D} \mathbf{H}_k + \alpha \mathbf{L})\mathbf{f} = \sum_{k=1}^{K} \mathbf{H}_k^T \mathbf{D}^T \mathbf{g}_k \,, \qquad (8.29)$$

step 2) $\mathbf{h}^{m+1} = \arg\min_{\mathbf{h}} E(\mathbf{f}^m, \mathbf{h})$

$$\Leftrightarrow ([\mathbf{I}_K \otimes \mathbf{F}^T \mathbf{D}^T \mathbf{D} \mathbf{F}] + \beta_1 \mathcal{N}^T \mathcal{N} + \beta_2 \mathbf{L})\mathbf{h} = [\mathbf{I}_K \otimes \mathbf{F}^T \mathbf{D}^T]\mathbf{g} \,, \qquad (8.30)$$

where $\mathbf{F} := \mathbf{C}_H^v\{f\}$, $\mathbf{g} := [\mathbf{g}_1^T, \ldots, \mathbf{g}_K^T]^T$ and m is the iteration step. Note that both steps consist of simple linear equations.

Energy E as a function of both variables $\mathbf{f}$ and $\mathbf{h}$ is not convex due to the coupling of the variables via convolution in the first term of (8.28). Therefore, it is not guaranteed that the BSR algorithm reaches the global minimum. In our experience, convergence properties improve significantly if we add feasible regions for the HR image and PSFs specified as lower and upper bound constraints. To solve step 1, we use the method of conjugate gradients (function *cgs* in MATLAB®) and then adjust the solution $\mathbf{f}^m$ to contain values in the admissible range, typically, the range of values of $\mathbf{g}$. It is common to assume that PSF is positive ($h_k \geq 0$) and that it preserves image brightness. We can therefore write the lower and upper bound constraints for PSFs as $\mathbf{h}_k \in \langle 0, 1 \rangle^{H^2}$. In order to enforce the bounds in step 2, we solve (8.30) as a constrained minimization problem (function *fmincon* in MATLAB®) rather than using the projection as in step 1. Constrained minimization problems are more computationally demanding, but we can afford it in this case since the size of $\mathbf{h}$ is much smaller than the size of $\mathbf{f}$.

The weighting constants α and β_i depend on the level of noise. If noise increases, α and β_2 should increase, and β_1 should decrease. One can use parameter estimation techniques, such as cross-validation [21] or expectation maximization [27], to determine the correct weights. However, in our experiments we set the values manually according to a visual assessment. If the iterative algorithm begins to amplify noise, we have underestimated the noise level. On the contrary, if the algorithm begins to segment the image, we have overestimated the noise level.

8.6   Experiments

This section consists of two parts. In the first one, a set of experiments on synthetic data evaluate performance of the BSR algorithm[1] with respect to noise and different regularization terms $R(\mathbf{h})$. The second part demonstrates the applicability of the proposed method to real data and evaluates performance under different input scenarios. Moreover, we compare the reconstruction quality with two other methods: one interpolation technique and one state-of-the-art SR approach. A brief description of these methods follows later.

In all experiments, the sensor blur is fixed and set to a Gaussian function of standard deviation $\sigma = 0.34$ (relative to the scale of LR images). One should underline that the proposed BSR method is fairly robust to the choice of the Gaussian variance, since it can compensate for the insufficient variance by automatically including the missing factor of Gaussian functions in the volatile blurs.

Another potential pitfall that we have taken into consideration is the feasible range of SR factors. Clearly, as the SR factor ε increases we need more LR images and the stability of BSR decreases. In addition, rational SR factors p/q, where p and q are incommensurable and large regardless of the effective value of ε, also make the BSR algorithm unstable. It is the numerator p that determines the internal SR factor used in the algorithm; see Section 8.3.2. Hence we limit ourselves to ε between 1 and 2.5, such as 3/2, 5/3, 2, and so on, which is sufficient in most practical applications.

8.6.1   Simulated Data

First, let us demonstrate the BSR performance with a simple experiment. A 270×200 image in Figure 8.5(a) blurred with six masks in Figure 8.5(b) and downsampled with factor 2 generated six LR images. Using the LR images as an input, we estimated the original HR image with the proposed BSR algorithm for $\varepsilon = 1.25$ and 1.75. In Figure 8.6 one can compare the results printed in their original size. The HR image for $\varepsilon = 1.25$ (Figure 8.6(b)) has improved significantly on the LR images due to deconvolution; however, some details, such as the shirt texture, exhibit artifacts. For the SR factor 1.75, the reconstructed image in Figure 8.6(c) is almost perfect.

Next, we would like to compare performance of different matrices $\mathcal{N}$ inside the blur regularization term $R(\mathbf{h})$ and robustness of the BSR algorithm to noise. Section 8.4 has shown that two distinct approaches exist for blur estimation. Either we use the naive approach in (8.14) that directly utilizes

---

[1]The BSR algorithm is implemented in MATLAB® v7.1 and is available on request.

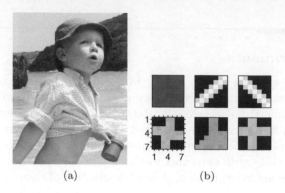

(a)   (b)

FIGURE 8.5:   Simulated data: (a) original 270×200 image; (b) six 7×7 volatile PSFs used to blur the original image.

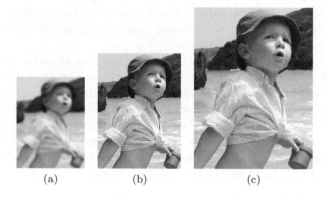

(a)   (b)   (c)

FIGURE 8.6:   BSR of simulated data: (a) one of six LR images with the downsampling factor 2; (b) BSR for $\varepsilon = 1.25$; (c) BSR for $\varepsilon = 1.75$. The shirt texture shows interference for the SR factor 1.25 but becomes well reconstructed for the SR factor 1.75.

the MBD formulation, or we apply the intrinsically SR approach (polyphase formulation) proposed in (8.20) and (8.24) for integer and rational factors, respectively. Depending on the approach, the nullity of $\mathcal{N}$ varies, which influences the shape of the blur regularization functional. We considered three distinct scenarios with the original image and PSFs in Figure 8.5. In the first one, we downsampled the blurred images with the integer factor 2 and performed BSR for $\varepsilon = 2$ using the naive approach inside blur regularization $R(\mathbf{h})$. In the second scenario, we again downsampled the images with the factor 2 and performed BSR for $\varepsilon = 2$ but utilized the polyphase approach for integer factors. The last scenario simulates a situation with rational SR factors. We downsampled the images with the rational factor $7/4 \; (= 1.75)$ and applied BSR with $\varepsilon = 7/4$ and with the polyphase approach for rational factors. In order to evaluate also the noise robustness, we added white Gaussian noise to the LR images with SNR 50dB and 30dB. Note that the signal-to-noise ratio is defined as $\text{SNR} = 10\log(\sigma_f^2/\sigma_n^2)$, where σ_f and σ_n are the image and noise standard deviations, respectively. The BSR algorithm ran without the smoothing term on volatile blurs ($\beta_2 = 0$, refer to discussion in Section 8.5) to avoid artificial enhancement of blurs and to study solely the effect of different matrices $\mathcal{N}$.

Results are not evaluated with any measure of reconstruction quality, such as mean square errors or peak signal-to-noise ratios. Instead we print the results and leave the comparison to a human eye as we believe that in this case the visual assessment is the only reasonable method. Estimated HR images and volatile blurs for three scenarios and two levels of noise are in Figures 8.7 and 8.8. For 50dB (low noise), the performance strongly depends on the applied regularization. If we use the naive approach in the first scenario, the estimated PSFs are inaccurate and hence the reconstructed HR image contains many artifacts as one can see in Figure 8.7(a). In the second scenario (Figure 8.7(b)), the reconstructed PSFs resemble the original ones but exhibit patch-like patterns due to higher nullity of $\mathcal{N}$, which is 16 for the integer SR factor 2. The patch-like pattern emanates from our inability to determine the mixing matrix for polyphase components. However, the reconstructed HR images display quality comparable to the original image. The third scenario (Figure 8.7(c)) with the rational SR factor 1.75 provides the most accurate estimates of the PSFs (the nullity is 1) but for the HR image the improvement on the SR factor 2 is negligible. Clearly, more strict and accurate regularization terms improve HR images, yet as the noise level increases the performance boost diminishes. In the case of 30dB in Figure 8.8, the reconstructed HR images are very similar for all three scenarios, though the PSFs still differ a lot and, e.g., the naive approach gives totally erroneous estimates. The reason for this higher tolerance to inaccurate PSFs resides in constants α and β (weights). To prevent amplification of noise, the weight α of the image smoothing term must be set higher and the other terms in the energy function become less important. Consequently, the estimated HR images tend to piecewise constant functions (general behavior of the total variation (TV) seminorm) and discrepancies in the volatile blurs become less important. The BSR algorithm flattens the interiors of objects to nearly homogeneous regions. Notice, e.g., that the shirt texture well reconstructed in 50dB is removed in 30dB reconstruction.

8.6.2 Real Data

We tested the BSR method on real photos acquired with three different acquisition devices: mobile phone, webcamera, and standard digital camera. The mobile phone of Nokia brand was equipped with a 1-Mpixel camera. The webcam was a Logitech QuickCam for Notebooks Pro with the maximum video resolution at 640×480 and the minimum shutter speed at 1/10s. The

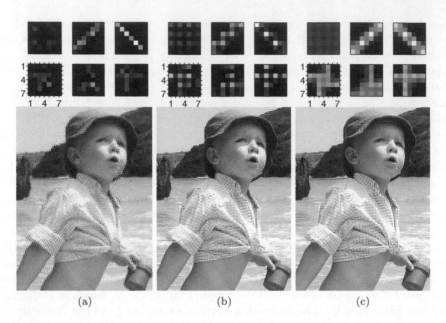

FIGURE 8.7: BSR of simulated data with SNR = 50dB: estimated HR images and PSFs for three different regularization matrices; (a) naive approach for the SR factor 2; (b) proposed polyphase approach for the integer SR factor 2; (c) proposed polyphase approach for the rational SR factor 7/4 (1.75). Note artifacts in (a) and the accurate PSF reconstruction in (c).

last and most advanced device was a 5-Mpixel color digital camera (Olympus C5050Z) equipped with an optical 3×zoom. Since this work considers gray-level images, LR images correspond either to green channels or to gray-level images converted from color photos. To compare the quality of SR reconstruction, we provide results of two additional methods: an interpolation technique and the state-of-the-art SR method. The former technique consists of the MBD method proposed in [13] followed by standard bilinear interpolation (BI) resampling. The MBD method first removes volatile blurs and then BI of the deconvolved image achieves the desired spatial resolution. The latter method, which we will call herein a "standard SR algorithm," is a MAP formulation of the SR problem proposed, e.g., in [16,17]. This method uses a MAP framework for the joint estimation of image registration parameters (in our case only translation) and the HR image, assuming only the sensor blur (**U**) and no volatile blurs. For an image prior, we use edge-preserving Huber Markov Random Fields [26].

First, the performance of the proposed BSR method was tested on data with negligible volatile blurs. Using the mobile phone, we took eight images of a sheet of paper with text. Figure 8.9(a) shows one part (70 × 80) of

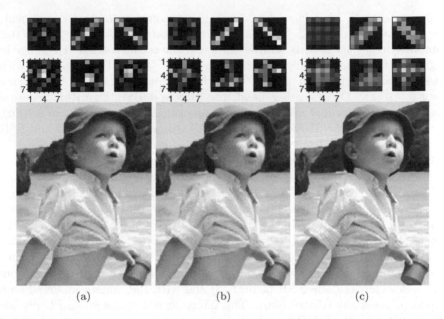

(a)   (b)   (c)

FIGURE 8.8: BSR of simulated data with SNR = 30dB: estimated HR images and PSFs for three different regularization matrices; (a) naive approach for the SR factor 2; (b) proposed polyphase approach for the integer SR factor 2; (c) proposed polyphase approach for the rational SR factor 7/4 (1.75). Due to noise, the image smoothing regularization term takes over and the HR images are smoother than for SNR = 50dB in Figure 8.7.

the image (zero-order interpolation) considered in this experiment. Since the light conditions were good, the shutter speed of the mobile was short, which minimized any possible occurrence of volatile blurs. We set the desired SR factor to 2 and applied the standard SR method, MBD with BI, and BSR with outcomes in Figures 8.9(b), (c), and (d), respectively. The standard SR technique gave results equivalent to those obtained by the BSR algorithm. In both cases the text is legible and the PSFs are almost identical, which indicates that the volatile blurs estimated by the BSR method were close to Dirac pulses. Consequently, the MBD method achieved only little improvement as there was no blurring, and bilinear interpolation does not create any new information.

The next two experiments demonstrate the true power of the BSR algorithm as we now consider LR images with substantial blurring. In the first one, with the webcam handheld, we captured a short video sequence under poor light conditions. Then we extracted ten consecutive frames and considered a small region of size 80 × 60; see one frame with zero-order interpolation in Figure 8.10(a). The long shutter speed (1/10s), together with the inevitable motion of hands, introduced blurring into the images. In this experiment, the SR factor was again 2. The standard SR algorithm could not handle this complicated situation with volatile blurs and the reconstructed HR image in Figure 8.10(b) shows many disturbing artifacts. The MBD combined with BI removed blurring, but subtle details in the image remained hidden; see Figure 8.10(c). On the other hand, the proposed BSR algorithm removed blurring and performed SR correctly, as one can compare in Figure 8.10(d); note, e.g., the word "BŘEZEN." The PSFs estimated by the MBD and BSR look similar, as expected, but the BSR blurs contain more details.

In the second experiment, we compared the three reconstruction techniques on blurred photos of a car front. With the digital camera, we took four shots in a row and cropped a 120×125 rectangle from each. All four cuttings printed in their original size (no interpolation) are in Figure 8.11(a). Similar to the previous experiment, the camera was handheld, and due to the longer shutter speed, the LR images exhibit blurring. We set the SR factor to $5/3$. Again we applied all three techniques as before. In order to better assess the obtained results we took one additional image with optical zoom $1.7\times$ (close to the desired SR factor $5/3$) and with the camera mounted on a tripod to avoid any volatile blurs. This image served as the ground truth; see Figure 8.11(e). Both MBD with BI in Figure 8.11(b) and the standard SR approach in Figure 8.11(c) failed to provide sharp HR images. The proposed BSR method outperformed both techniques and returned a well-reconstructed HR image (Figure 8.11(d)), which is comparable to the "ground-true" image acquired with the optical zoom and tripod. The PSFs estimated with BSR are in Figure 8.11(f). To better evaluate the results, refer to the four close-ups in Figure 8.11(g).

8.6.3 Performance Experiments

When dealing with real data, one cannot expect that the performance will increase indefinitely as the number of available LR images increases. At a certain point, possible discrepancies between the measured data and our mathematical model take over, and the estimated HR image does not improve any more or it can even get worse. We conducted several experiments on real data (short shutter speed and still shooting objects) with different SR factors and number of LR images K. See the results of one such experiment in Figure 8.12 for $\varepsilon = 7/4$ and the number of LR images ranging from four to ten. Note that at the end of Section 8.4.2 we concluded that the minimum number of LR images necessary to construct the blur regularization $R(\mathbf{h})$ for $\varepsilon = 7/4$ is four. A certain improvement is apparent in using six instead of four LR images; see Figure 8.12(c). However, results obtained with more images (eight and ten) show almost no improvement. We deduce that for each SR factor exists an optimal number of LR images that is close to the minimum necessary number. Therefore, in practice, we recommend to use the minimum or close to minimum number of LR images for the given SR factor.

The last experiment of this chapter demonstrates that the BSR method truly reconstructs high-frequency information, which is otherwise unavailable in the single LR image. With the digital camera, we took eight images of an ISO test pattern "Chart." The original chart is in Figure 8.13(b) and one of the acquired LR images with zero-order interpolation is in Figure 8.13(a). A riveting feature of this chart is that if BSR can recover high frequencies, then we should be able to distinguish the rays closer to the center and thus eliminate the Moire effect apparent on the LR images. We applied the BSR algorithm with six different SR factors from $\varepsilon = 1$ to 2.5; see the estimated HR images in Figure 8.13(c). From the obtained results it is clear that the amount of high-frequency information estimated by BSR depends on the SR factor. The Moire effect, still visible for $\varepsilon = 1.25$, disappears for $\varepsilon = 1.5$. As the SR factor increases, the rays become better outlined closer to the center of the chart. However, this does not continue to infinity for real noisy data. In this case, we did not see any objective improvement beyond $\varepsilon = 2$.

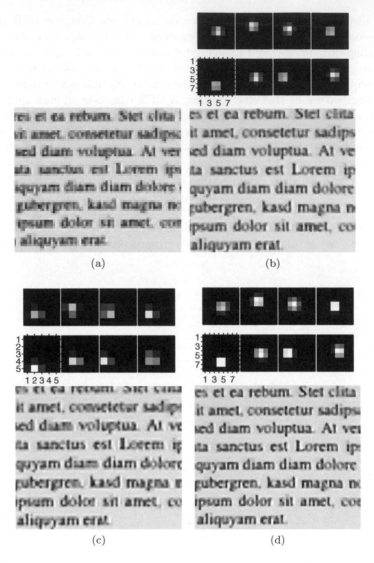

FIGURE 8.9: Reconstruction of images acquired with a mobile phone ($\varepsilon = 2$): (a) one of eight LR images shot with the mobile phone, zero-order interpolation; (b) HR image and blur shifts estimated by the standard SR algorithm; (c) HR image and blurs estimated by MBD with bilinear interpolation; (d) HR image and blurs estimated by the BSR algorithm. Volatile blurs were negligible in this case and the main source of degradation in (a) was the insufficient resolution of the device. Both the standard SR (b) and BSR (d) methods give similar results that improve legibility of the text significantly. MBD (c) can hardly achieve any improvement. The LR images in (a) are provided courtesy of Janne Heikkila from the University of Oulu, Finland.

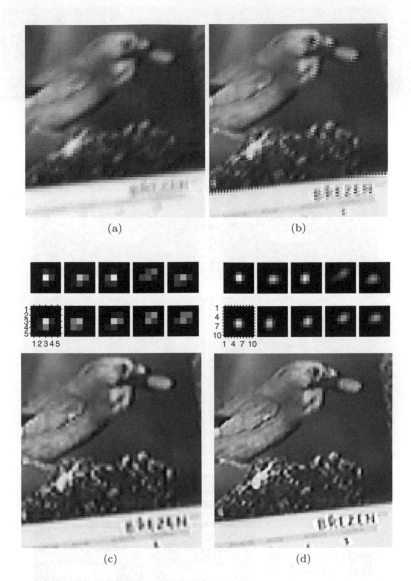

FIGURE 8.10: Reconstruction of images acquired with a webcam ($\varepsilon = 2$): (a) one of ten LR frames extracted from a short video sequence captured with the webcam, zero-order interpolation; (b) standard SR algorithm; (c) HR image and blurs estimated by MBD with bilinear interpolation; (d) HR image and blurs estimated by the BSR algorithm. Due to blurring in the LR frames, the standard SR method (b) gives unsatisfactory results. MBD (c) improves the image slightly, but the true enhancement is achieved only with the BSR method (d).

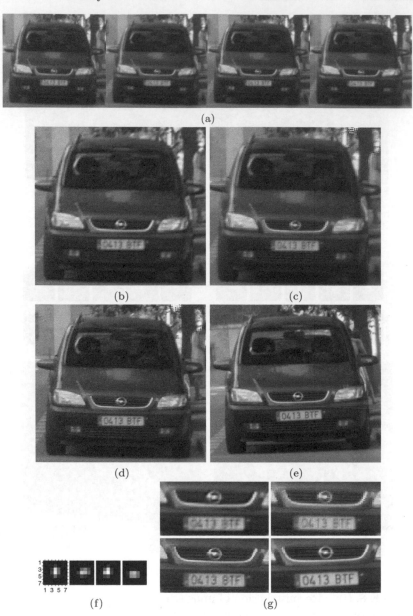

FIGURE 8.11: Reconstruction of images acquired with a digital camera ($\varepsilon = 5/3$): (a) four LR images used in the reconstruction; (b) MBD followed by bilinear interpolation; (c) standard SR algorithm; (d–f) BSR algorithm showing the HR image together with recovered blurs; (e) image acquired with the camera mounted on a tripod and with optical zoom 1.7×; (g) close-ups of the results (b), (c) on top and (d), (e) on bottom. Due to blurring and insufficient resolution of the LR images, both MBD (b) and the standard SR method (c) give unsatisfactory results. Only the BSR algorithm (d) achieves reconstruction comparable to the image with optical zoom (e).

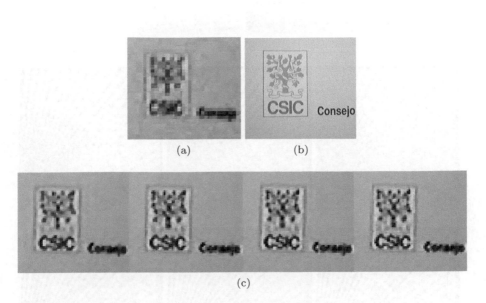

(a) (b)

(c)

FIGURE 8.12: Performance of the BSR algorithm with respect to the number of LR images ($\varepsilon = 7/4 = 1.75$): (a) one of ten LR images, zero-order interpolation; (b) original image; (c) HR images estimated by the BSR algorithm using four, six, eight, and ten LR images (from left to right). A small improvement is visible between four and six images (compare letter "S" and details of the tree images). However, any further increase of the number of LR images proves fruitless. (**See color insert following page 140.**)

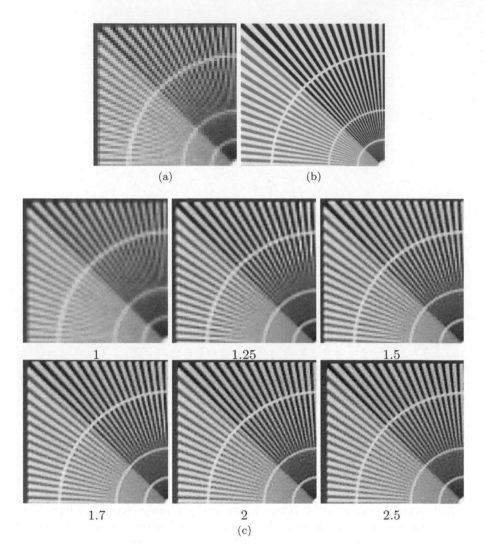

(a)   (b)

1   1.25   1.5

1.7   2   2.5

(c)

FIGURE 8.13: Performance of the BSR algorithm with respect to the SR factor: (a) one of eight LR images acquired with a camera, zero-order interpolation; (b) original cropped "Chart" ISO image, courtesy of B. Brower (ITT Industries); (c) HR images estimated by the BSR algorithm with the SR factor 1, 1.25, 1.5, 1.7, 2, and 2.5. The HR images were bilinearly interpolated to have the same size. The BSR algorithm truly reconstructs high-frequency information. With the increasing SR factor, we can distinguish chart rays closer to the center. The Moire effect visible on the LR images (a) disappears completely for the SR factor 1.5 and more.

8.7 Conclusions

This chapter described a general method for blind deconvolution and resolution enhancement. We have shown that the SR problem permits a stable solution even in the case of unknown blurs. The fundamental idea is to split radiometric deformations into sensor and volatile parts and assume that only the sensor part is known. We can then construct a convex functional using the LR images and observe that the volatile part minimizes this functional. Due to resolution decimation, the functional is not strictly convex and reaches its minimum on a subspace that depends on the integer SR factor. We have also extended our conclusions to rational factors by means of polyphase decomposition. To achieve a robust solution, the regularized energy minimization approach was adopted. The proposed BSR method goes far beyond the standard SR techniques. The introduction of volatile blurs makes the method particularly appealing to real situations. While reconstructing the blurs, we estimate not only subpixel shifts but also any possible blurs imposed by the acquisition process. To our knowledge, this is the only method that can perform deconvolution and resolution enhancement simultaneously. Several experiments with promising results give the reader a precise notion of the quality of the BSR methodology and wide applicability of the proposed algorithm to all sorts of real problems.

Acknowledgment

This work has been supported by the Czech Ministry of Education and by the Spanish Ministries of Education and Health under the projects TEC2004-00834, TEC2005-24739-E,2004SOE184, No. 1M0572 (Research Center DAR), the bilateral project 2004CZ0009 of CSIC and AS CR, and by the Grant Agency of the Czech Republic under the projects No. 102/04/0155. F. Šroubek was also supported by the Spanish States Secretary of Education and Universities fellowship.

References

[1] B. Zitová and J. Flusser. Image registration methods: A survey. *Image and Vision Computing*, 21: pp. 977–1000, 2003.

[2] R.L. Lagendijk, J. Biemond, and D.E. Boekee. Identification and restoration of noisy blurred images using the expectation-maximization algorithm. *IEEE Transactions on Acoustic, Speech, and Signal Processing*, 38(7): pp. 1180–1191, July 1990.

[3] S.J. Reeves and R.M. Mersereau. Blur identification by the method of generalized cross-validation. *IEEE Transactions on Image Processing*, 1(3): pp. 301–311, July 1992.

[4] T.F. Chan and C.K. Wong. Total variation blind deconvolution. *IEEE Transactions on Image Processing*, 7(3): pp. 370–375, March 1998.

[5] M. Haindl. Recursive model-based image restoration. In *Proceedings of the 15th International Conference on Pattern Recognition*, vol. III, pp. 346–349. IEEE Press, 2000.

[6] D. Kundur and D. Hatzinakos. Blind image deconvolution. *IEEE Signal Processing Magazine*, 13(3): pp. 43–64, May 1996.

[7] D. Kundur and D. Hatzinakos. Blind image deconvolution revisited. *IEEE Signal Processing Magazine*, 13(6): pp. 61–63, November 1996.

[8] G. Harikumar and Y. Bresler. Perfect blind restoration of images blurred by multiple filters: Theory and efficient algorithms. *IEEE Transactions on Image Processing*, 8(2): pp. 202–219, February 1999.

[9] G.B. Giannakis and R.W. Heath. Blind identification of multichannel FIR blurs and perfect image restoration. *IEEE Transactions on Image Processing*, 9(11): pp. 1877–1896, November 2000.

[10] Hung-Ta Pai and A.C. Bovik. On eigenstructure-based direct multichannel blind image restoration. *IEEE Transactions on Image Processing*, 10(10): pp. 1434–1446, October 2001.

[11] G. Panci, P. Campisi, S. Colonnese, and G. Scarano. Multichannel blind image deconvolution using the bussgang algorithm: Spatial and multiresolution approaches. *IEEE Transactions on Image Processing*, 12(11): pp. 1324–1337, November 2003.

[12] F. Šroubek and J. Flusser. Multichannel blind iterative image restoration. *IEEE Transactions on Image Processing*, 12(9): pp. 1094–1106, September 2003.

[13] F. Šroubek and J. Flusser. Multichannel blind deconvolution of spatially misaligned images. *IEEE Transactions on Image Processing*, 14(7): pp. 874–883, July 2005.

[14] S.C. Park, M.K. Park, and M.G. Kang. Super-resolution image reconstruction: A technical overview. *IEEE Signal Processing Magazine*, 20(3): pp. 21–36, 2003.

[15] S. Farsui, D. Robinson, M. Elad, and P. Milanfar. Advances and challenges in super-resolution. *International Journal of Imaging Systems and Technology*, 14(2): pp. 47–57, August 2004.

[16] R.C. Hardie, K.J. Barnard, and E.E. Armstrong. Joint map registration and high-resolution image estimation using a sequence of undersampled images. *IEEE Transactions on Image Processing*, 6(12): pp. 1621–1633, December 1997.

[17] C.A. Segall, A.K. Katsaggelos, R. Molina, and J. Mateos. Bayesian resolution enhancement of compressed video. *IEEE Transactions on Image Processing*, 13(7): pp. 898–911, July 2004.

[18] N.A. Woods, N.P. Galatsanos, and A.K. Katsaggelos. Stochastic methods for joint registration, restoration, and interpolation of multiple undersampled images. *IEEE Transactions on Image Processing*, 15(1): pp. 201–213, January 2006.

[19] S. Farsiu, M.D. Robinson, M. Elad, and P. Milanfar. Fast and robust multiframe super resolution. *IEEE Transactions on Image Processing*, 13(10): pp. 1327–1344, October 2004.

[20] E. Shechtman, Y. Caspi, and M. Irani. Space-time super-resolution. *IEEE Transactions on Pattern Analysis and Machine Intelligence*, 27(4): pp. 531–545, April 2005.

[21] N. Nguyen, P. Milanfar, and G. Golub. Efficient generalized cross-validation with applications to parametric image restoration and resolution enhancement. *IEEE Transactions on Image Processing*, 10(9): pp. 1299–1308, September 2001.

[22] N.A. Woods, N.P. Galatsanos, and A.K. Katsaggelos. EM-based simultaneous registration, restoration, and interpolation of super-resolved images. In *Proceedings IEEE ICIP*, vol. 2, pp. 303–306, 2003.

[23] W. Wirawan, P. Duhamel, and H. Maitre. Multi-channel high resolution blind image restoration. In *Proceedings IEEE ICASSP*, pp. 3229–3232, 1999.

[24] A.E. Yagle. Blind superresolution from undersampled blurred measurements. In *Advanced Signal Processing Algorithms, Architectures, and Implementations XIII*, vol. 5205, pp. 299–309, Bellingham, 2003. SPIE.

[25] D.S. Biggs, C.L. Wang, T.J. Holmes, and A. Khodjakov. Subpixel deconvolution of 3D optical microscope imagery. In *Proceedings SPIE*, vol. 5559, pp. 369–380, October 2004.

[26] D. Capel. *Image Mosaicing and Super-Resolution*. Springer-Verlag, New York, 2004.

[27] R. Molina, M. Vega, J. Abad, and A.K. Katsaggelos. Parameter estimation in Bayesian high-resolution image reconstruction with multisensors. *IEEE Transactions on Image Processing*, 12(12): pp. 1655–1667, December 2003.

[28] F. Šroubek and J. Flusser. Resolution enhancement via probabilistic deconvolution of multiple degraded images. *Pattern Recognition Letters*, 27: pp. 287–293, March 2006.

[29] Y. Chen, Y. Luo, and D. Hu. A general approach to blind image super-resolution using a PDE framework. In *Proceedings SPIE*, vol. 5960, pp. 1819–1830, 2005.

[30] T.J. Moore, B.M. Sadler, and R.J. Kozick. Regularity and strict identifiability in MIMO systems. *IEEE Transactions on Signal Processing*, 50(8): pp. 1831–1842, August 2002.

[31] G. Aubert and P. Kornprobst. *Mathematical Problems in Image Processing*. Springer-Verlag, New York, 2002.

9

Blind Reconstruction of Multiframe Imagery Based on Fusion and Classification

Dimitrios Hatzinakos, Alexia Giannoula, Jianxin Han

Department of Electrical and Computer Engineering, University of Toronto, Toronto, Ontario, M5S 3G4, Canada

e-mail: dimitris@comm.utoronto.ca

Abstract

This chapter deals with the problem of restoring an unknown image scene when multiple degraded acquisitions are available. To address this problem, a recursive filtering framework relying on classification and fusion of the observed image frames is considered and two blind multiframe restoration algorithms are described. Experiments with simulated and real data clearly indicate the feasibility and promise of the methods.

9.1 Introduction

Processing of multiframe imagery, i.e., capturing of the same scene by different sensors and fusion of the gathered data for achieving a better understanding of a given situation, is widely performed in satellite remote sensing, computer vision, and military and surveillance applications. In such cases, the objective is, usually, to obtain an enhanced representation of the true scene from a sequence of possibly degraded acquisitions. Similar cases can be encountered in medical imaging. For example, in magnetic resonance imaging (MRI), the availability of multiple fast scans of the same organ imposes the challenging task of their combination into a higher-quality version of the true original image. In many other image capturing applications, the imaging sensors generate poor-quality and possibly poor-resolution scene representations. Usually, multiple low-resolution frames of the scene are captured through sub-pixel motion of the camera. In addition, these images suffer from sensor and optical blurring (motion-induced or out-of-focus) and noise (quantization errors, sensor measurement, model errors, and so on). Therefore, image restoration techniques are required to be applied on the degraded data.

Several iterative or noniterative blind restoration algorithms have appeared in the literature [1–10]. The problem of restoring the original image from a degraded observation and incomplete information about the blur is called *blind deconvolution*. In many practical applications, limited knowledge of the blurring process is available and thus, techniques of this kind are required. Most of the methods presented in the above works, however, deal with the single-frame restoration problem or, at most, combine individual restorations (resulting from single-frame algorithms) to provide a more regularized global result. Particularly for the latter case, it was pointed out in [1], that if blind image restoration is performed individually on each frame followed by fusion and classification, then a more regularized solution can be produced. This technique is limited, though, by constraints imposed due to its single-frame nature, e.g., the constraint of finite support. In general, even though the problem of multiframe blind image restoration has been theoretically defined and dealt with, nevertheless, from a practical point of view, it is only recently that promising approaches have appeared in the literature, [11–17].

A number of *superresolution restoration* methods have also appeared lately, but they aim at reconstructing a higher-resolution image from a sequence of undersampled (low-resolution) and degraded frames, and they commonly involve a three-step process, i.e.: registration, interpolation, and restoration [18–22]. In this chapter, we will be focusing on the restoration process itself, which, in fact, can be simply considered a special case of superresolution imaging, and therefore we do not require the image resolution (dimensions) to increase. Two novel blind restoration approaches are proposed, where recursive inverse filtering is performed with multiple variable finite impulse response (FIR) filters. In both algorithms classification of the recursive filter outputs plays an important role in producing "desired" features for blind optimization and filter coefficient adjustment. Nonlinear clustering methods of signal values such as classification have been found to possess strong blind properties in general. On the other hand, in a multiframe scenario, fusion of multiple frames is another natural mechanism to extract salient features for blind restoration. In our treatment these two mechanisms, that is, classification and fusion, are merged, either directly or indirectly in order to produce a more regularized result.

Our first approach, the Recursive Inverse Filtering with Normal-density Mixtures (RIF-FNM), utilizes classification as a means to effectively fuse multiple frames and then let the differences between the fused and individual frames drive the reconstruction filter parameters. On the other hand, the second approach, the Fusion and Classification with Recursive Inverse Filtering (FAC-RIF), utilizes the differences between classified characteristics of the individual frames and classified characteristics of a fused image to drive the filter adaptation. Even though specific classification and fusion methodologies are proposed, nevertheless, different possible implementations of the proposed general architectures are possible.

9.2 System Overview

In most practical cases, each of the obtained image frames are subject to different distortion and noise. In general, we may assume that we obtain M frames $y_m[n_1, n_2]$ of size $N_{m,1}xN_{m,2}$ where $m = 0, 1, 2, \ldots, M - 1$. The degradation image model is depicted as follows:

$$y_m[n_1, n_2] = (x_m * h_m)[n_1, n_2] + w_m[n_1, n_2] \qquad (9.1)$$

where * denotes two-dimensional linear convolution, $x_m[n_1, n_2]$ is the true frame, $h_m[n_1,n_2]$ is the distorting Point Spread Function(PDF), and $w_m[n_1,n_2]$ stands for the additive zero-mean background noise in the m-th obtained image. The PDF and noise are supposed to be different and statistically independent between different frames; however, in the sequel we will assume that either

$$x_0[n_1, n_2] = \cdots = x_{M-1}[n_1, n_2] = x[n_1, n_2] \qquad (9.2)$$

or

$$x_0[n_1, n_2] \simeq x_1[n_1, n_2] \simeq \cdots \simeq x_{M-1}[n_1, n_2]. \qquad (9.3)$$

In other words, the observed frames are distorted versions of either the same original scene, or slightly different scenes, which, however, will provide useful complimentary information by means of fusion. The objective is: given

$y_m[n_1, n_2]$ of size $N_{m,1} x N_{m,2}$ to obtain an estimate $\hat{x}_m[n_1, n_2]$ of $x_m[n_1, n_2]$ of the same size $N_1 x N_2$ where $N_1 \geq N_{m,1}$, $N_2 \geq N_{m,2}$ and $m = 0, 1 \ldots, M - 1$ assuming no information of the distorting parameters is available. In many practical applications, this ill-conditioned problem [8] is generally addressed by minimizing a cost function of the error between the true (or desired) image and a linear or nonlinear transformation (filtering) of the observed image. In those cases where the true images are not available, estimates based on *a priori* partial known or otherwise acquired information is utilized instead. For example in a multiframe scenario, the fusion of different frames may provide reliable characteristics of the desired image, and the following optimization problem may be constructed:

$$\hat{x}_m = \operatorname*{argmin}_{\{u_m\}} J(u_m) = \operatorname*{argmin}_{\{u_m\}} Q\left[\hat{x}_d - P(y_m, u_m)\right] \qquad (9.4)$$

where Q and P are appropriate linear or nonlinear transformations of the observed image y_m and the desired image characteristics $\hat{x}_d$ are parameterized by $\{u_m\}$. In this chapter, a recursive implementation is considered to obtain $\{u_m\}$ and then

$$\hat{x}_m = y_m * u_m \qquad (9.5)$$

by means of 2-D FIR inverse filtering. Two algorithms are presented based on classification and fusion of the recursive FIR filter outputs. In both cases the outputs of the recursive filters are first classified to produce an L-level image $\hat{x}_m^c$, $m = 0, 1, \ldots, M - 1$. The first algorithm uses the image classification as a means of fusing the M filter outputs $\hat{x}_m$, $m = 0, 1, \ldots, M - 1$ and thus producing an estimate of the desired image. Then, the error between $\hat{x}_d$ and $\hat{x}_m$ is used to update the corresponding recursive filters by means of a pseudo-Newton optimization procedure. The second algorithm utilizes a joint fusion–classification process to obtain a K-level desired image $\hat{x}_d^c$. Then, the error between $\hat{x}_d^c$ and $\hat{x}_m^c$ is used to update the corresponding recursive filters by means of the pseudo-Newton optimization procedure. Block diagrams of the two schemes are depicted in Figure 9.1 and Figure 9.2.

9.3 Recursive Inverse Filtering with Finite Normal-Density Mixtures (RIF-FNM)

9.3.1 Image Modeling Using Finite Mixture Distributions

The implicit goal of the proposed recursive algorithm is to effectively classify the filtered images $\hat{x}_m$, $m = 0, \ldots, M - 1$ in a fashion that the subsequent

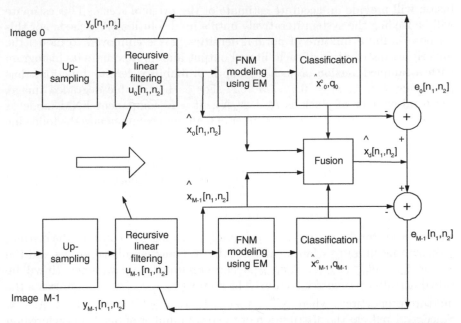

FIGURE 9.1: Block diagram of image restoration scheme 1 (RIF-FNM).

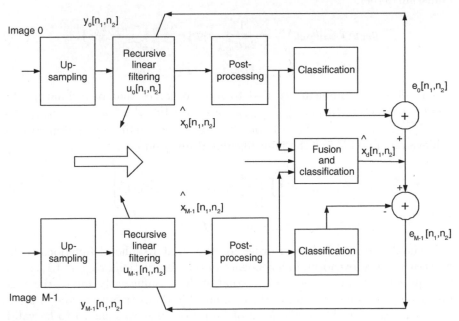

FIGURE 9.2: Block diagram of image restoration scheme 2 (FAC-RIF).

fusion will provide an accurate estimate of the original scene. This estimate will be feeding the system iteratively until a final solution is reached. For this purpose, a finite mixture of normal densities [23] is employed to model the underlying distribution of each filtered output and approximate its histogram with a sampled version of such a smooth probability density function, thus producing x_m^{FNM}, $m = 0, \ldots, M - 1$. For each available degraded image, the total number of pixels is $N = N_1 N_2$. If a K-component FNM model is assumed, where K denotes the number of Gaussian components, the following sum is formed for each filtered image (given any pixel $\hat{x}_{m,i}$):

$$x_m^{FNM}(\hat{x}_{m,i}|\psi_{\mathbf{m}}) = \sum_{k=1}^{K} \pi_{m,k} g_{m,k}(\hat{x}_{m,i}|\theta_{\mathbf{m,k}}), \qquad (9.6)$$

$$m = 0, \ldots, M - 1, \quad i = 1, \ldots, N,$$

where the vector $\theta_{\mathbf{m}} = [\mu_{m,1}, \ldots, \mu_{m,K}, \sigma_{m,1}^2, \ldots, \sigma_{m,K}^2]^T$ contains the distinct unknown mean and variance values of the K Gaussian distributions and the vector $\psi_{\mathbf{m}} = [\theta_{\mathbf{m}}^T, \pi_{m,1}, \ldots, \pi_{m,K}]$ denotes all unknown parameters (it will be called hereafter *parameter vector*). The vector $(\pi_{m,1}, \ldots, \pi_{m,K})$ represents the mixture proportions, where $\sum_{k=1}^{K} \pi_{m,k} = 1$, $\pi_{m,k} \geq 0$ for $m = 0, \ldots, M - 1$. Since $\pi_{m,k}$ reflects the distribution of the total number of pixels in each region (class or component), it can be interpreted as a prior probability of the pixel classes. The k-th component density of the mixture can be described by the Gaussian kernel:

$$g_{m,k}(\hat{x}_{m,i}|\theta_{\mathbf{m,k}}) = \frac{1}{\sqrt{2\pi}\sigma_{m,k}} exp\{-\frac{(\hat{x}_{m,i} - \mu_{m,k})^2}{2\sigma_{m,k}^2}\}, \qquad (9.7)$$

$$m = 0, \ldots, M - 1, \quad i = 1, \ldots, N.$$

Therefore, the problem we need to solve becomes that of estimating the model parameters $(\mu_{m,k}, \sigma_{m,k}^2, \pi_{m,k})$, based on the observations $\{\hat{x}_{m,i}\}$, for all x_m^{FNM}, $m = 0, \ldots, M - 1$ and $k = 1, \ldots, K$. This estimation can be achieved by maximizing the joint likelihood function:

$$L_m(\psi_{\mathbf{m}}) = \prod_{i=1}^{N} \left[\sum_{k=1}^{K} \pi_{m,k} g_{m,k}(\hat{x}_{m,i}|\theta_{\mathbf{m,k}}) \right], \qquad (9.8)$$

$$m = 0, \ldots, M - 1.$$

It should be noted that in order for the joint likelihood functions to be written in the above form, independence of the observations $\hat{x}_{m,i}$, $i = 1, \ldots, N$ is required. It was shown in [24] that if N is sufficiently large, the true image distribution converges to the estimated FNM distribution, and in many practical situations [3, 24] the form of Equation (9.8) was shown to be valid. In our problem, the validity of the above assumption is also experimentally verified in Section 9.7.

Explicit calculation of maximum likelihood (ML) estimates is usually not possible. However, it does make available the general class of iterative procedures, widely known as *expectation maximization* (EM) algorithms [25]. In its general form, the EM algorithm first calculates the posterior Bayesian probabilities of the data through the observations and the current parameter estimates (*E*-step) and then updates the parameter estimates using generalized mean ergodic theorems [26] (*M*-step). The successive iterations increase the likelihood of the model parameters. ML estimation of the FNM parameter vector has been efficiently applied in [27, 28] using expectation maximization. Alternative numerical techniques for performing ML estimation have appeared in the literature, such as the classification maximization (CM) algorithm, competitive learning, and the K-means algorithm [29]. However, the cost to be paid for an efficient method is usually biased estimates [23]. Although the EM algorithm may converge slowly, it is numerically stable and it was thus incorporated in the proposed scheme for estimating the Gaussian parameter vector.

Application of the EM algorithm on the logarithm of Equation (9.8), i.e., the log-likelihood function, yields the following updating rules for the ML estimates of the FNM model [25]:

$$\mu_{m,k}^{r+1} = \frac{\sum_{i=1}^{N} \hat{x}_{m,i} p(\theta_{\mathbf{m,k}}^{\mathbf{r}}|\hat{x}_{m,i})}{\sum_{i=1}^{N} p(\theta_{\mathbf{m,k}}^{\mathbf{r}}|\hat{x}_{m,i})} \tag{9.9}$$

$$var_{m,k}^{r+1} = \frac{\sum_{i=1}^{N} p(\theta_{\mathbf{m,k}}^{\mathbf{r}}|\hat{x}_{m,i})(\hat{x}_{m,i} - \mu_{m,k}^{r+1})(\hat{x}_{m,i} - \mu_{m,k}^{r+1})^T}{\sum_{i=1}^{N} p(\theta_{\mathbf{m,k}}^{\mathbf{r}}|\hat{x}_{m,i})} \tag{9.10}$$

$$\pi_{m,k}^{r+1} = \frac{1}{N} \sum_{i=1}^{N} p(\theta_{\mathbf{m,k}}^{\mathbf{r}}|\hat{x}_{m,i}) \tag{9.11}$$

$$m = 0, \ldots, M - 1, \quad k = 1, \ldots, K, \quad r = 0, 1, \ldots$$

where r denotes the current iteration, $\mu_{m,k}$, $var_{m,k}=\sigma_{m,k}^2$ and $\pi_{m,k}$ represent the mean value, the variance value, and the mixing factor, correspondingly, for the k-th Gaussian component in image $\hat{x}_m$. The term $p(\theta_{\mathbf{m,k}}|\hat{x}_{m,i})$ denotes the posterior probability, which, by using the Bayes rule, can be written as follows:

$$p(\theta_{\mathbf{m,k}}^{\mathbf{r}}|\hat{x}_{m,i}) = \frac{g_{m,k}(\hat{x}_{m,i}|\theta_{\mathbf{m,k}}^{\mathbf{r}})\pi_{m,k}^r}{x_m^{FNM}(\hat{x}_{m,i}|\psi_{\mathbf{m}}^{\mathbf{r}})}. \tag{9.12}$$

The formulas defined in (9.9) through (9.11) provide the updating rules for the computation of the FNM parameter vector (mean, variance values, and mixing proportions of the local Gaussian components). Initial values $\mu_{m,k}^0, var_{m,k}^0$, and $\pi_{m,k}^0$ (iteration $r = 0$) are approximated using the K-means algorithm.

9.3.2 Pixel Classification

The process of *classification* is to assign the pixels of each frame $\hat{x}_m$, $m = 0, \ldots, M-1$ to its appropriate category, i.e., to the appropriate Gaussian component of the mixture model, based on the estimated FNM parameters described in the previous subsection. Classification will be shown to constitute an essential part in each iteration of the restoration algorithm, by enabling efficient fusion of the filtered frames. In other terms, a more accurately classified image will contribute to a greater extent to the formation of the final fused output. A measure of the accuracy of the classification decision will be described in the following section. Pixel classification of the individual frames involves two stages:

- First, a single-pass (soft) maximum likelihood (ML) classification is performed, where the image pixels $\hat{x}_{m,i}$, $m = 0, \ldots, M-1$, $i = 1, \ldots, N$ are assigned to the k-th Gaussian component with the highest individual likelihood $g_{m,k}(\hat{x}_{m,i}; \theta_{\mathbf{m,k}})$, described by Equation (9.7). Maximization of the logarithm of (9.7) is equivalent to assigning a pixel $\hat{x}_{m,i}$ to the class (Gaussian component) k, such that:

$$log(\sigma_{m,k}) + (\hat{x}_{m,i} - \mu_{m,k})^2 / 2\sigma_{m,k}^2 \qquad (9.13)$$

  yields its minimum value for $k = 1, \ldots, K$. This way of classifying pixels is known to be susceptible to noise, resulting in high classification errors.

- A local-based classification follows that will refine the soft-classification decision of the previous step. An approach similar to [24] is undertaken, where the motivation lies in the concept of *relaxation labeling*[1] [30]. A pixel $\hat{x}_{m,i}$, $i = 1, \ldots, N$ is randomly visited and is classified in class k that minimizes

$$log(\sigma_{m,k}) - log(p_m(c_{m,i} = k|nbh_i)) + (\hat{x}_{m,i} - \mu_{m,k})^2 / 2\sigma_{m,k}^2 \qquad (9.14)$$

  where $p_m(c_{m,i} = k|nbh_{m,i})$ represents the conditional prior of the class (label) $c_{m,i}$ of a pixel $\hat{x}_{m,i}$, given the classification of its neighboring pixels,[2] e.g., a 3×3 neighborhood $nbh_{m,i}$. This probability is analogous to the cluster prior probability $\pi_{m,k}$. In fact, it can be simply described by the proportion of the pixels in this neighborhood that have been assigned to the same class as that of the pixel $\hat{x}_{m,i}$.

Minimizing Equation (9.14) is equivalent to maximizing the joint likelihood of a pixel $\hat{x}_{m,i}$ and its class label $c_{m,i}$, conditioned by the local classification

---

[1]The structure of relaxation labeling relies on two basic considerations: 1) decomposition of a global computation scheme into a network performing simple local computations, and 2) utilization of appropriate local class regularities for resolving ambiguities.

[2]It has been assumed that pixels can be decomposed into pixel images $\hat{x}_{m,i}$ and context images $c_{m,i}$ (i.e., pixel classes).

information nbh_i, i.e.:

$$p_{m,k}(\hat{x}_{m,i}, c_{m,i}|nbh_{m,i}) = p_m(c_{m,i} = k|nbh_{m,i})p_m(\hat{x}_{m,i}|c_{m,i}, nbh_{m,i}) =$$

$$p_m(c_{m,i} = k|nbh_{m,i})g_{m,k}(\hat{x}_{m,i}; \theta_{\mathbf{m,k}}) \tag{9.15}$$

similarly to the ML classification defined in Equation (9.13). It has been shown in [30] that when the number of pixels that change their class labels becomes very small, relaxation labeling converges to a steady classification point, where the classified images $\hat{x}_m^c$, $m = 0, \ldots, M - 1$ have been finally formed and no further pixels need to update their class labels.

9.3.3 ML-Based Image Fusion

In the subsequent fusion of the filtered images $\hat{x}_m$, $m = 0, \ldots, M - 1$, the most significant features are expected to be extracted from the individual restorations, in order to produce a more consistent global result, with respect to the normal mixture model that characterizes the original scene. For determining the appropriate weights that will be employed in the fusion process, the final classification output of the previous subsection will be taken into account. For this reason, an *average local consistency (ALC) measure* is employed for each classified image $\hat{x}_m^c$, $m = 0, \ldots, M - 1$ to link consistent classification (labeling) to global optimization [30]. It was shown that when this spatial compatibility measure is symmetric and attains a local maximum for some specific classification assignment, then this is considered a consistent labeling [30,31]. In particular, a weight q_m is formed for each classified image, as the summation of the joint likelihoods $p_{m,k}(\hat{x}_{m,i}, c_{m,i}|nbh_{m,i})$ of Equation (9.15), i.e.:

$$q_m = ALC_m = \sum_{i=1}^{N}\left[\sum_{k=1}^{K}\delta(c_{m,i}, k)p_{m,k}(\hat{x}_{m,i}, c_{m,i}|nbh_{m,i})\right], \tag{9.16}$$

$$m = 0, \ldots, M - 1$$

where $\delta(c_{m,i}, k)$ is the *Kronecker delta*. In other terms, the ALC_m metric has a local nature (by considering classification information in a local neighborhood), but overall provides a global assessment of the classification decision (by summing over all pixels).

By choosing to fuse the filtered images based on probabilistic compatibility constraints, we expect that the most accurately classified images (high q_m values) will contribute in the formation of the fused result to a greater extent, while images that possess low consistency values will not affect significantly the final output. Subsequently, the desired reconstructed fused image $\hat{x}_d$ is formed as the following weighted sum:

$$\hat{x}_d = \frac{\sum_{m=1}^{M} q_m \cdot \hat{x}_m}{\sum_{m=0}^{M-1} q_m}. \tag{9.17}$$

In the proposed recursive algorithm, the weights q_m, hence the fused image $\hat{x}_d$, are iteratively modified according to the updated FNM approximation and classification of the current (outer-loop) iteration (see the block diagram of Figure 9.1). It can be also readily concluded based on the above analysis, that appropriate FNM modeling and pixel classification are intimately related and consist of essential processes for more reliably fusing the filtered frames.

9.4   Optimal Filter Adaptation

Finally, the proposed methodology involves the adaptation of each FIR filter u_m, $m = 0, \ldots, M-1$ that will be fed back into the recursive algorithm. The cost functions J_m that will be used in the blind restoration process are defined as:

$$J_m = \sum_{i=1}^{N}(\hat{x}_{d,i} - \hat{x}_{m,i})^2, \quad m = 0, \ldots, M-1. \tag{9.18}$$

Taking into account that $\hat{x}_m = y_m * u_m$ (see Equation (9.5)) and its corresponding FNM modeling, each filtered output $\hat{x}_m$ can be approximated as follows:

$$\hat{x}_{m,i} \simeq \sum_{k=1}^{K} \pi_{m,k} g_{m,k}(\hat{x}_{m,i}; \theta_{\mathbf{m},\mathbf{k}}) = \sum_{k=1}^{K} \pi_{m,k}\{u_m * g_{m,k}^{p}(\hat{x}_{m,i}|\theta_{\mathbf{m},\mathbf{k}})\}, \tag{9.19}$$

$$i = 1, \ldots, N$$

where $g_{m,k}^{p}$ denotes the k-th normal component of the mixture approximation for the degraded images prior to filtering (note also that the term $\psi_{\mathbf{m}}$ has been dropped for convenience of notation). Similarly, by combining (9.5), (9.6), and (9.17), the fused image $\hat{x}_d$ can be expressed by the following formula:

$$\hat{x}_{d,i} = \sum_{m=0}^{M-1} q'_m \left[\sum_{k=1}^{K} \pi_{m,k}\{u_m * g_{m,k}^{p}(\hat{x}_{m,i}|\theta_{\mathbf{m},\mathbf{k}})\} \right], \quad i = 1, \ldots, N \tag{9.20}$$

where q'_m denotes the normalized fusion weights q_m, $m = 0, \ldots, M-1$, such that $\sum_{m=0}^{M-1} q'_m = 1$.

Differentiating the above equations (9.19) and (9.20) with respect to the filter coefficients $u_m(l)$, the corresponding partial derivatives are obtained:

$$\frac{\partial \hat{x}_{m,i}}{\partial u_m(l)} = \sum_{k=1}^{K} \pi_{m,k} g_{m,k}^{p}(\hat{x}_{m,i-l}; \theta_{\mathbf{m},\mathbf{k}}) \tag{9.21}$$

$$\frac{\partial \hat{x}_{d,i}}{\partial u_m(l)} = \sum_{m=0}^{M-1}\sum_{k=1}^{K} q'_m \pi_{m,k} g_{m,k}(\hat{x}_{m,i-l}; \theta_{\mathbf{m},\mathbf{k}}) \tag{9.22}$$

$$m = 0, \ldots, M - 1, \quad i = 1, \ldots, N.$$

The class of quasi-Newton gradient optimization methods was selected for minimizing the cost function corresponding to each individual frame [32], i.e.:

$$u_m^{t+1} = u_m^t - \mu S_m^t \cdot \nabla J_m^t[u_m], \tag{9.23}$$

$$m = 0, \ldots, M - 1, \quad t = 0, 1, \ldots$$

where t denotes the current iteration, $\nabla J_m[u_m]$ is a vector containing the gradient of the cost function J_m with respect to the filter coefficients $\{u_m(l)\}$ described in (9.21) and (9.22), and μ is a suitable update step-size that may be chosen experimentally or derived using a line-search algorithm along the Newton direction [33]. The matrices S_m are appropriately constructed to approximate the inverse *Hessian* H_m^{-1} of each J_m [34], as in practical applications it may be computationally inefficient to explicitly derive and invert the Hessian H_m. In fact, the motivation behind the adoption of a quasi-Newton optimization technique in the proposed restoration algorithm was to reduce computational complexities (by avoiding the evaluation of second derivatives and matrix inversions), while maintaining good convergence properties. Specifically, we can write:

$$\delta_m^t = -S_m^t \nabla J_m^t[u_m] \quad (Newton \; direction) \tag{9.24}$$

$$u_m^{t+1} = u_m^t + \mu \, \delta_m^t \tag{9.25}$$

and

$$\gamma_m^t = \nabla J_m^{t+1}[u_m] - \nabla J_m^t[u_m] \tag{9.26}$$

for $m = 0, \ldots, M - 1$ and then make an additive correction to S_m^t of the form:

$$S_m^{t+1} = S_m^t + C_m^t \tag{9.27}$$

for $t = 0, 1, \ldots$, where C_m^t represents a correction matrix that satisfies the conditions $S_m^{t+1}\gamma_m^i = \delta_m^i$, for $0 \leq i \leq t$. Initially, the approximate matrices S_m^0 for $m = 0, \ldots, M - 1$ and $t = 0$ are set equal to the $N_u \times N_u$ identity matrix I_{N_u}. For updating the matrices S_m^{t+1}, the Davidon-Fletcher-Powell (DFP) or the Broyden-Fletcher-Goldfarb-Shanno (BFGS) formulas may be employed [32, 33].

The filters are initialized with a value of unity in the center of the FIR window and zero elsewhere (discrete unit impulses) and are normalized at each iteration in order to sum up to unity. Finally, the deviation in the classification decision of each image, from one iteration to the next, has been set as the stopping criterion for the iterative restoration algorithm. In particular, if less than 5% of the image pixels change their class assignments with respect to the previous iteration, then the corresponding frame is considered converged and restored. However, it should be noted that if excessive noise amplification starts being observed visually (see Section 9.5), then the algorithm may be terminated before the aforementioned criterion is reached.

9.5   Effects of Noise

The effects of noise for classical linear and nonlinear image restoration problems have been studied in [2, 26, 35]. Blind deconvolution is well known to be an ill-posed problem, as small perturbations of the given data produce large deviations in the resulting solution. Direct inverse filtering attempts to restore the image by inverting the PSF. However, due to the presence of noise the problem becomes ill-posed. Specifically, the direct inverse of the PSF transfer function often has a large magnitude at high frequencies and therefore, excessive amplification of the noise at these frequencies appears. Analysis of the effects of noise on the cost functions (defined in the previous section) would provide useful information about the behavior of the proposed restoration algorithm in practical situations.

Because the cost functions J_m, $0 = 1, \ldots, M-1$ are nonlinear (see Equation (9.18)), their minima in the presence of noise $u^*_{m,noise}$, $m = 0, \ldots, M-1$ are hard to represent in terms of the corresponding minima u^*_m in the noiseless scenario. However, if the continuity of J_m with respect to u_m is considered, it can be concluded that the value of the cost functions in noisy conditions at the ideal points u^*_m reveals significant information about the degree of bias that has been introduced in the restored images. For the case of zero-mean stationary additive white Gaussian noise w_m, $m = 0, \ldots, M-1$ with variance σ^2_w, the bias can be calculated as follows: the cost functions of Equation (9.18) are first expanded using Equation (9.17), i.e.,

$$J_m = \sum_{i=1}^{N} (\hat{x}^2_{d,i} - 2\hat{x}_{d,i}\hat{x}_{m,i} + x^2_{m,i}) \Rightarrow$$

$$J_m = \sum_{i=1}^{N} [\sum_k \sum_{j \neq k} q'_k q'_j \hat{x}_{k,i} \hat{x}_{j,i} + \sum_k q'^2_k \hat{x}^2_{k,i} - 2\hat{x}_{m,i} \sum_k q'_k \hat{x}_{k,i} + \hat{x}^2_{m,i}] \quad (9.28)$$

for $m = 0, \ldots, M-1$. Each filtered image $\hat{x}_m$ is given by:

$$\hat{x}_m = y_m * u_m = \tilde{x}_m + \tilde{w}_m, \quad m = 0, \ldots, M-1 \quad (9.29)$$

where

$$y_m = \tilde{y}_m + w_m$$

$$\tilde{x}_m = \tilde{y}_m * u_m$$

$$\tilde{w}_m = \tilde{w}_m * u_m. \quad (9.30)$$

Rewriting the cost functions of (9.28) in terms of $\tilde{x}$ and $\tilde{w}$ yields:

$$J_m = \sum_{i=1}^{N} [\sum_k \sum_{j \neq k} q'_k q'_j (\tilde{x}_{k,i} + \tilde{w}_{k,i})(\tilde{x}_{j,i} + \tilde{w}_{j,i}) + \sum_k q'^2_k (\tilde{x}_{k,i} + \tilde{w}_{k,i})^2 -$$

$$-2(\tilde{x}_{m,i}+\tilde{w}_{m,i})\sum_k q'_k(\tilde{x}_{k,i}+\tilde{w}_{k,i})+(\tilde{x}_{m,i}+\tilde{w}_{m,i})^2], \quad m=0,\ldots,M-1. \quad (9.31)$$

Since w_m, $m=0,\ldots,M-1$ are zero-mean and Gaussian processes, it can be trivially shown that $\tilde{w}_m$ are also zero-mean and Gaussian processes (filtered versions of w_m) with variance

$$\sigma^2 = \sigma_w^2 \sum_{\forall k} u_k^2. \quad (9.32)$$

Using the above results and the fact that $E\{\tilde{w}_m\}=0$ (zero-mean noise), the expectation (bias) of the cost functions in the presence of noise, at the true inverse PSF u_m^*, is evaluated. It should be noted that $\tilde{x}_m = u_m^* * \tilde{y}_m = x_m$ for $m=0,\ldots,M-1$. Therefore:

$$E\{J_m(u^*)\} = \sum_{i=1}^{N}[(x_{m,i}-\sum_k q'_k x_{k,i})^2 + \sigma^2(1-\sum_k q_k^2)], \quad (9.33)$$

$$m=0,\ldots,M-1.$$

It can be clearly seen that the bias is a function of the true (undistorted) images x_k, $k=0,\ldots,M-1$, the fusion weights q'_k, $k=0,\ldots,M-1$, the variance of the noise σ_w^2, and the energy of the optimal coefficients u_m^*. By observing Equation (9.33) one may conclude that the effect of noise is small when each filtered frame does not deviate significantly from the fused image. The contribution of the classification-based fusion weights in an accurate fused result is therefore essential and is analytically established using (9.33). Furthermore, it can be easily seen that the bias is proportional to the variance of the noise σ_w^2. For a fixed value of σ_w^2, the noise has little effect on the restoration if the variance of the inverse, of the ideal PSF is small, i.e., if the inverse PSF is lowpass. Practical PSFs are generally lowpass and their associated inverse transfer functions are highpass, such that noise amplification is expected in the solution.

9.6 The Fusion and Classification Recursive Inverse Filtering Algorithm (FAC-RIF)

A multiframe ad-hoc blind image restoration approach is proposed in this section. The proposed algorithm which is called the Fusion and Classification Recursive Inverse Filtering (FAC-RIF) algorithm requires the availability of at least two different frames of the degraded image all having the same spatial resolution. Recursive FIR filtering is applied separately to each of the image frames. The outputs of the filters are then fused and classified jointly and separately to a predetermined number of support region levels. Finally, the error between the classified output of each filter and their fusion–classification output is used to control the adaptation of the corresponding FIR filter parameters. By fusing the images in the classification stage we hope to extract the salient features from each of the different restorations to produce a more regularized overall result. The block diagram of the proposed reconstruction scheme is depicted in Figure 9.2. A description of the basic processing stages of the proposed algorithm follows.

9.6.1 The Iterative Algorithm

At iteration $t + 1$, the adjustment of the coefficients $u_m^{t+1}[n_1, n_2]$ or equivalently, by adopting a 1-D notation, of the pixel $u_m^{t+1}[i]$, $i = 1, \ldots, N_u$ of the recursive filters $m = 0, \ldots, M - 1$ is carried out using a gradient optimization according to the modified Newton method [36]:

$$u_m^{t+1} = u_m^t - \mu S_m^t \cdot \nabla J_m^t[u_m], \qquad (9.34)$$

$$m = 0, \ldots, M - 1, \;\; t = 0, 1, \ldots$$

where t denotes the current iteration, $\nabla J_m^t[u_m]$ is the vector containing the gradients of the cost function J_m with respect to the filter coefficients at iteration t. The gradient of J_m with respect to the coefficient $u_m(i)$ is defined as follows:

$$\nabla J_m^t[u_m(i)] = \frac{\partial J_m^t(u_m)}{\partial u_m(i)} = \left\{ \sum_l e_m^t(l)\hat{y}_m(l - n - 1) \right\} \qquad (9.35)$$

and

$$e_m^t(i) = C_{\text{fusion-class}}^t(i) - C_{m,\text{class}}^t(i) \qquad (9.36)$$

where, given K classification levels, $e_m^t(i)$ is the error between the classified–fused image pixel $C_{\text{fusion-class}}^t(i)$ and the individually classified image pixel $C_{m,\text{class}}^t(i)$, $m = 0, \ldots, M - 1$ at iteration t. As before, the computation of the inverse Hessian matrix can be obtained by means of the Davidon-Fletcher-Powell algorithm described which proceeds as follows, [36]:

$$v^t = u^{t+1} - u^t$$

$$z^t = \partial J^{t+1}(u) - J_m^t(u)$$

$$S^{t+1} = S^t + \frac{v^t v^{t^T}}{z^{t^T} v^t} - \frac{(S^t z^t) - (S^t z^t)^T}{z^{t^T} S^t z^t}. \qquad (9.37)$$

An approximation $S^0 = I$, the identity matrix, is usually taken for initialization.

The selection of a good stopping criterion for the recursive algorithm is very important, and it can guarantee that we obtain best results. Due to the inverse filtering nature of the FAC-RIF, we know that noise and artifact amplification such as ringing effects will be introduced, and this will in turn affect both the quality of the results as well as the classification process sooner or later at some iteration. Since each frame is subject to different distortion in general, the stopping criterion must be applied individually to each branch. Adaptation will continue until all frames have been sufficiently restored.

Our motivation for using the error of Equation (9.36) stems from the strong blind convergence properties of the image support constraint used in algorithms such as the NAS-RIF [1, 10]. There, it is assumed that the original image is of finite support against a uniform background, and blind image deconvolution is achieved by penalizing image values outside the image support which needs to be either *a priori* known or estimated. Assuming two classification levels (K=2) with binary finite support images, our error definition is equivalent to the application of the finite support constraint. In other words, the assumption made here is that the two-level classification of the fused image is equivalent to determining the support of the original image and therefore the difference between the classified–fused and individually classified images form an error equivalent to the finite support constraint. Some advantages of our approach compared to existing support determination approaches are i) it allows the automatic determination of arbitrarily shaped finite support images; ii) it is applicable to nonfinite support images as well, in which case the dispersive effect of blurring is viewed as a support leaking effect; iii) it refines recursively the estimated image support as the individual images' quality improves; iv) it takes into account not only blurring, but other image degradations as well (e.g., noise). It is important to note that our approach corresponds to the minimization of a nonconvex cost function in general and therefore local convergence of the algorithm may be experienced.

9.6.2 Prefiltering and Postfiltering Processing

Prior to filtering the low-resolution images may be interpolated to the same desired higher resolution. Postfiltering operations include image scaling and conditioning operations needed for the later stages of Fusion and Classification. This involves hard-limiting of each image between 0 and 255, and adjustment of the image mean and energy. Also, in case of major misalignment it may involve registration of the images (note, however, that small to moderate shifts among frames can be corrected by the recursive algorithm itself).

9.6.3 Classification

Classification is basically the assignment of a label to the pixels in an image. After some experimentation, we have decided to utilize a Markov Random Field (MRF)-based classification method which has been proposed in [37, 38] and has been previously used successfully in [1]. Accordingly, classifications in the image are determined by maximizing the *a posteriori* (MAP) distribution of a likelihood function which is based on a Gaussian observational model. This becomes equivalent to the minimization of the following energy function:

$$U(\hat{x}_{m,i}, C(i)) = U_{\text{obs}}(\hat{x}_{m,i}, C(i)) - U_{\text{prior}}(C(i)) \tag{9.38}$$

where

$$U_{\text{obs}}(\hat{x}_{m,i}, C(i)) = \frac{1}{2} ln\sigma^2_{C(i)} + \frac{1}{2\sigma^2_{C(i)}} \left(\hat{x}_{,i} - \mu_{C(i)}\right)^2 \tag{9.39}$$

and

$$U_{\text{prior}}(C(i)) = \beta n(C(i)) \tag{9.40}$$

and $C(i) = 1, 2, \ldots, K$ is the index of K classes and $n(C(i))$ is the number of neighbors of the image pixel $\hat{x}_{m,i}$ that are equal to the class level $C(i)$. The β is a user-specified nonnegative scalar parameter that controls the degree of spatial contextual information. $\sigma^2_{C(i)}$ and $\mu_{C(i)}$ are the variance and mean of class $C(i)$, respectively, which need to either be defined or estimated prior to classification. This is accomplished simply by applying a K-means clustering algorithm where the following average squared distance is minimized:

$$\sum_i ||\hat{x}_{m,i} - \mu_{C(i)}||^2, \; i = 1, \ldots, N. \tag{9.41}$$

9.6.4   Fusion-Based Classification

An assumption made at this point is that prior to fusion-based classification, the images from different sources have to be registered and have the same spatial solution. For the pixels $\{\hat{x}_{m,i}\}$, $m = 0, \ldots, M - 1$ we obtain a MAP classification estimate [1, 37] by minimizing the following energy function:

$$U(\hat{x}_{0,i}, \hat{x}_{1,i}, \ldots, \hat{x}_{M-1,i}, C(i)) = \sum_{m=0}^{M-1} U_{\text{obs}}(\hat{x}_{m,i}, C(i)) + U_{\text{prior}}(C(i)) \tag{9.42}$$

where $U_{\text{obs}}(\hat{x}_{m,i}, C(i))$ and $U_{\text{prior}}(C(i))$ are given by (9.39) and (9.40), respectively. The fusion–classification algorithm becomes simply a classification algorithm when only one image is considered.

9.6.5   Fusion of Reconstructed Images

It should be noted that step D does not provide a fused image, but only a classification based on fusion of the individual image frames. If a single fused image is desired then a classical fusion technique such as simple averaging or minimum variance fusion [1, 39] can be applied. In many applications, however, such as medical imaging, even small differences in the original frames may be important, and therefore observation of all reconstructed frames may be important.

FIGURE 9.3: The original image and the blurring function.

9.7 Experimental Results

In this section, experiments were performed to illustrate the efficiency of the two described restoration schemes with simulated and real cardiac data (CD).

With simulated data, to create a multiframe scenario, we take a high-resolution image which is then blurred, downsampled, and corrupted by noise to create a set of low-resolution frames of the same scene with different blur, noise, and relative shift. For example, the boat image (512×512), depicted in Figure 9.3, was first bilinearly interpolated to get a (1024×1024) size image. The image was then blurred and downsampled to create eight (256×256) images. Finally, random noise was added to the eight images, and four of them were used in the simulations.

Real multiframe images in the form of a sequence of eight (128×128) real cardiac CT images were provided by Canamet Inc. [40]. Due to the activity of the heart and the influence of electric noise, the images have been subjected to blurring and artifacts.

The iterative blind restoration scheme, RIF-FMN, was implemented by employing four out of the eight frames in both cases. Four (7×7) variable 2-D FIR filters were utilized. A four-component FNM model (see histogram matching in Figure 9.4) was found to be most appropriate for approximating the true histogram of the simulated individual frames, and a five-component FNM for the real data, based on a relative entropy minimization criterion (GRE). The iterative restoration process converged in less than 20 iterations, in both cases generating four reconstructed outputs. In the simulated data, shown in Figures 9.5 and 9.6, a final fused image is provided; however, with the real data, shown in Figure 9.7, only the individual restorations are depicted separately, as the originally observed frames may correspond to relatively different scenes. Therefore, the final fused image is not shown, although its role in feeding the recursive algorithm is essential. One can readily observe the sharpened details of each restored frame. The corresponding classified outputs after 29 iterations are also presented in the third column of Figure 9.7.

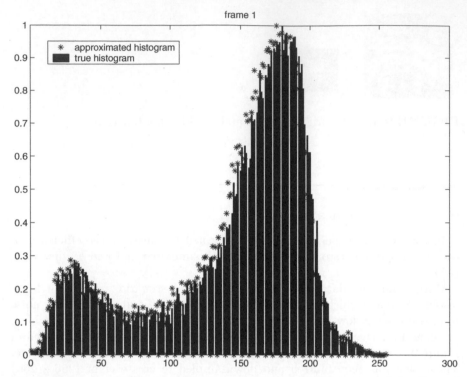

FIGURE 9.4: True histogram and approximated histogram of the "boats" image based on 4-FNN model.

The performance of the second algorithm, FAC-RIF, with simulated and real cardiac CT images, is depicted in Figures 9.5, 9.6, and 9.8. With the real data only frames 2 and 4 out of the four frames used by the FAC-RIF are illustrated; however, the performance with the other frames has been similar. Clearly, the restored images are sharper and show more detail. All the filters were again of size 7×7 pixels and with an initial value of 1 in the center and 0 elsewhere. The step size was taken constant between 0.08 and 0.2. Finally, in this implementation, without loss of generality, we have set the number K of classification levels at 2 while four neighbor pixels were used in forming Equation (9.40). Experiments with more than two classification levels have indicated that the performance can improve by using a higher number of levels; however, it is strongly dependent on the frame content under investigation.

In comparing the performance of the RIF-FNM and FAC-RIF algorithms, we observe that when the fused–reconstructed images are compared, as in Figures 9.5 and 9.6, the RIF-FNM produces a slightly better visual result than the FAC-RIF. On the other hand, the FAC-RIF shows superior performance when comparing individual reconstructed frames as those depicted in Figures 9.7 and 9.8. This observation is clearly compatible with the structure and

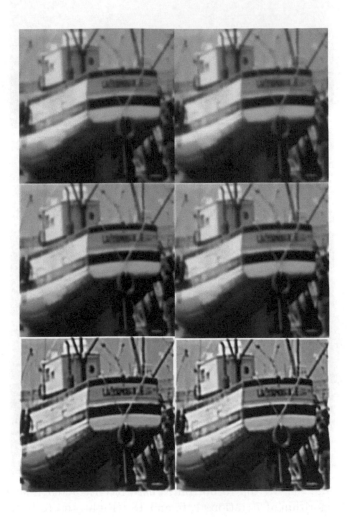

FIGURE 9.5: (Left, top to bottom) Fused-reconstructed image with the RIF-FNM algorithm at iterations 1, 6, and 18. (Right, top to bottom) Minimum variance fusion of the four image frames reconstructed by the FAC-RIF at iterations 1, 6, and 18, respectively. Noise-free case.

design of these two algorithms. From a complexity point of view, the Gaussian mixture modeling section of the RIF-FNM makes this algorithm more complex compared to the NAS-RIF. The complexity is strongly dependent on the number of Gaussian mixture components required by the assumed model. Finally, we have observed, based on experimentation, that the NAS-RIF ex-

FIGURE 9.6: (Left, top to bottom) Fused-reconstructed image with the RIF-FNM algorithm at iterations 1, 6, and 18. (Right, top to bottom) Minimum variance fusion of the four image frames reconstructed by the FAC-RIF at iterations 1, 6, and 18, respectively. Signal-to-noise ratio is 70dB.

hibits in general a faster convergence rate than the RIF-FNM; however, due to the ad-hoc nature of the algorithm, we are not able to provide an analytic proof of this behavior.

To further demonstrate the promise of the FAC-RIF algorithm, despite its ad-hoc nature, we have designed another experiment where real data were obtained using a standard ultrasound phantom. The phantom was illuminated with an adaptive ultrasound beam former developed by Canamet Inc. [40] The profile of the phantom and the illuminated area are shown in Figure 9.9.

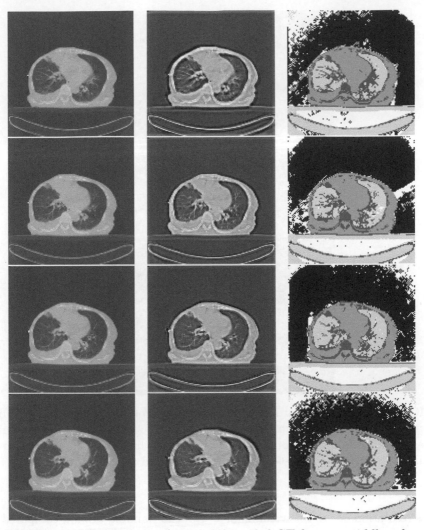

FIGURE 9.7: Left column: original degraded CT frames; middle column: individual restorations with the RIF-FNM algorithm at iteration 19; right column: pixel classification of each frame ($K = 5$).

To create a multiframe scenario we have applied a multiscanning process of the same phantom area. One frame was obtained by applying a rectangular window weighting to the beam former and a second frame was obtained by applying a Hanning window weighting to the beam former. The original and reconstructed images by applying the FAC-RIF are shown in Figure 9.10. A different multiscanning of the phantom by introducing different initial phase shifts could also provide multiframe images with complimentary information. One can observe the difference between the two frames with the left exhibiting

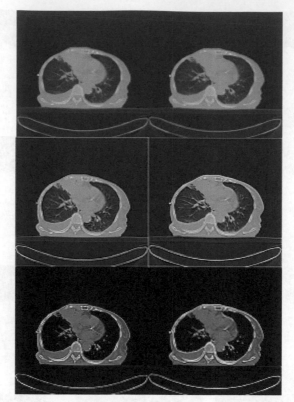

FIGURE 9.8: Real cardiac CT frames. Left column: frame 2 out of four frames processed. Right column: frame 4 out of four frames processed. Original frames (top row), reconstructed frames with the FAC-RIF at iteration 12 (middle row), and contrast-adjusted reconstructed frames at iteration 12.

less resolution due to the wider main lobe of the Hanning window compared to the rectangular window, while the right frames exhibit more evident side-lobe activity. In either case, however, the reconstructed images by the FAC-RIF are once again sharper and of higher quality compared to the original ones.

We should mention here that the selection of a good stopping criterion of the recursive algorithms described above is very important, and it can guarantee that we obtain best results. Due to the inverse filtering nature of the RIF-FNM and FAC-RIF algorithms, we know that noise and artifact amplification such as ringing effects will be introduced, and this will in turn affect both the quality of the results as well as the classification process sooner or later at some iteration. Since each frame is subject to different distortion in general, the stopping criterion must be applied individually to each branch. While adaptation in one or more branches may stop, it may still continue to the rest of the branches until all frames have been sufficiently restored. In our

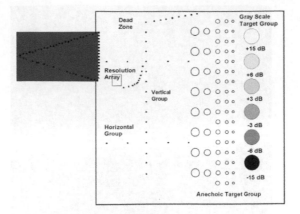

FIGURE 9.9: Illumination of phantom data.

experiments we have monitored the variations of classification error in each branch in order to determine the termination of the recursive operations.

9.8 Final Remarks

Two blind multiframe recursive restoration methods, the RIF-FNM and FAC-RIF, were described in this chapter. The objective is to restore an unknown image scene when multiple degraded acquisitions are available. Both methods rely on classification and fusion of the available multiple frames in order to adaptively drive the adjustment of a set of linear FIR reconstruction filters. Due to the highly nonlinear nature of the fusion–classification-based optimization criterion, it is difficult to provide an analytic characterization of their convergence behavior. However, experiments with simulated and real data clearly indicate the feasibility and promise of the methods.

References

[1] D. Kundur, D. Hatzinakos, and H. Leung, "Robust Classification of Blurred Imagery," in *IEEE Transactions on Image Processing*, Vol. 9(2), pages 243–255, Feb. 2000.

[2] D. Kundur and D. Hatzinakos, "A Novel Blind Deconvolution Scheme for Image Restoration Using Recursive Filtering," in *IEEE Transactions*

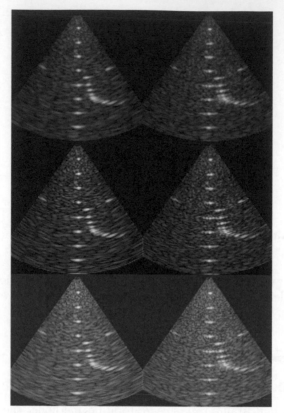

FIGURE 9.10:   Phantom frames obtained with adaptive ultrasound beam-former. Left column with Hanning window. Right column with rectangular window. Original frames (top row), reconstructed frames at iteration 12 (middle row), and contrast adjusted reconstructed frames at iteration 12 (bottom row).

on Signal Processing, Vol. 46(2), pages 375–389, Feb. 1998.

[3] R. Yoakeim and D. Taubman, "Quantitative analysis of resolution synthesis," in *Procedings of the International Conference on Image Processing (ICIP'04),* pages 1645–1648, Singapore, Oct. 2004.

[4] B. C. Tom and A. K. Katsaggelos, "Reconstruction of a high-resolution image by simultaneous registration, restoration, and interpolation of low-resolution images," in *Procedings of the International Conference on Image Processing (ICIP'95),* Vol. 2, pages 539–542, Oct. 1995.

[5] A. K. Katsaggelos, *Digital Image Restoration,* New York: Springer-Verlag, 1991.

[6] L. Brown, "A survey of image registration techniques," in *ACM Computing Surveys*, Vol. 24(4), pages 325–376, Dec. 1992.

[7] S. P. Kim and W.-Y. Su, "Recursive high-resolution reconstruction of blurred multiframe images," in *IEEE Signal Processing Magazine*, Vol. 2(40, pages 534–539, Oct. 1993.

[8] A. K. Katsaggelos and K. T. Lay, "Image identification and image restoration based on the expectation-maximization algorithm," in *Optical Engineering*, Vol. 29, pages 436–445, 1990.

[9] A. S. Carasso, "Direct blind deconvolution," in *SIAM Journal on Applied Mathematics*, Vol. 61, pages 1980–2007, 2001

[10] D. Kundur and D. Hatzinakos, "Blind image deconvolution," in *IEEE Signal Processing Magazine*, Vol. 13(3), pages 43–64, 1996.

[11] G. Harikumar and Y. Bresler, "Perfect blind restoration of images blurred by multiple filters: theory and efficient algorithms," in *IEEE Transactions on Image Processing*, Vol. 8(2), pages 202–219, Feb. 1999.

[12] G. Harikumar and Y. Bresler, "Exact image deconvolution from multiple FIR blurs," in *IEEE Transactions on Image Processing*, Vol. 8(6), pages 846–862, June 1999.

[13] G. Harikumar and Y. Bresler, "FIR perfect signal reconstruction from multiple convolutions: minimum deconvolver orders," in *IEEE Transaction on Acoustic, Speech, and Signal Processing*, Vol. 46(1), pages 215–218, Jan. 1998.

[14] J. Han and D. Hatzinakos, "Joint blind deconvolution, fusion and classification of multi-frame imagery," in *Proceedings of Canadian Conference on Electrical and Computer Engineering (CCECE'04)*, Vol. 4, pages 2061–2064, Niagara Falls, Canada, 2004.

[15] A. Giannoula and D. Hatzinakos, "Recursive deconvolution of multisensor imagery using finite mixture distributions," in *Proceedings ICASSP'05*, Philadelphia, USA, March 19–23, 2005.

[16] F. Sroubek and J. Flusser, "Multichannel blind iterative image restoration," in *IEEE Transactions on Image Processing*, Vol. 12(9), pages 1094–1106, Sept. 2003.

[17] F. Sroubek and J. Flusser, "Multichannel blind deconvolution of spatially misaligned images," in *IEEE Transactions on Image Processing*, Vol. 14(7), pages 874–883, July 2005.

[18] C. Srinivas and M. D. Srinath, "Stochastic model-based approach for simultaneous restoration of multiple misregistered images," in *Proceedings SPIE* Vol. 1360, *Visual Communications and Image Processing '90: Fifth in a Series*, Murat Kunt, Ed., pages 1416–1427, Sept. 1990.

[19] N. Nguyen, P. Milanfar, and G. Golub, "Efficient generalized cross-validation with applications to parametric image restoration and resolution enhancement," in *IEEE Transactions on Image Processing*, Vol. 10(9), pages 1299–1308, Sept. 2001.

[20] S. C. Park, M. K. Park, and M. G. Kang, "Super-resolution image reconstruction: a technical overview," in *IEEE Signal Processing Magazine*, Vol. 20(3), pages 21–36, May 2003.

[21] S. Chaudhuri, "Super-Resolution Imaging," Kluwer Academic Publishers Norwell, MA, 2001.

[22] S. Farsiu, D. Robinson, N. Elad, and P. Milanfar, "Fast and robust multiframe super resolution," in *IEEE Transactions on Image Processing*, Vol. 13(10), pages 1327–1344, Oct. 2004.

[23] D. M. Titterington, A. F. M. Smith, and U. E. Makov, "Statistical Analysis of Finite Mixture Distributions," New York: John Wiley, 1985.

[24] Y. Wang, H. Lin, Li, Y. Kung, "Data mapping by probabilistic modular networks and information-theoretic criteria," in *IEEE Transactions on Signal Processing*, Vol. 46(12), pages 3378–3397, Dec. 1998.

[25] G. J. McLachlan and T. Krishnan, "The EM Algorithm and Extensions," New York: John Wiley, 1997.

[26] A. Papoulis, "Probability & Statistics," Prentice Hall, Upper Saddle River, NJ, 1991.

[27] T. Lei, W. Sewchand, "Statistical approach to X-ray CT imaging and its applications in image analysis. II. A new stochastic model-based image segmentation technique for X-ray CT image," in *IEEE Transactions on Medical Imaging*, Vol. 11(1), pages 62–69, March 1992.

[28] L. Zhengrong, J. R. MacFall, and D. P. Harrington, "Parameter estimation and tissue segmentation from multispectral MR images," in *IEEE Transactions on Medical Imaging*, Vol. 13(3), pages 441–449, Sept. 1994.

[29] J. L. Marroquin and F. Girosi, "Some extensions of the K-means algorithm for image segmentation and pattern classification," *Tech. Rep. MIT Artif. Intell. Lab.*, Cambridge, MA, Jan. 1993.

[30] R.A. Hummel and S.W. Zucker, "On the Foundations of Relaxation Labeling Processes," *McGill Univ. TR*, 1980.

[31] W. C. Lin, C. K. Tsao, and T. Chen, "Constraint satisfaction neural networks for image segmentation," in *Pattern Recognition*, Vol. 25, pages 679–693, Dec. 1992.

[32] W. Press, B. Flannery, S. Teukolsky, and W. Vetterling, "Numerical Recipes in C," Cambridge University Press, 1993.

[33] D. G. Luenberger, "Introduction to Linear and Nonlinear Programming," Addison-Wesley, Reading, MA, 1973.

[34] T. K. Moon and W. C. Stirling, "Mathematical Methods and Algorithms for Signal Processing," Prentice-Hall, Upper Saddle River, NJ, 2000.

[35] M. E. Zervakis and A. N. Venetsanopoulos, "Resolution-to-noise tradeoff in linear image restoration," in *IEEE Transactions on Circuits and Systems*, Vol. 38, pages 1206–1212, Oct. 1991.

[36] J. Borwein and A. S. Lewis, "Convex Analysis and Nonlinear Optimization," Springer-Verlag, 2000.

[37] A. H. S. Solberg, T. Text, and A. K. Jain, "Multisource classification of remotely sensed data: Fusion of landsat tm and sar images," in *IEEE Transactions on Geoscience and Remote Sensing*, Vol. 34, pages 100–113, Jan. 1996.

[38] K. Held and E. R. Kops, et al., "Markov Random Field Segmentation of Brain MR Image," in *IEEE Transactions on Medical Imaging*, Vol. 16(6), pages 878–886, 1997.

[39] S. Stergiopoulos, "Advanced Signal Processing Handbook: Theory and Implementation for Radar, Sonar, and Medical Imaging Real Time Systems," CRC Press, 2001.

[40] "Canamet Inc., Canadian National Medical Technologies," *Toronto, Ontario, http://www.canamet.com*

10

Blind Deconvolution and Structured Matrix Computations with Applications to Array Imaging

Michael K. Ng

Department of Mathematics, Hong Kong Baptist University, Kowloon Tong, Hong Kong

e-mail: mng@math.hkbu.edu.hk

Robert J. Plemmons

Department of Mathematics and Computer Science, Wake Forest University, Winston-Salem, NC 27109, USA

e-mail: plemmons@wfu.edu

Abstract

In this chapter, we study using total least squares (TLS) methods for solving blind deconvolution problems arising in image recovery. Here, the true image is to be estimated using only partial information about the blurring operator, or point spread function, which is also subject to error and noise. Iterative, regularized, and constrained TLS methods are discussed and analyzed. As an application, we study TLS methods for the reconstruction of high-resolution images from multiple undersampled images of a scene that is obtained by using a charge-coupled device (CCD) or a CMOS detector array of sensors which are shifted relative to each other by subpixel displacements. The objective is improving the performance of the signal-processing algorithms in the presence of the ubiquitous perturbations of displacements around the ideal subpixel locations because of imperfections in fabrication and so on, or because of shifts designed to enable superresolution reconstructions in array imaging. The target architecture consists of a regular array of identical lenslets whose images are grouped, combined, and then digitally processed. Such a system will have the resolution, field of view, and sensitivity of a camera with an effective aperture that would be considerably larger than the single-lenslet aperture, yet with a short focal length typical of each lenslet. As a means for solving the resulting blind deconvolution problems, the errors-in-variables (or the TLS) method is applied.

10.1   Introduction

The fundamental issue in image enhancement or restoration is blur removal in the presence of observation noise. Recorded images almost always represent a degraded version of the original scene. A primary example is images taken by an optical instrument recording light that has passed through a turbulent medium, such as the atmosphere. Here, changes in the refractive index at different positions in the atmosphere result in a nonplanar wave front [1, Chapter 3]. In general, the degradation by noise and blur is caused by fluctuations in both the imaging system and the environment.

In the important case where the blurring operation is *spatially invariant*, the basic restoration computation involved is a deconvolution process that faces the usual difficulties associated with ill-conditioning in the presence of noise [2, Chapter 2]. The image observed from a shift-invariant linear blurring process, such as an optical system, is described by how the system blurs a point source of light into a larger image. The image of a point source is called the *point spread function*, PSF, which we denote by $\mathbf{h}$. The observed image $\mathbf{g}$ is then the result of convolving the PSF $\mathbf{h}$ with the "true" image $\mathbf{f}$, and with noise present in $\mathbf{g}$.

The *standard deconvolution problem* is to recover the image $\mathbf{f}$ given the observed image $\mathbf{g}$ and the point spread function $\mathbf{h}$. See, e.g., the survey paper on standard deconvolution written by Banham and Katsaggelos [3]. If the PSF $\mathbf{h}$ is not known, then the problem becomes one of *blind deconvolution*, sometimes called myopic deconvolution if $\mathbf{h}$ is partially known (see, e.g., the survey paper on blind or myopic deconvolution written by Kundur and Hatzinakos [4]).

We also mention the recent work of Bardsley, Jefferies, Nagy and Plemmons [5] on the blind restoration of images with an unknown, spatially varying PSF.

Their algorithm uses a combination of techniques. First, they section the image, and then treat the sections as a sequence of frames whose unknown PSFs are correlated and approximately spatially invariant. To estimate the PSFs in each section phase diversity (see, e.g., [6]), is used. With these PSF estimates in hand, they then use an interpolation technique to restore the image globally.

In the sections to follow, we concentrate primarily on blind deconvolution. Emphasis is given to the use of TLS methods where the blurring operator has a natural, structured form determined by boundary conditions. Applications to the construction of high-resolution images using low-resolution images from thin array imaging cameras with millimeter-diameter lenslets and focal lengths are provided to illustrate the techniques discussed in this chapter.

10.2 One-Dimensional Deconvolution Formulation

For simplicity of notation we begin with the one-dimensional deblurring problem. Consider the original signal

$$\mathbf{f} = (\cdots, f_{-n}, f_{-n+1}, \cdots, f_0, f_1, \cdots, f_n, f_{n+1}, \cdots, f_{2n}, f_{2n+1}, \cdots)^t$$

and the *discrete point spread function* given by

$$\mathbf{h} = (\cdots, 0, 0, h_{-n+1}, \cdots, h_0, \cdots, h_{n-1}, 0, 0, \cdots)^t.$$

Here "t" denotes transposition. The blurred signal is the convolution of $\mathbf{h}$ and $\mathbf{f}$, i.e., the i-th entry $\bar{g}_i$ of the blurred signal is given by

$$\bar{g}_i = \sum_{j=-\infty}^{\infty} h_{i-j} f_j. \tag{10.1}$$

Therefore, the blurred signal vector is given by

$$\bar{\mathbf{g}} = [\bar{g}_1, \cdots, \bar{g}_n]^t.$$

For a detailed discussion of digitizing images, see Castleman [2, Chapter 2].

From (10.1), we have

$$
\begin{bmatrix} \bar{g}_1 \\ \bar{g}_2 \\ \vdots \\ \bar{g}_{n-1} \\ \bar{g}_n \end{bmatrix} = \left[\begin{array}{ccc|ccc|ccc} \bar{h}_{n-1} & \cdots & \bar{h}_1 & \bar{h}_0 & \cdots & \cdots & \bar{h}_{-n+1} & 0 & \cdots & 0 \\ 0 & \ddots & & \vdots & & & \vdots & \bar{h}_{-n+1} & 0 & \vdots \\ \vdots & 0 & \bar{h}_{n-1} & \vdots & & & \vdots & & \ddots & 0 \\ 0 & \cdots & 0 & \bar{h}_{n-1} & \cdots & \cdots & \bar{h}_0 & \bar{h}_{-1} & \cdots & \bar{h}_{-n+1} \end{array}\right] \begin{bmatrix} f_{-n+2} \\ \vdots \\ f_0 \\ \hline f_1 \\ \vdots \\ f_n \\ \hline f_{n+1} \\ \vdots \\ f_{2n-1} \end{bmatrix}
$$

$$
= [\bar{H}_l | \bar{H}_c | \bar{H}_r] \begin{bmatrix} \mathbf{f}_l \\ \hline \mathbf{f}_c \\ \hline \mathbf{f}_r \end{bmatrix} = [\bar{H}_l | \bar{H}_c | \bar{H}_r] \mathbf{f}. \tag{10.2}
$$

Here $[A|B]$ denotes an m-by-(n_1+n_2) matrix where A and B are m-by-n_1 and m-by-n_2 matrices, respectively. The matrices $\bar{H}_l$ and $\bar{H}_r$ refer to the parts of the blurring matrix affecting the boundary of the signal. The matrix $\bar{H}_c$ refers to the main part of the blurring matrix affecting the signal inside.

For a given n, the deconvolution problem is to recover the vector $[f_1, \cdots, f_n]^t$ given the point spread function $\bar{\mathbf{h}}$ and a blurred signal $\bar{\mathbf{g}} = [\bar{g}_1, \cdots, \bar{g}_n]^t$ of finite length n. Notice that the blurred signal $\bar{\mathbf{g}}$ is determined not only by $\mathbf{f}_c = [f_1, \cdots, f_n]^t$, but by $\mathbf{f} = [\mathbf{f}_l \ \mathbf{f}_c \ \mathbf{f}_r]^t$.

The linear system (10.2) is underdetermined. To recover the vector $\mathbf{f}_c$, we assume the data outside $\mathbf{f}_c$ are reflections of the data inside $\mathbf{f}_c$, i.e.,

$$
\left\{ \begin{array}{ccc} f_0 & = & f_1 \\ \vdots & \vdots & \vdots \\ f_{-n+2} & = & f_{n-1} \end{array} \right. \quad \text{and} \quad \left\{ \begin{array}{ccc} f_{n+1} & = & f_n \\ \vdots & \vdots & \vdots \\ f_{2n-1} & = & f_2. \end{array} \right. \tag{10.3}
$$

In [7], it has been shown that the use of this (Neumann) boundary condition can reduce the boundary artifacts and that solving the resulting systems is much better and faster than using zero and periodic boundary conditions. For example, reverting to the two-dimensional case for illustration purposes, see Figures 10.1–10.4. The detailed discussion of using the Neumann boundary condition can be found in [7].

In classical image restoration, the point spread function is assumed to be known or adequately sampled [8]. However, in practice, one is often faced with imprecise knowledge of the PSF. For instance, in two-dimensional deconvolution problems arising in ground-based atmospheric imaging, the problem

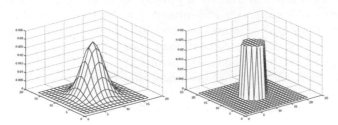

FIGURE 10.1: Gaussian (atmospheric turbulence) blur (left) and out-of-focus blur (right).

FIGURE 10.2: Original image (left), noisy and blurred satellite image by Gaussian (middle), and out-of-focus blur (right).

rel. error = 1.24×10^{-1} rel. error = 1.15×10^{-1} rel. error = 6.59×10^{-2}

FIGURE 10.3: Restoring Gaussian blur with zero boundary (left), periodic boundary (middle), and Neumann boundary (right) conditions.

rel. error = 1.20×10^{-1} rel. error = 1.09×10^{-1} rel. error = 4.00×10^{-2}

FIGURE 10.4: Restoring out-of-focus blur with zero boundary (left), periodic boundary (middle), and Neumann boundary (right) conditions.

consists of an image received by a ground-based imaging system, together with an image of a guide star PSF observed under the effects of atmospheric turbulence. Empirical estimates of the PSF can sometimes be obtained by imaging a relatively bright, isolated point source. The point source might be a natural guide star or a guide star artificially generated using range-gated laser backscatter, e.g., [9–11]. Notice here that the PSF as well as the image are degraded by blur and noise.

Because the spatial invariance of h translates into the spatial invariance of the noise in the blurring matrix, we assume that the "true" PSF can be represented by the following formula:

$$\bar{\mathbf{h}} = \mathbf{h} + \delta\mathbf{h}, \tag{10.4}$$

where

$$\mathbf{h} = [h_{-n+1}, \ldots, h_{n-1}]^t$$

is the estimated (or measured) point spread function and

$$\delta\mathbf{h} = [\delta h_{-n+1}, \delta h_{-n+2}, \cdots, \delta h_0, \cdots, \delta h_{n-2}, \delta h_{n-1}]^t,$$

is the error component of the PSF. Each δh_i is modeled as independent uniformly distributed noise, with zero-mean and variance σ_h^2; see, for instance, [12]. The blurred signal $\bar{\mathbf{g}}$ is also subject to errors. We assume that the observed signal $\mathbf{g} = [g_1, \ldots, g_n]^t$ can be represented by

$$\bar{\mathbf{g}} = \mathbf{g} + \delta\mathbf{g}, \tag{10.5}$$

where

$$\delta\mathbf{g} = [\delta g_1, \delta g_2, \cdots, \delta g_n]^t$$

and δg_i is independent, uniformly distributed noise with zero-mean and variance σ_g^2. Here the noise in the PSF and in the observed signal are assumed to be uncorrelated. Thus our image restoration problem is to recover the vector $\mathbf{f}$ from the given inexact point spread function $\mathbf{h}$ and a blurred and noisy signal $\mathbf{g}$.

10.3   Regularized and Constrained TLS Formulation

In the literature, blind deconvolution methods (see, e.g., [4, 13–17]), have been developed to estimate both the true image $\mathbf{f}$ and the point spread function $\mathbf{h}$ from the degraded image $\mathbf{g}$. In order to obtain a reasonable restored image, these methods require one to impose suitable constraints on the PSF and the image. In our image restoration applications, the PSF is *not* known exactly (e.g., it is corrupted by errors resulting from blur and/or noise). A review of optimization models for blind deconvolution can be found in the survey paper by Kundur and Hatzinakos [4].

Recently there have been growing interest and progress in using total least squares (TLS) methods for solving these blind deconvolution problems arising in image restoration and reconstruction (see, e.g., [12,18–24]). It is well known that the total least squares (TLS) is an effective technique for solving a set of error-contaminated equations [25, 26]. The TLS method is an appropriate method for consideration in astro-imaging applications. In [20], Kamm and Nagy have proposed the use of the TLS method for solving Toeplitz systems arising from image restoration problems. They applied Newton and Rayleigh quotient iterations to solve the Toeplitz total least squares problems. A possible drawback of their approach is that the point spread function in the TLS formulation is not constrained to be spatially invariant. Mesarović, Galatsanos, and Katsaggelos [12] have shown that formulating the TLS problem for image restoration with the spatially invariant constraint improves the restored image greatly (see the numerical results in [12]).

The determination of $\mathbf{f}$ given the recorded data $\mathbf{g}$ and knowledge of the PSF $\mathbf{h}$ is an inverse problem [27]. Deconvolution algorithms can be extremely sensitive to noise. It is necessary to incorporate regularization into deconvolution to stabilize the computations. Regarding the regularization, Golub, Hansen, and O'Leary [19] have shown how Tikhonov regularization methods, for regularized least squares computations, can be recast in a total least squares framework, suited for problems in which both the coefficient matrix and the right-hand side are known only approximately. However, their results do not hold for the constrained total least squares formulation [12]. Therefore, we cannot use the algorithm in [19].

Here we develop a constrained TLS approach to solving the image restoration problem, using the one-dimensional case for simplicity of presentation.

By using (10.4) and (10.5), the convolution equation (10.2) can be reformulated as follows:

$$Hf - \mathbf{g} + \delta Hf - \delta \mathbf{g} = 0, \tag{10.6}$$

where

$$H = \begin{bmatrix} h_{n-1} \cdots & h_1 & h_0 & \cdots & h_{-n+1} & & 0 \\ & \ddots & & \vdots & & \vdots & h_{-n+1} & \\ & & h_{n-1} & \vdots & & \vdots & & \ddots \\ 0 & & & h_{n-1} & \cdots & h_0 & h_{-1} & \cdots & h_{-n+1} \end{bmatrix} = [H_l | H_c | H_r] \tag{10.7}$$

and

$$
\delta H =
\begin{bmatrix}
\delta h_{n-1} & \cdots & \delta h_1 & \delta h_0 & \cdots & \delta h_{-n+1} & & & 0 \\
& \ddots & & \vdots & & \vdots & \delta h_{-n+1} & & \\
& & \delta h_{n-1} & \vdots & & \vdots & & \ddots & \\
0 & & & \delta h_{n-1} & \cdots & \delta h_0 & \delta h_{-1} & \cdots & \delta h_{-n+1}
\end{bmatrix}
$$
$$
= [\delta H_l | \delta H_c | \delta H_r]. \tag{10.8}
$$

Correspondingly, we can define the Toeplitz matrices H_l, H_c, and H_r and δH_l, δH_c, and δH_r similar to $\bar{H}_l$, $\bar{H}_c$, and $\bar{H}_r$ in (10.2), respectively. The constrained total least squares formulation amounts to determining the necessary "minimum" quantities δH and $\delta \mathbf{g}$ such that (10.6) is satisfied.

Mathematically, the constrained total least squares formulation can be expressed as

$$
\min_{\mathbf{f}_c} \| [\delta H | \delta g] \|_F^2 \quad \text{subject to} \quad H\mathbf{f} - \mathbf{g} + \delta H\mathbf{f} - \delta \mathbf{g} = 0,
$$

where $\mathbf{f}$ satisfies (10.3).

Recall that image restoration problems are in general ill-conditioned inverse problems, and restoration algorithms can be extremely sensitive to noise [28, p. 282]. Regularization can be used to achieve stability. Using classical Tikhonov regularization [27, p. 117], stability is attained by introducing a regularization operator D and a regularization parameter μ to restrict the set of admissible solutions. More specifically, the regularized solution $\mathbf{f}_c$ is computed as the solution to

$$
\min_{\mathbf{f}_c} \{ \| [\delta H | \delta g] \|_F^2 + \mu \| D\mathbf{f}_c \|_2^2 \} \tag{10.9}
$$

subject to

$$
H\mathbf{f} - \mathbf{g} + \delta H\mathbf{f} - \delta \mathbf{g} = 0, \tag{10.10}
$$

and $\mathbf{f}$ satisfies (10.3). The term $\| D\mathbf{f}_c \|_2^2$ is added in order to regularize the solution. The regularization parameter μ controls the degree of regularity (i.e., degree of bias) of the solution. In many applications [28–30], $\| D\mathbf{f}_c \|_2$ is chosen to be the L_2 norm $\| \mathbf{f}_c \|_2$ or the H_1 norm $\| L\mathbf{f}_c \|_2$ where L is a first-order difference operator matrix. In this chapter, we only consider the L_2 and H_1 regularization functionals.

In [12], the authors addressed the problem of restoring images from noisy measurements in both the PSF and the observed data as a regularized and constrained total least squares problem. It was shown in [12] that the regularized minimization problem obtained is nonlinear and nonconvex. Thus fast algorithms for solving this nonlinear optimization problem are required. In [12], circulant, or periodic, approximations are used to replace the convolution matrices in subsequent computations. In the Fourier domain, the system of nonlinear equations is decoupled into a set of simplified equations and therefore the computational cost can be reduced significantly. However, practical signals and images often do not satisfy these periodic assumptions, and ringing effects will appear on the boundary [31]. In the image processing literature, various methods have been proposed to assign boundary values (see Lagendijk and Biemond [32, p. 22] and the references therein). For instance, the boundary values may be fixed at a local image mean, or they can be obtained by a model-based extrapolation. In this chapter, we consider the image formulation model for the regularized constrained TLS problem using the *Neumann boundary condition* for the image, i.e., we assume that the scene immediately outside is a reflection of the original scene near the boundary. This Neumann boundary condition has been studied in image restoration [7, 31, 32] and in image compression [33, 34]. Results in [7] show that the boundary artifacts resulting from the deconvolution computations are much less prominent than those under the assumption of zero [20] or periodic [12] boundary conditions.

The theorem below characterizes the constrained, regularized TLS formulation of the one-dimensional deconvolution problem.

THEOREM 10.1 *Under the Neumann boundary condition (10.3), the regularized constrained total least squares solution can be obtained as the $\mathbf{f}_c$ that minimizes the functional:*

$$P(\mathbf{f}_c) = (A\mathbf{f}_c - \mathbf{g})^t Q(\mathbf{f}_c)(A\mathbf{f}_c - \mathbf{g}) + \mu \mathbf{f}_c^t D^t D \mathbf{f}_c, \qquad (10.11)$$

where A is an n-by-n Toeplitz-plus-Hankel matrix,

$$A = H_c + [0|H_l]J + [H_r|0]J, \qquad (10.12)$$

J is the n-by-n reversal matrix,

$$Q(\mathbf{f}_c) = ([T(\mathbf{f}_c)|I][T(\mathbf{f}_c)|I]^t)^{-1} \equiv [T(\mathbf{f}_c)T(\mathbf{f}_c)^t + I]^{-1},$$

$T(\mathbf{f}_c)$ is an n-by-$(2n-1)$ Toeplitz matrix,

$$T(\mathbf{f}_c) = \frac{1}{\sqrt{n}} \begin{bmatrix} f_n & f_{n-1} & \cdots & f_2 & f_1 & f_1 & \cdots & f_{n-2} & f_{n-1} \\ f_n & f_n & \ddots & \ddots & f_2 & f_1 & \ddots & f_{n-3} & f_{n-2} \\ \vdots & \ddots & \ddots & \ddots & \ddots & \ddots & \ddots & \ddots & \vdots \\ f_3 & f_4 & \ddots & f_n & f_{n-1} & \ddots & \ddots & f_1 & f_1 \\ f_2 & f_3 & \cdots & f_n & f_n & f_{n-1} & \cdots & f_2 & f_1 \end{bmatrix}, \qquad (10.13)$$

and I is the n-by-n identity matrix.

PROOF From (10.10), we have

$$[T(\mathbf{f}_c)|I] \begin{bmatrix} \sqrt{n}\delta\mathbf{h} \\ -\delta\mathbf{g} \end{bmatrix} = \sqrt{n}T(\mathbf{f}_c)\delta\mathbf{h} - \delta\mathbf{g} = \delta H\mathbf{f} - \delta\mathbf{g} = \mathbf{g} - H\mathbf{f} = \mathbf{g} - A\mathbf{f}_c. \quad (10.14)$$

We note that

$$\|[\delta H|\delta \mathbf{g}]\|_F^2 = n\|\delta \mathbf{h}\|_2^2 + \|\delta \mathbf{g}\|_2^2 = \left\|\begin{bmatrix} \sqrt{n}\delta \mathbf{h} \\ -\delta \mathbf{g} \end{bmatrix}\right\|_2^2.$$

Therefore, we obtain the minimum 2-norm solution of the underdetermined system in (10.14) (see for instance [25]). Since the rank of the matrix $[T(\mathbf{f}_c)|I]$ is n, we have

$$\begin{bmatrix} \sqrt{n}\delta \mathbf{h} \\ -\delta \mathbf{g} \end{bmatrix} = [T(\mathbf{f}_c)|I]^t Q(\mathbf{f}_c)(\mathbf{g} - A\mathbf{f}_c)$$

or

$$\left\|\begin{bmatrix} \sqrt{n}\delta \mathbf{h} \\ -\delta \mathbf{g} \end{bmatrix}\right\|_2^2 = (A\mathbf{f}_c - \mathbf{g})^t Q(\mathbf{f}_c)(A\mathbf{f}_c - \mathbf{g}). \tag{10.15}$$

By inserting (10.15) into (10.9), we obtain (10.11). $\qquad\qquad\qquad$ ☐

10.3.1   Symmetric Point Spread Functions

The estimates of the discrete blurring function may not be unique, in the absence of any additional constraints, mainly because blurs may have any kind of Fourier phase (see [32]). Nonuniqueness of the discrete blurring function can in general be avoided by enforcing a set of constraints. In many papers dealing with blur identification [13,35,36], the point spread function is assumed to be symmetric, i.e.,

$$\bar{h}_k = \bar{h}_{-k}, \quad k = 1, 2, \cdots, n-1.$$

We remark that point spread functions are often symmetric (see [30, p. 269]), for instance, the Gaussian point spread function arising in atmospheric turbulence induced blur is symmetric with respect to the origin. For guide star images [35] this is usually not the case. However, they often appear to be fairly symmetric, which can be observed by measuring their distance to a nearest symmetric point spread function. In [35], Hanke and Nagy use the symmetric part of the measured point spread function to restore atmospherically blurred images.

Similarly, we thus incorporate the following symmetry constraints into the total least squares formulation of the problem:

$$\bar{h}_k = \bar{h}_{-k} \quad \text{and} \quad \delta h_k = \delta h_{-k}, \quad k = 1, 2, \cdots, n-1. \tag{10.16}$$

Then, using Neumann boundary conditions (10.3), the convolution equation (10.6) becomes

$$A\mathbf{f}_c - \mathbf{g} + \delta A\mathbf{f}_c - \delta \mathbf{g} = 0,$$

where A is defined in (10.12) and δA is defined similarly. It was shown in [7] that these Toeplitz-plus-Hankel matrices A and δA can be diagonalized by an

n-by-n discrete cosine transform matrix C with entries

$$C_{ij} = \sqrt{\frac{2 - \delta_{i1}}{n}} \cos\left(\frac{(i-1)(2j-1)\pi}{2n}\right), \quad 1 \le i, j, \le n,$$

where δ_{ij} is the Kronecker delta (see [30, p. 150]). We note that C is orthogonal, i.e., $C^t C = I$. Also, for any n-vector $\mathbf{v}$, the matrix–vector multiplications $C\mathbf{v}$ and $C^t\mathbf{v}$ can be computed in $O(n \log n)$ real operations by fast cosine transforms (FCTs); see [37, pp. 59–60].

In the following discussion, we write

$$A = C^t \operatorname{diag}(\mathbf{w}) \, C \quad \text{and} \quad \delta A = C^t \operatorname{diag}(\delta\mathbf{w}) \, C. \tag{10.17}$$

Here, for a general vector $\mathbf{v}$, $\operatorname{diag}(\mathbf{v})$ is a diagonal matrix with its diagonal entries given by

$$[\operatorname{diag}(\mathbf{v})]_{i,i} = \mathbf{v}_i, \quad i = 1, 2, \cdots, n.$$

Using (10.17), we can give a new regularized constrained total least squares formulation to this symmetric case as follows:

THEOREM 10.2 *Under the Neumann boundary condition (10.3) and the symmetry constraint (10.16), the regularized constrained total least squares solution can be obtained as the $\hat{\mathbf{f}}_c$ that minimizes the functional*

$$P(\hat{\mathbf{f}}_c) = [\operatorname{diag}(\mathbf{w})\hat{\mathbf{f}}_c - \hat{\mathbf{g}}]^t \{[\operatorname{diag}(\hat{\mathbf{f}}_c)|I][\operatorname{diag}(\hat{\mathbf{f}}_c)|I]^t\}^{-1}[\operatorname{diag}(\mathbf{w})\hat{\mathbf{f}}_c - \hat{\mathbf{g}}] + \mu \hat{\mathbf{f}}_c^t \Lambda \hat{\mathbf{f}}_c, \tag{10.18}$$

where

$$\hat{\mathbf{f}} = C\mathbf{f}, \quad \hat{\mathbf{g}} = C\mathbf{g},$$

and Λ is an n-by-n diagonal matrix given by

$$\Lambda = C D^t D C^t,$$

and D is the regularization operator.

PROOF In this regularized total least squares formulation, we minimize $\|[\delta A|\delta g]\|_F^2 + \mu\|D\mathbf{f}_c\|_2^2$ subject to $A\mathbf{f}_c - \mathbf{g} + \delta A\mathbf{f}_c - \delta\mathbf{g} = 0$. Since A and δA can be diagonalized by C, the constraint now becomes

$$\operatorname{diag}(\mathbf{w})\hat{\mathbf{f}}_c - \hat{\mathbf{g}} + \operatorname{diag}(\delta\mathbf{w})\hat{\mathbf{f}}_c - \hat{\delta\mathbf{g}} = 0.$$

Let us define $\mathbf{y} = [\operatorname{diag}(\hat{\mathbf{f}}_c)]^{-1}\operatorname{diag}(\delta\mathbf{w})\hat{\mathbf{f}}_c$. It is easy to show that

$$[\operatorname{diag}(\hat{\mathbf{f}}_c)|I]\begin{bmatrix} \mathbf{y} \\ -\hat{\delta\mathbf{g}} \end{bmatrix} = \operatorname{diag}(\hat{\mathbf{f}}_c)\mathbf{y} - \hat{\delta\mathbf{g}} = \operatorname{diag}(\delta\mathbf{w})\hat{\mathbf{f}}_c - \hat{\delta\mathbf{g}} = \hat{\mathbf{g}} - \operatorname{diag}(\mathbf{w})\hat{\mathbf{f}}_c.$$

Hence we have

$$\begin{bmatrix} \mathbf{y} \\ -\hat{\delta\mathbf{g}} \end{bmatrix} = [\operatorname{diag}(\hat{\mathbf{f}}_c)|I]^t \{[\operatorname{diag}(\hat{\mathbf{f}}_c)|I][\operatorname{diag}(\hat{\mathbf{f}}_c)|I]^t\}^{-1}(\hat{\mathbf{g}} - \operatorname{diag}(\mathbf{w})\hat{\mathbf{f}}_c) \tag{10.19}$$

The diagonalization of A by C implies that

$$\|[\delta A|\delta\mathbf{g}]\|_F^2$$
$$= \|C^t\mathrm{diag}(\delta\mathbf{w})C\|_F^2 + \|C^t\delta\hat{\mathbf{g}}\|_2^2 \;=\; \|\mathrm{diag}(\delta\mathbf{w})\|_F^2 + \|\delta\hat{\mathbf{g}}\|_2^2 \;=\; \left\|\begin{bmatrix} \mathbf{y} \\ -\delta\hat{\mathbf{g}} \end{bmatrix}\right\|_2^2$$

Now, by using (10.19), it is easy to verify that $\|[\delta A|\delta\mathbf{g}]\|_F^2 + \mu\|D\mathbf{f}_c\|_2^2$ is equal to the second member of (10.18).   ∐

We recall that when the L_2 and H_1 regularization functionals are used in the restoration process, the main diagonal entries of Λ are just given by

$$\Lambda_{ii} = 1 \quad\text{and}\quad \Lambda_{ii} = 4\cos^2\left(\frac{(i-1)\pi}{2n}\right), \quad 1 \le i \le n,$$

respectively.

10.4   Numerical Algorithms

In this section, we introduce an approach to minimizing (10.11). For simplicity, we let

$$\frac{\partial Q(\mathbf{f}_c)}{\partial f_i} = \begin{bmatrix} \frac{\partial Q_{11}(\mathbf{f}_c)}{\partial f_i} & \cdots & \frac{\partial Q_{1n}(\mathbf{f}_c)}{\partial f_i} \\ \vdots & & \vdots \\ \frac{\partial Q_{n1}(\mathbf{f}_c)}{\partial f_i} & \cdots & \frac{\partial Q_{nn}(\mathbf{f}_c)}{\partial f_i} \end{bmatrix}.$$

Here $\partial Q_{jk}(\mathbf{f}_c)/\partial f_i$ is the derivative of the (j,k)-th entry of $Q(\mathbf{f}_c)$ with respect to f_i. By applying the product rule to the matrix equality

$$Q(\mathbf{f}_c)[T(\mathbf{f}_c)T(\mathbf{f}_c)^t + I] = I,$$

we obtain

$$\frac{\partial Q(\mathbf{f}_c)}{\partial f_i} = -Q(\mathbf{f}_c)\frac{\partial\{T(\mathbf{f}_c)T(\mathbf{f}_c)^t + I\}}{\partial f_i}Q(\mathbf{f}_c)$$

or equivalently

$$\frac{\partial Q(\mathbf{f}_c)}{\partial f_i} = -Q(\mathbf{f}_c)\left\{\frac{\partial T(\mathbf{f}_c)}{\partial f_i}T(\mathbf{f}_c)^t + T(\mathbf{f}_c)\frac{\partial T(\mathbf{f}_c)^t}{\partial f_i}\right\}Q(\mathbf{f}_c).$$

The gradient $G(\mathbf{f}_c)$ (derivative with respect to $\mathbf{f}_c$) of the functional (10.11) is given by

$$G(\mathbf{f}_c) = 2A^tQ(\mathbf{f}_c)(A\mathbf{f}_c - \mathbf{g}) + 2\mu D^tD\mathbf{f}_c + \mathbf{u}(\mathbf{f}_c),$$

where

$$\mathbf{u}(\mathbf{f}_c) = \begin{bmatrix} (A\mathbf{f}_c - \mathbf{g})^t \frac{\partial Q(\mathbf{f}_c)}{\partial f_1}(A\mathbf{f}_c - \mathbf{g}) \\ (A\mathbf{f}_c - \mathbf{g})^t \frac{\partial Q(\mathbf{f}_c)}{\partial f_2}(A\mathbf{f}_c - \mathbf{g}) \\ \vdots \\ (A\mathbf{f}_c - \mathbf{g})^t \frac{\partial Q(\mathbf{f}_c)}{\partial f_n}(A\mathbf{f}_c - \mathbf{g}) \end{bmatrix}.$$

The gradient descent scheme yields

$$\mathbf{f}_c^{(k+1)} = \mathbf{f}_c^{(k)} - \tau_k G(\mathbf{f}_c^{(k)}), \quad k = 0, 1, \cdots.$$

A line search can be added to select the step size τ_k in a manner which gives sufficient decrease in the objective functional in (10.11) to guarantee convergence to a minimizer. This gives the method of steepest descent (see [38–40]). While numerical implementation is straightforward, steepest descent has rather undesirable asymptotic convergence properties which can make it very inefficient. Obviously, one can apply other standard unconstrained optimization methods with better convergence properties, like the nonlinear conjugate gradient method or Newton's method. These methods converge rapidly near a minimizer provided the objective functional depends on smoothly on $\mathbf{f}_c$. Since the objective function in (10.11) is nonconvex, this results in a loss of robustness and efficiency for higher-order methods like Newton's method. Moreover, implementing Newton's method requires the inversion of an n-by-n unconstructed matrix, clearly an overwhelming task for any reasonable-sized image, for instance, $n = 65,536$ for a 256×256 image. Thus, these approaches may all be unsuitable for the image restoration problem.

Here we develop an alternative approach to minimizing (10.11). At a minimizer, we know that $G(\mathbf{f}_c) = 0$, or equivalently,

$$2A^t Q(\mathbf{f}_c)(A\mathbf{f}_c - \mathbf{g}) + 2\mu D^t D\mathbf{f}_c - \mathbf{u}(\mathbf{f}_c) = 0.$$

The iteration can be expressed as

$$\left[A^t Q(\mathbf{f}_c^{(k)})A + \mu D^t D \right] \mathbf{f}_c^{(k+1)} = \frac{\mathbf{u}(\mathbf{f}_c^{(k)})}{2} + A^t Q(\mathbf{f}_c^{(k)})\mathbf{g}. \tag{10.20}$$

Note that at each iteration, one must solve a linear system depending on the previous iterate $\mathbf{f}_c^{(k)}$, to obtain the new iterate $\mathbf{f}_c^{(k+1)}$. We also find that

$$\mathbf{d}^{(k)} = \mathbf{f}_c^{(k+1)} - \mathbf{f}_c^{(k)} = -\frac{1}{2} \left[A^t Q(\mathbf{f}_c^{(k)})A + \mu D^t D \right]^{-1} G(\mathbf{f}_c^{(k)}).$$

Hence the iteration is of quasi-Newton form, and an existing convergence theory can be applied (see, for instance, [38, 39]). Since the matrix $Q(\mathbf{f}_c^{(k)})$ is symmetric positive definite, and therefore $(A^t Q(\mathbf{f}_c^{(k)})A + \mu D^t D)$ is symmetric positive definite with its eigenvalues bounded away from zero (because of the regularization), each step computes the descent direction $\mathbf{d}^{(k)}$, and global convergence can be guaranteed by using the appropriate step size, i.e.,

$$\mathbf{f}_c^{(k+1)} = \mathbf{f}_c^{(k)} - \frac{\tau_k}{2}\left[A^t Q(\mathbf{f}_c^{(k)})A + \mu D^t D\right]^{-1} G(\mathbf{f}_c^{(k)}),$$

where

$$\tau_k = \operatorname{argmin}_{\tau^k > 0} P(\mathbf{f}_c^{(k)} + \tau^k \mathbf{d}^{(k)}).$$

With this proposed iterative scheme, one must solve a symmetric positive definite linear system

$$\left[A^t Q(\mathbf{f}_c^{(k)})A + \mu D^t D\right]\mathbf{x} = \mathbf{b} \qquad (10.21)$$

for some $\mathbf{b}$ at each iteration. Of course these systems are dense in general, but have structures that can be utilized. We apply a preconditioned conjugate gradient (CG) method to solving these linear systems.

10.4.1   The Preconditioned Conjugate Gradient Method

In this subsection, we introduce the conjugate gradient (CG) method for solving linear systems of equations. For a more in-depth treatment of other Krylov space methods see [41]. The CG method was invented in the 1950s [42] (Hestenes and Steifel, 1952) as a direct method for solving Hermitian positive definite systems. It has come into wide use over the last 20 years as an iterative method.

Let us consider $A\mathbf{x} = \mathbf{b}$ where $A \in \mathbb{C}^{n \times n}$ is a nonsingular Hermitian positive definite matrix and $\mathbf{b} \in \mathbb{C}^n$. Given an initial guess $\mathbf{x}_0$ and the corresponding initial residual $\mathbf{r}_0 = \mathbf{b} - A\mathbf{x}_0$, the kth iterate $\mathbf{x}_k$ of CG minimizes the functional

$$\phi(\mathbf{x}) \equiv \frac{1}{2}\mathbf{x}^* A\mathbf{x} - \mathbf{x}^* \mathbf{b}$$

over $\mathbf{x}_0 + \mathcal{K}_k$, where $\mathcal{K}_k$ is the kth Krylov subspace

$$\mathcal{K}_k \equiv \operatorname{span}(\mathbf{r}_0, A\mathbf{r}_0, \ldots, A^{k-1}\mathbf{r}_0), \quad k = 1, 2, \ldots.$$

Note that if $\mathbf{x}$ minimizes $\phi(\mathbf{x})$, then $\nabla\phi(\mathbf{x}) = A\mathbf{x} - \mathbf{b} = 0$ and hence $\mathbf{x}$ is the solution.

Denote $\mathbf{x}_t$, the true solution of the system, and define the norm

$$\|\mathbf{x}\|_A \equiv \sqrt{\mathbf{x}^* A\mathbf{x}}.$$

One can show that minimizing $\phi(\mathbf{x})$ over $\mathbf{x}_0 + \mathcal{K}_k$ is the same as minimizing $\|\mathbf{x} - \mathbf{x}_t\|_A$ over $\mathbf{x}_0 + \mathcal{K}_k$, i.e.,

$$\|\mathbf{x}_t - \mathbf{x}_k\|_A = \min_{\mathbf{y} \in \mathbf{x}_0 + \mathcal{K}_k} \|\mathbf{x}_t - \mathbf{y}\|_A.$$

Since any $\mathbf{y} \in \mathbf{x}_0 + \mathcal{K}_k$ can be written as

$$\mathbf{y} = \mathbf{x}_0 + \sum_{i=0}^{k-1} \alpha_i A^i \mathbf{r}_0$$

for some coefficients $\{\alpha_i\}_{i=0}^{k-1}$, we can express $\mathbf{x}_t - \mathbf{y}$ as

$$\mathbf{x}_t - \mathbf{y} = \mathbf{x}_t - \mathbf{x}_0 - \sum_{i=0}^{k-1} \alpha_i A^i \mathbf{r}_0.$$

As $\mathbf{r}_0 = \mathbf{b} - A\mathbf{x}_0 = A(\mathbf{x}_t - \mathbf{x}_0)$, we have

$$\mathbf{x}_t - \mathbf{y} = \mathbf{x}_t - \mathbf{x}_0 - \sum_{i=0}^{k-1} \alpha_i A^{i+1}(\mathbf{x}_t - \mathbf{x}_0) = p(A)(\mathbf{x}_t - \mathbf{x}_0),$$

where the polynomial

$$p(z) = 1 - \sum_{i=0}^{k-1} \alpha_i z^{i+1}$$

has degree k and satisfies $p(0) = 1$. Hence

$$\|\mathbf{x}_t - \mathbf{x}_k\|_A = \min_{p \in \mathcal{P}_k, p(0)=1} \|p(A)(\mathbf{x}_t - \mathbf{x}_0)\|_A, \qquad (10.22)$$

where $\mathcal{P}_k$ is the set of polynomials of degree k.

Hermitian positive definite matrices asserts that $A = U\Lambda U^*$, where U is a unitary matrix whose columns are the eigenvectors of A and Λ is the diagonal matrix with the positive eigenvalues of A on the diagonal. Since $UU^* = U^*U = I$, we have $A^k = U\Lambda^k U^*$. Hence $p(A) = Up(\Lambda)U^*$. Defining $A^{\frac{1}{2}} = U\Lambda^{\frac{1}{2}}U^*$, we have

$$\|p(A)\mathbf{x}\|_A = \|A^{\frac{1}{2}}p(A)\mathbf{x}\|_2 \leq \|p(A)\|_2 \|\mathbf{x}\|_A.$$

Together with (10.22), this implies that

$$\|\mathbf{x}_t - \mathbf{x}_k\|_A \leq \|\mathbf{x}_t - \mathbf{x}_0\|_A \min_{p \in \mathcal{P}_k, p(0)=1} \max_{\lambda \in \sigma(A)} |p(\lambda)|. \qquad (10.23)$$

Clearly, if $k = n$, we can choose p to be the nth-degree polynomial that passes through all the eigenvalues $\lambda \in \sigma(A)$ with $p(0) = 1$. Then the maximum in the right-hand side of (10.23) is zero and we have the following theorem given by Hestenes and Steifel [42]:

THEOREM 10.3 *Let A be a Hermitian positive definite matrix of size n. Then the CG algorithm finds the solution of $A\mathbf{x} = \mathbf{b}$ within n iterations in the absence of roundoff errors.*

In most applications, the number of unknowns n is very large. It is better to consider CG as an iterative method and terminate the iteration when some specified error tolerance is reached. The usual implementation of the CG method is to find, for a given ϵ, a vector $\mathbf{x}$ such that $\|\mathbf{b} - A\mathbf{x}\|_2 \leq \epsilon\|\mathbf{b}\|_2$. Algorithm CG is a typical implementation of the method. Its inputs are the right-hand side $\mathbf{b}$, a routine which computes the action of A on a vector, and the initial guess $\mathbf{x}_0$ which will be overwritten by the subsequent iterates $\mathbf{x}_k$. We limit the number of iterations to $k_{\max}$ and return the solution $\mathbf{x}_k$ and the residual norm ρ_k.

Algorithm $\mathrm{CG}(\mathbf{x}, \mathbf{b}, A, \epsilon, k_{\max})$

1. $\mathbf{r} = \mathbf{b} - A\mathbf{x}$, $\rho_0 = \|\mathbf{r}\|_2^2$, $k = 1$

2. Do while $\sqrt{\rho_{k-1}} > \epsilon\|\mathbf{b}\|_2$ and $k < k_{\max}$

   > if $k = 1$ then $\mathbf{p} = \mathbf{r}$
   > else
   > $\beta = \rho_{k-1}/\rho_{k-2}$ and $\mathbf{p} = \mathbf{r} + \beta\mathbf{p}$
   > $\mathbf{w} = A\mathbf{p}$
   > $\alpha = \rho_{k-1}/\mathbf{p}^*\mathbf{w}$
   > $\mathbf{x} = \mathbf{x} + \alpha\mathbf{p}$
   > $\mathbf{r} = \mathbf{r} - \alpha\mathbf{w}$
   > $\rho_k = \|\mathbf{r}\|_2^2$
   > $k = k + 1$

3. End Do

Note that the matrix A itself need not be formed or stored — only a routine for matrix–vector products $A\mathbf{p}$ is required.

Next we consider the cost. We need to store only four vectors: $\mathbf{x}, \mathbf{w}, \mathbf{p}$, and $\mathbf{r}$. Each iteration requires a single matrix–vector product to compute $\mathbf{w} = A\mathbf{p}$, two scalar products (one for $\mathbf{p}^*\mathbf{w}$ and one to compute $\rho_k = \|\mathbf{r}\|_2^2$), and three operations of the form $\alpha\mathbf{x} + \mathbf{y}$, where $\mathbf{x}$ and $\mathbf{y}$ are vectors and α is a scalar. Thus, besides the matrix–vector multiplication, each iteration of Algorithm CG requires $O(n)$ operations[1], where n is the size of the matrix A.

The convergence rate of the conjugate gradient method can be bounded using the condition number $\kappa(A)$ of the matrix A, where

$$\kappa(A) \equiv \|A\|_2 \cdot \|A^{-1}\|_2 = \frac{\lambda_{\max}(A)}{\lambda_{\min}(A)}.$$

---

[1] Let f and g be two functions defined on the set of integers. We say that $f(n) = O(g(n))$ as $n \to \infty$ if there exists a constant K such that $|f(n)/g(n)| \leq K$ as $n \to \infty$.

In fact, by choosing p in (10.23) to be a kth-degree Chebyshev polynomial, one can derive the following theorem in Luenberger [43, p. 187].

THEOREM 10.4 *Let A be a Hermitian positive definite matrix with condition number $\kappa(A)$. Then the kth iterate $\mathbf{x}_k$ of the conjugate gradient method satisfies*

$$\frac{\|\mathbf{x}_t - \mathbf{x}_k\|_A}{\|\mathbf{x}_t - \mathbf{x}_0\|_A} \leq 2 \left(\frac{\sqrt{\kappa(A)} - 1}{\sqrt{\kappa(A)} + 1} \right)^k. \tag{10.24}$$

In particular, for any given tolerance $\tau > 0$, $\|\mathbf{x}_t - \mathbf{x}_k\|_A / \|\mathbf{x}_t - \mathbf{x}_0\|_A \leq \tau$ if

$$k \geq \frac{1}{2} \sqrt{\kappa(A)} \log \left(\frac{2}{\tau} \right) + 1 = O(\sqrt{\kappa(A)}). \tag{10.25}$$

This shows that the convergence rate is at least linear:

$$\frac{\|\mathbf{x}_t - \mathbf{x}_k\|_A}{\|\mathbf{x}_t - \mathbf{x}_0\|_A} \leq 2r^k$$

where $r < 1$. However, if we have more information about the spectrum of A, then we can have a better bound of the error (see [44]).

When the condition number or the distribution of the eigenvalues of a matrix is large, one can improve the performance of the CG iteration by preconditioning. In effect, one tries to replace the given system $A\mathbf{x} = \mathbf{b}$ by another Hermitian positive definite system with the same solution $\mathbf{x}$, but with the new coefficient matrix having a more favorable spectrum.

Suppose that P is a Hermitian positive definite matrix such that either the condition number of $P^{-1}A$ is close to 1 or the eigenvalues of $P^{-1}A$ are clustered around 1. Then, the CG method, when applied to the preconditioned system

$$P^{-1}A\mathbf{x} = P^{-1}\mathbf{b},$$

will converge very fast. We will call P the *preconditioner* of the system $A\mathbf{x} = \mathbf{b}$ or of the matrix A. The following algorithm is a typical implementation of the preconditioned conjugate gradient (PCG) method. The input of this algorithm is the same as that for Algorithm CG, except that we now have an extra routine to compute the action of the inverse of the preconditioner on a vector.

Algorithm PCG$(\mathbf{x}, \mathbf{b}, A, P, \epsilon, k_{\max})$

1. $\mathbf{r} = \mathbf{b} - A\mathbf{x}$, $\rho_0 = \|\mathbf{r}\|_2^2$, $k = 1$

2. Do while $\sqrt{\rho_{k-1}} > \epsilon\|\mathbf{b}\|_2$ and $k < k_{\max}$

   $\mathbf{z} = P^{-1}\mathbf{r}$ (or solve $P\mathbf{z} = \mathbf{r}$)

$$\tau_{k-1} = \mathbf{z}^*\mathbf{r}$$

if $k = 1$ then $\beta = 0$ and $\mathbf{p} = \mathbf{z}$

else

$\beta = \tau_{k-1}/\tau_{k-2}$ and $\mathbf{p} = \mathbf{z} + \beta\mathbf{p}$

$\mathbf{w} = A\mathbf{p}$

$\alpha = \tau_{k-1}/\mathbf{p}^*\mathbf{w}$

$\mathbf{x} = \mathbf{x} + \alpha\mathbf{p}$

$\mathbf{r} = \mathbf{r} - \alpha\mathbf{w}$

$\rho_k = \|\mathbf{r}\|_2^2$

$k = k + 1$

3. **End Do**

Note that the cost of Algorithm PCG is identical to that of Algorithm CG with the addition of the solution of the preconditioner system $P\mathbf{z} = \mathbf{r}$ and the inner product to compute τ_{k-1} in Step 2. Thus the criteria for choosing a good preconditioner P are:

• The system $P\mathbf{z} = \mathbf{r}$ for any given $\mathbf{r}$ should be solved efficiently.

• The spectrum of $P^{-1}A$ should be clustered and its condition number should be close to 1.

10.4.2   Cosine Transform-Based Preconditioners

In this application, we need to compute a matrix–vector product

$$(A^t Q(\mathbf{f}_c^{(k)})A + \mu D^t D)\mathbf{v}$$

for some vector $\mathbf{v}$ in each CG iteration. Since the matrices A and A^t are Toeplitz-plus-Hankel matrices, their matrix–vector multiplications can be done in $O(n \log n)$ operations for any n-vector (see, for instance, [45]). However, for the matrix–vector product

$$Q(\mathbf{f}_c^{(k)})\mathbf{v} \equiv \{[T(\mathbf{f}_c^{(k)})|I][T(\mathbf{f}_c^{(k)})|I]^t\}^{-1}\mathbf{v},$$

we need to solve another linear system

$$\left\{[T(\mathbf{f}_c^{(k)})|I][T(\mathbf{f}_c^{(k)})|I]^t\right\}\mathbf{z} = \mathbf{v}. \tag{10.26}$$

Notice that the matrix–vector multiplications $T(\mathbf{f}_c^{(k)})\mathbf{y}$ and $T(\mathbf{f}_c^{(k)})^t\mathbf{v}$ can also be computed $O(n \log n)$ operations for any n-vector $\mathbf{y}$. A preconditioned conjugate gradient method will also be used for solving this symmetric positive definite linear system.

We remark that all matrices that can be diagonalized by the discrete cosine transform matrix C must be symmetric [7], so C above can only diagonalize matrices with symmetric point spread functions for this problem. On the other hand, for nonsymmetric point spread functions, we can construct cosine transform-based preconditioners to speed up the convergence of the conjugate gradient method.

Given a matrix X, we define the optimal cosine transform preconditioner $c(X)$ to be the minimizer of $\|X - Q\|_F^2$ over all Q that can diagonalized by C (see [7]). In this case, the cosine transform preconditioner $c(A)$ of A in (10.12) is defined to be the matrix $C^t \Lambda C$ such that Λ minimizes

$$\|C^t \Lambda C - A\|_F^2.$$

Here Λ is any nonsingular diagonal matrix. Clearly, the cost of computing $c(A)^{-1}\mathbf{y}$ for any n-vector $\mathbf{y}$ is $O(n \log n)$ operations. In [7], Ng et al. gave a simple approach for finding $c(A)$. The cosine transform preconditioner $c(A)$ is just the blurring matrix (cf. (10.12)) corresponding to the symmetric point spread function $s_i \equiv (h_i + h_{-i})/2$ with the Neumann boundary condition imposed. This approach allows us to precondition the symmetric positive definite linear system (10.21).

Next we construct the cosine transform preconditioner for

$$\{[T(\mathbf{f}_c^{(k)})|I][T(\mathbf{f}_c^{(k)})|I]^t\}$$

which exploits the Toeplitz structure of the matrix. We approximate $T(\mathbf{f}_c)$ by

$$\tilde{T}(\mathbf{f}_c) = \frac{1}{2n-1} \begin{bmatrix} f_n & f_{n-1} & \cdots & f_2 & f_1 & 0 & \cdots \cdots & 0 \\ 0 & f_n & \ddots & \ddots & f_2 & f_1 & \ddots & 0 \\ \vdots & \ddots & \ddots & \ddots & \ddots & \ddots & \ddots & \vdots \\ 0 & & \ddots & f_n & f_{n-1} & \ddots & \ddots & f_1 & 0 \\ 0 & \cdots & \cdots & 0 & f_n & f_{n-1} & \cdots & f_2 & f_1 \end{bmatrix}.$$

In [46], Ng and Plemmons have proved that if $\mathbf{f}_c$ is a stationary stochastic process, then the expected value of $T(\mathbf{f}_c)T(\mathbf{f}_c)^t - \tilde{T}(\mathbf{f}_c)\tilde{T}(\mathbf{f}_c)^t$ is close to zero. Since $\tilde{T}(\mathbf{f}_c)\tilde{T}(\mathbf{f}_c)^t$ is a Toeplitz matrix, $c(\tilde{T}(\mathbf{f}_c)\tilde{T}(\mathbf{f}_c)^t)$ can be found in $O(n)$ operations (see [45]). However, the original matrix $T(\mathbf{f}_c)T(\mathbf{f}_c)^t$ is much more complicated and thus the construction cost of $c(\tilde{T}(\mathbf{f}_c)\tilde{T}(\mathbf{f}_c)^t)$ is cheaper than that of $c(T(\mathbf{f}_c)T(\mathbf{f}_c)^t)$. It is clear that the cost of computing $c(\tilde{T}(\mathbf{f}_c))^{-1}\mathbf{y}$ for any n-vector $\mathbf{y}$ is again $O(n \log n)$ operations. It follows that the cost per each iteration in solving the linear systems (10.21) and (10.26) are $O(n \log n)$ operations.

Finally, we remark that the objective function is simplified in the cosine transform domain when the symmetry constraints are incorporated into the total least squares formulation. In accordance with Theorem 2, the minimization of $P(\hat{\mathbf{f}}_c)$ in (10.18) is equivalent to

$$\min_{\hat{\mathbf{f}}_i} \left[\frac{(\mathbf{w}_i\hat{\mathbf{f}}_i - \hat{\mathbf{g}}_i)^2}{\hat{\mathbf{f}}_i^2 + 1} + \mu\Lambda_{ii}\hat{\mathbf{f}}_i^2 \right], \quad 1 \leq i \leq n.$$

We note that the objective function is decoupled into n equations, each to be minimized independently with respect to one DCT coefficient of $\hat{\mathbf{f}}_c$. It follows that each minimizer can be determined by a one-dimensional search method (see [40]).

10.5   Two-Dimensional Deconvolution Problems

The results of the previous section extend in a natural way to image deconvolution. The main interest concerns optical image enhancement. Applications of image deconvolution in optics can be found in many areas of science and engineering, e.g., see the book by Roggemann and Welsh [1]. For example, work to enhance the quality of optical images has important applications in astronomical imaging [10]. Only partial *a priori* knowledge about the degradation phenomena or point spread function in aero-optics is generally known, so here the use of constrained total least squares methods is appropriate. In addition, the estimated point spread function is generally degraded in a manner similar to that of the observed image [1].

Let $f(x, y)$ and $\bar{g}(x, y)$ be the functions of the original and the blurred images, respectively. The image restoration problem can be expressed as a linear integral equation

$$\bar{g}(x, y) = \int \int \bar{h}(x - y, u - v) f(y, v) dy dv. \tag{10.27}$$

The convolution operation, as is often the case in optical imaging, acts uniformly (i.e., in a spatially invariant manner) on f. We consider numerical methods for approximating the solution to the linear restoration problem in discretized (matrix) form obtained from (10.27). For notation purposes we assume that the image is n-by-n, and thus contains n^2 pixels. Typically, n is chosen to be a power of 2, such as 256 or larger. Then the number of unknowns grows to at least 65,536. The vectors $\mathbf{f}$ and $\bar{\mathbf{g}}$ represent the "true" and observed image pixel values, respectively, unstacked by rows. After discretization of (10.27), the blurring matrix $\bar{H}$ defined by $\bar{h}$ is given by

$$
\bar{H} = \left[\begin{array}{ccc|cccc|ccc}
H^{(n-1)} & \cdots & H^{(1)} & H^{(0)} & \cdots & \cdots & H^{(-n+1)} & 0 & \cdots & 0 \\
0 & \ddots & & \vdots & & & \vdots & H^{(-n+1)} & 0 & \vdots \\
\vdots & & 0 \; H^{(n-1)} & \vdots & & & \vdots & & \ddots & 0 \\
0 & \cdots & 0 & H^{(n-1)} & \cdots & \cdots & H^{(0)} & H^{(-1)} & \cdots & H_{(-n+1)}
\end{array}\right]
$$
(10.28)

with each subblock $H^{(j)}$ being an n-by-$(2n-1)$ matrix of the form given by (10.7). The dimensions of the discrete point spread function h are $2n-1$ and $2n-1$ in the x-direction and y-direction, respectively. In the above partitioned matrix, the left- and the right-hand side matrices refer to the parts of the blurring matrix affecting the image boundary. The middle one refers to the main part of the blurring matrix affecting the inside image.

Applying Neumann boundary conditions, the resulting matrix A is a block-Toeplitz-plus-Hankel matrix with Toeplitz-plus-Hankel blocks. More precisely,

$$
A = \left[\begin{array}{ccccc}
A^{(0)} & A^{(-1)} & \cdots & \cdots & A^{(-n+1)} \\
A^{(1)} & A^{(0)} & \ddots & \ddots & \vdots \\
\vdots & \ddots & \ddots & \ddots & \vdots \\
\vdots & & \ddots & \ddots & A^{(0)} \; A^{(-1)} \\
A^{(n-1)} & \cdots & \cdots & A^{(1)} & A^{(0)}
\end{array}\right] + \left[\begin{array}{ccccc}
A^{(1)} & A^{(2)} & \cdots & A^{(n-1)} & 0 \\
A^{(2)} & & & \iddots & A^{(-n+1)} \\
\vdots & & \iddots & \iddots & \vdots \\
A^{(n-1)} & \iddots & \iddots & & A^{(-2)} \\
0 & A^{(-n+1)} & \cdots & A^{(-2)} & A^{(-1)}
\end{array}\right]
$$
(10.29)

with each block $A^{(j)}$ being an n-by-n matrix of the form given in (10.12). We note that the $A^{(j)}$ in (10.29) and the $H^{(j)}$ in (10.28) are related by (10.12). A detailed discussion of using the Neumann boundary conditions for two-dimensional problems can be found in [7].

Using a similar argument, we can formulate the regularized constrained total least squares problems under the Neumann boundary conditions.

THEOREM 10.5 *Under the Neumann boundary condition (10.3), the regularized constrained total least squares solution can be obtained as the $\mathbf{f}_c$ that minimizes the functional:*

$$
P(\mathbf{f}_c) = (A\mathbf{f}_c - g)^t([T|I][T|I]^t)^{-1}(A\mathbf{f}_c - g) + \mu\mathbf{f}_c^t D^t D\mathbf{f}_c,
$$

where T is an n^2-by-$(2n-1)^2$ block-Toeplitz-Toeplitz-block matrix

$$T = \frac{1}{n} \begin{bmatrix} T_n & T_{n-1} & \cdots & T_2 & T_1 & T_1 & \cdots & T_{n-2} & T_{n-1} \\ T_n & T_n & \ddots & \ddots & T_2 & T_1 & \ddots & T_{n-3} & T_{n-2} \\ \vdots & \ddots & \ddots & \ddots & \ddots & \ddots & \ddots & \ddots & \vdots \\ T_3 & T_4 & \ddots & T_n & T_{n-1} & \ddots & \ddots & T_1 & T_1 \\ T_2 & T_3 & \cdots & T_n & T_n & T_{n-1} & \cdots & T_2 & T_1 \end{bmatrix}$$

and each subblock T_j is a n-by-$(2n-1)$ matrix of the form given by (10.7).

For a symmetric point spread function, we have the following theorem.

THEOREM 10.6   *Under the Neumann boundary condition (10.3) and the symmetry constraint*

$$h_{i,j} = h_{i,-j} = h_{-i,j} = h_{-i,-j},$$

the regularized constrained total least squares solution can be obtained as the $\hat{\mathbf{f}}_c$ that minimizes the functional

$$P(\mathbf{f}_c) = [\mathrm{diag}(\mathbf{w})\hat{\mathbf{f}}_c - \hat{\mathbf{g}}]^t \{[\mathrm{diag}(\hat{\mathbf{f}}_c)|I][\mathrm{diag}(\hat{\mathbf{f}}_c)|I]^t\}^{-1}[\mathrm{diag}(\mathbf{w})\hat{\mathbf{f}}_c - \hat{\mathbf{g}}] + \mu\hat{\mathbf{f}}_c^t\Lambda\hat{\mathbf{f}}_c,$$

where $\hat{\mathbf{f}} = C\mathbf{f}$, $\hat{\mathbf{g}} = C\mathbf{g}$, and Λ is an n-by-n diagonal matrix given by $\Lambda = CD^t DC^t$ and D is the regularization operator.

10.6   Numerical Examples

In this section, we illustrate that the quality of a restored image given by using the regularized, constrained TLS method with the Neumann boundary conditions is generally superior to that obtained by the least squares method. The data source is a photo from the 1964 Gatlinburg Conference on Numerical Algebra taken from Matlab. From (10.2), we see that to construct the right-hand side vector $\bar{\mathbf{g}}$ correctly, we need the vectors $\mathbf{f}_l$ and $\mathbf{f}_r$, i.e., we need to know the image outside the given domain. Thus we start with the 480×640 image of the photo and cut out a 256×256 portion from the image. Figure 10.5 gives the 256×256 image of this picture.

We consider restoring the "Gatlinburg Conference" image blurred by a truncated (band-limited) Gaussian point spread function,

$$h_{i,j} = \begin{cases} ce^{-0.1(i^2+j^2)}, & \text{if } |i-j| \le 8, \\ 0, & \text{otherwise,} \end{cases}$$

FIGURE 10.5: "Gatlinburg Conference" test image.

see [30, p. 269], where $h_{i,j}$ is the jth entry of the first column of $A^{(i)}$ in (10.29) and c is the normalization constant such that $\sum_{i,j} h_{i,j} = 1$. We remark that the Gaussian point spread function is symmetric, and is often used in the literature to simulate the blurring effects of imaging through the atmosphere [1, 3]. Gaussian noise with signal-to-noise ratios of 20dB, 30dB, and 40dB is then added to the blurred images and the point spread functions to produce test images. Noisy, blurred images are shown in Figure 10.6(a). We note that after the blurring, the cigarette held by Prof. Householder (the rightmost person) is not clearly shown.

In Table 10.1, we present results for the regularized constrained total least squares method with PSF symmetry constraints. We denote this method by RCTLS in the table. As a comparison, the results in solving the regularized least squares (RLS) problems with the exact and noisy point spread functions are also listed. For all methods tested, the Neumann boundary conditions are employed and the corresponding blurring matrices can be diagonalized by a discrete cosine transform matrix. Therefore, the image restoration can be done efficiently in the transform domain. We remark that Tikhonov regularization of the least squares method can be recast as in a total least squares framework (see [19]). In the tests, we used the L_2 norm as the regularization functional. The corresponding regularization parameters μ are chosen to minimize the relative error of the restored image which is defined as

$$\frac{\|\mathbf{f}_c - \mathbf{f}_c(\mu)\|_2}{\|\mathbf{f}_c\|_2}, \tag{10.30}$$

where $\mathbf{f}_c$ is the original image. In the tests, the regularization parameters are obtained by trial and error.

We see from Table 10.1 that the relative errors in using the RCTLS method are less than that of using the RLS method with the noisy PSF, except in the case where the SNR of noises added to the blurred image and PSF are 20dB and 40dB, respectively. However, for some cases, the improvement of using the RCTLS method is not significant when the SNR ratio is low, that is, the noise

TABLE 10.1: The relative errors for different methods.

Blurred image Noise added SNR (dB)	PSF Noise added SNR (dB)	Exact PSF RLS method	Noisy PSF RLS method	Noisy PSF RCTLS method
40	20	7.78×10^{-2}	1.07×10^{-1}	8.94×10^{-2}
40	30	7.78×10^{-2}	8.72×10^{-2}	8.48×10^{-2}
40	40	7.78×10^{-2}	8.07×10^{-2}	8.03×10^{-2}
30	20	8.66×10^{-2}	1.07×10^{-1}	9.98×10^{-2}
30	30	8.66×10^{-2}	9.17×10^{-2}	8.88×10^{-2}
30	40	8.66×10^{-2}	8.75×10^{-2}	8.71×10^{-2}
20	20	9.68×10^{-2}	1.13×10^{-1}	1.09×10^{-1}
20	30	9.68×10^{-2}	1.00×10^{-1}	9.99×10^{-2}
20	40	9.68×10^{-2}	9.76×10^{-2}	9.77×10^{-2}

level to the blurred image is very high. In Figure 10.6, we present the restored images for different methods. We see from Figure 10.6 that the cigarette is better restored by using the RCTLS method than that by using the RLS method. We remark that the corresponding relative error is also significantly smaller than that obtained by using the RLS method. When noise with a low SNR is added to the blurred image and point spread function, visually the restored images look similar. In Figure 10.6(e), we present the restored images for the periodic boundary condition using the RCTLS method. We see from all the figures that by using the RCTLS method and imposing the Neumann boundary condition, the relative errors and the ringing effects in the restorations are significantly reduced.

10.7 Application: High-Resolution Image Reconstruction

Processing methods for single images often provide unacceptable results because of the ill-conditioned nature of associated inverse problems and the lack of diversity in the data. Therefore, image processing using a sequence of images, as well as images captured simultaneously using a camera with multiple lenslets, has developed into an active research area because multiple deconvolution operators can be used to make the problem better posed. Rapid progress in computer and semiconductor technology is making it possible to implement such image processing tasks reasonably quickly, but the need for processing in real time requires attention to design of efficient and robust algorithms for implementation on current and future generations of

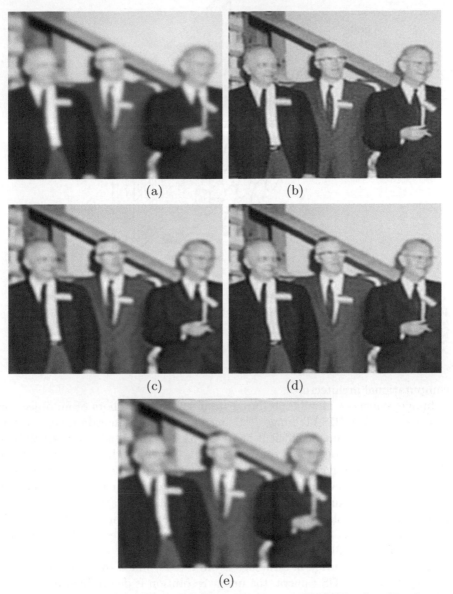

(a)

(b)

(c)

(d)

(e)

FIGURE 10.6: (a) Noisy and blurred images with SNR of 40dB, the restored images by using (b) the exact PSF (rel. err. $= 7.78 \times 10^{-2}$), (c) the RLS method for the noisy PSF with SNR of 20dB (rel. err. $= 1.07 \times 10^{-1}$), (d) the RCTLS method for a noisy PSF with SNR of 20dB under Neumann boundary conditions (rel. err. $= 8.94 \times 10^{-2}$), and (e) the RCTLS method for a noisy PSF with SNR of 20dB under periodic boundary conditions (rel. err. $= 1.14 \times 10^{-1}$).

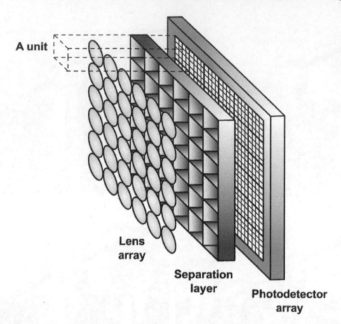

A unit

**Lens
array**

**Separation
layer**

**Photodetector
array**

FIGURE 10.7: A depiction of the basic components of an array imaging camera system [55].

computational architectures.

Image sequences may be produced from several snapshots of an object or a scene (e.g., LANDSAT images). Using sequential estimation theory in the wave number domain, an efficient method was developed [47] for interpolating and recursively updating to provide filtering provided the displacements of the frames with respect to a reference frame were either known or estimated. It was observed that the performance deteriorated when the blur produced zeros in the wave number domain, and theoretical justification for this can be provided. The problem of reconstruction in the wave number domain with errors present both in the observation and data was tackled by the total least squares method [48].

The spatial resolution of an image is often determined by imaging sensors. In a CCD or a CMOS camera, the image resolution is determined by the size of its photo-detector. An ensemble of several shifted images could be collected by a prefabricated planar array of CCD or CMOS sensors and one may reconstruct with higher resolution which is equivalent to an effective increase of the sampling rate by interpolation. Fabrication limitations are known to cause subpixel displacement errors, which, coupled with observation noise, limit the deployment of least squares techniques in this scenario. TLS is an effective technique for solving a set of such error-contaminated equations [48] and therefore is an appropriate method for consideration in high-resolution image reconstruction applications. However, a possible drawback of using the conventional TLS approach is that the formulation is not constrained to handle point spread functions obtained from multisensors. In this section, an image processing technique that leads to the deployment of constrained total least squares (CTLS) theory is described. Then a computational procedure is advanced and the role of regularization is considered.

10.7.1 Mathematical Model

A brief introduction to a mathematical model in high-resolution image reconstruction is provided first (see Figure 10.8). Details can be found in [49]. Consider a sensor array with $L_1 \times L_2$ sensors in which each sensor has $N_1 \times N_2$ sensing elements (pixels) and the size of each sensing element is $T_1 \times T_2$. The goal is to reconstruct an image of resolution $M_1 \times M_2$, where $M_1 = L_1 N_1$ and $M_2 = L_2 N_2$. To maintain the aspect ratio of the reconstructed image the case where $L_1 = L_2 = L$ is considered. For simplicity, L is assumed to be an even positive integer in the following discussion.

To generate enough information to resolve the high-resolution image, sub-pixel displacements between sensors are necessary. In the ideal case, the sensors are shifted from each other by a value proportional to $T_1/L \times T_2/L$. However, in practice there can be small perturbations around these ideal sub-pixel locations due to imperfections of the mechanical imaging system during fabrication. Thus, for $l_1, l_2 = 0, 1, \cdots, L - 1$ with $(l_1, l_2) \neq (0, 0)$, the horizontal and vertical displacements $d^x_{l_1 l_2}$ and $d^y_{l_1 l_2}$, respectively, of the $[l_1, l_2]$-th sensor with respect to the $[0, 0]$-th reference sensor are given by

$$d^x_{l_1 l_2} = \frac{T_1}{L}(l_1 + \bar{\epsilon}^x_{l_1 l_2}) \quad \text{and} \quad d^y_{l_1 l_2} = \frac{T_2}{L}(l_2 + \bar{\epsilon}^y_{l_1 l_2}),$$

where $\bar{\epsilon}^x_{l_1 l_2}$ and $\bar{\epsilon}^y_{l_1 l_2}$ denote, respectively, the actual normalized horizontal and vertical displacement errors. The estimates, $\epsilon^x_{l_1 l_2}$ and $\epsilon^y_{l_1 l_2}$, of these parameters, $\bar{\epsilon}^x_{l_1 l_2}$ and $\bar{\epsilon}^y_{l_1 l_2}$, can be obtained by manufacturers during camera calibration.

It is reasonable to assume that

$$|\bar{\epsilon}^x_{l_1 l_2}| < \frac{1}{2} \quad \text{and} \quad |\bar{\epsilon}^y_{l_1 l_2}| < \frac{1}{2},$$

because if that is not the case, then the low-resolution images acquired from two different sensors may have more than the desirable overlapping information for reconstructing satisfactorily the high-resolution image [49].

Let $f(x_1, x_2)$ denote the original bandlimited high-resolution scene, as a function of the continuous spatial variables, x_1, x_2. Then the observed low-resolution digital image $\bar{g}_{l_1 l_2}$ acquired from the (l_1, l_2)-th sensor, characterized

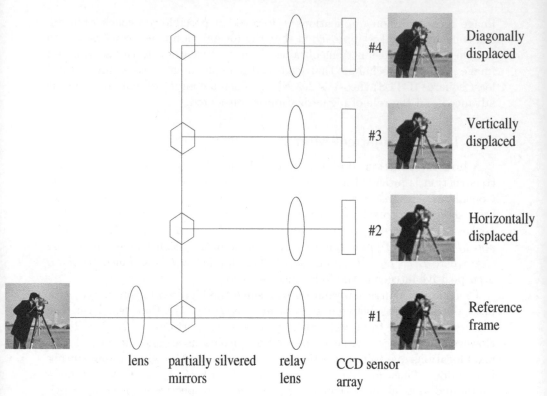

FIGURE 10.8: Example of low-resolution image formation.

by a point spread function, is modeled by

$$\bar{g}_{l_1 l_2}[n_1, n_2] = \int_{T_2(n_2-\frac{1}{2})+d^y_{l_1 l_2}}^{T_2(n_2+\frac{1}{2})+d^y_{l_1 l_2}} \int_{T_1(n_1-\frac{1}{2})+d^x_{l_1 l_2}}^{T_1(n_1+\frac{1}{2})+d^x_{l_1 l_2}} f(x_1, x_2) dx_1 dx_2, \qquad (10.31)$$

for $n_1 = 1, \ldots, N_1$ and $n_2 = 1, \ldots, N_2$. These low-resolution images are combined to yield the $M_1 \times M_2$ high-resolution image $\bar{g}$ by assigning its pixel values according to

$$\bar{g}[L(n_1 - 1) + l_1, L(n_2 - 1) + l_2] = \bar{g}_{l_1 l_2}[n_1, n_2], \qquad (10.32)$$

for $l_1, l_2 = 0, 1, \cdots, (L-1)$, $n_1 = 1, \ldots, N_1$ and $n_2 = 1, \ldots, N_2$.

The continuous image model $f(x_1, x_2)$ in (10.31) can be discretized by the rectangular rule and approximated by a discrete image model. Let **g** and **f** be, respectively, the vectors formed from discretization of $g(x_1, x_2)$ and $f(x_1, x_2)$ using a column ordering. The Neumann boundary condition is applied on the images. This assumes that the scene immediately outside is a reflection of the original scene at the boundary, i.e.,

$$f(i,j) = f(k,l) \quad \text{where} \quad \begin{cases} k = 1 - i, & i < 1, \\ k = 2M_1 + 1 - i, & i > M_1, \\ l = 1 - j, & j < 1, \\ l = 2M_2 + 1 - j, & j > M_2. \end{cases}$$

Under the Neumann boundary condition, the blurring matrices are banded matrices with bandwidth $L + 1$, but the entries at the upper left part and the lower right part of the matrices are changed. The resulting matrices, denoted by $H_{l_1 l_2}^x(\bar{\epsilon}_{l_1,l_2}^x)$ and $H_{l_1 l_2}^y(\bar{\epsilon}_{l_1,l_2}^y)$, have a Toeplitz-plus-Hankel structure,

$$H_{l_1 l_2}^x(\bar{\epsilon}_{l_1,l_2}^x) = \frac{1}{L} \overbrace{\begin{pmatrix} 1 & \cdots & 1 & \frac{1}{2} - \bar{\epsilon}_{l_1 l_2}^x & & 0 \\ \vdots & \ddots & \ddots & \ddots & \ddots & \\ 1 & & \ddots & \ddots & & \ddots & \frac{1}{2} - \bar{\epsilon}_{l_1 l_2}^x \\ \frac{1}{2} + \bar{\epsilon}_{l_1 l_2}^x & \ddots & & \ddots & & \ddots & 1 \\ & \ddots & & \ddots & \ddots & \ddots & \vdots \\ 0 & & \frac{1}{2} + \bar{\epsilon}_{l_1 l_2}^x & 1 & \cdots & 1 \end{pmatrix}}^{L/2 \text{ ones}} +$$

$$\frac{1}{L} \overbrace{\begin{pmatrix} 1 & \cdots & 1 & \frac{1}{2} + \bar{\epsilon}_{l_1 l_2}^x & & 0 \\ \vdots & \ddots & & \ddots & & \\ 1 & & \ddots & & & \frac{1}{2} - \bar{\epsilon}_{l_1 l_2}^x \\ \frac{1}{2} + \bar{\epsilon}_{l_1 l_2}^x & & & \ddots & 1 \\ & & \ddots & \ddots & \vdots \\ 0 & \frac{1}{2} - \bar{\epsilon}_{l_1 l_2}^x & 1 & \cdots & 1 \end{pmatrix}}^{L/2-1 \text{ ones}} . \quad (10.33)$$

The matrix $H_{l_1 l_2}^y(\bar{\epsilon}_{l_1,l_2}^y)$ is defined similarly. The blurring matrix corresponding to the (l_1, l_2)-th sensor under the Neumann boundary condition is given by the Kronecker product,

$$H_{l_1 l_2}(\bar{\epsilon}_{l_1,l_2}) = H_{l_1 l_2}^x(\bar{\epsilon}_{l_1,l_2}^x) \otimes H_{l_1 l_2}^y(\bar{\epsilon}_{l_1,l_2}^y),$$

where the 2×1 vector $\bar{\epsilon}_{l_1,l_2}$ is denoted by $(\bar{\epsilon}_{l_1,l_2}^x \ \bar{\epsilon}_{l_1,l_2}^y)^t$. The blurring matrix for the whole sensor array is made up of blurring matrices from each sensor:

$$H_L(\bar{\epsilon}) = \sum_{l_1=0}^{L-1} \sum_{l_2=0}^{L-1} D_{l_1 l_2} H_{l_1 l_2}(\bar{\epsilon}_{l_1,l_2}), \quad (10.34)$$

where the $2L^2 \times 1$ vector $\bar{\epsilon}$ is defined as

$$\bar{\epsilon} = [\bar{\epsilon}_{00}^x \ \bar{\epsilon}_{00}^y \ \bar{\epsilon}_{01}^x \ \bar{\epsilon}_{01}^y \ \cdots \ \bar{\epsilon}_{L-1L-2}^x \ \bar{\epsilon}_{L-1L-2}^y \ \bar{\epsilon}_{L-1L-1}^x \ \bar{\epsilon}_{L-1L-1}^y]^t.$$

Here $D_{l_1 l_2}$ are diagonal matrices with diagonal elements equal to one if the corresponding component of $\mathbf{g}$ comes from the (l_1, l_2)-th sensor and zero otherwise (see [49] for more details).

10.7.2   Image Reconstruction Formulation

In this subsection, the displacement errors are not known exactly. The spatial invariance of the blurring function translates into the spatial invariance of the displacement error in the blurring matrix. The "true" blur function is represented as follows. For each $l_1, l_2 \in \{0, 1, \cdots, L-1\}$,

$$\bar{\mathbf{h}}_{l_1 l_2}^z = \frac{1}{L} [\overbrace{\frac{1}{2} - \bar{\epsilon}_{l_1 l_2}^z \ \ 1 \ \cdots \ 1 \ \cdots \ 1 \ \ \frac{1}{2} + \bar{\epsilon}_{l_1 l_2}^z}^{\text{there are } L+1 \text{ entries}}]^t = \mathbf{h}_{l_1 l_2}^z + \delta \mathbf{h}_{l_1 l_2}^z, \quad \forall z \in \{x, y\}, \tag{10.35}$$

where the $(L+1) \times 1$ component vectors are

$$\mathbf{h}_{l_1 l_2}^z = \frac{1}{L} [\frac{1}{2} - \epsilon_{l_1 l_2}^z \ \ 1 \ \cdots \ 1 \ \cdots \ 1 \ \ \frac{1}{2} + \epsilon_{l_1 l_2}^z]^t, \quad z \in \{x, y\}$$

and

$$\delta \mathbf{h}_{l_1 l_2}^z = \frac{1}{L} [-\delta \epsilon_{l_1 l_2}^z \ \ 0 \ \cdots \ 0 \ \cdots \ 0 \ \ \delta \epsilon_{l_1 l_2}^z]^t, \quad z \in \{x, y\}.$$

Here $\mathbf{h}_{l_1 l_2}^z$ is the estimated (or measured) point spread function and $\delta \mathbf{h}_{l_1 l_2}^z$ is the error component of the point spread function at the (l_1, l_2)-th sensor in the z-direction ($z \in \{x, y\}$).

The observed signal vector $\bar{\mathbf{g}}$ is also subject to errors. It is assumed that this observed $M_1 M_2 \times 1$ signal vector $\bar{\mathbf{g}} = [g_1 \ \cdots \ g_{M_1 M_2}]^t$ can be represented by

$$\bar{\mathbf{g}} = \mathbf{g} + \delta \mathbf{g}, \tag{10.36}$$

where

$$\delta \mathbf{g} = [\delta g_1 \ \delta g_2 \ \cdots \ \delta g_{M_1 M_2}]^t$$

and $\{\delta g_i\}$ is independent and identically distributed Gaussian noise with zero mean and variance σ_g^2. Then the image reconstruction problem requires the recovery of the vector $\mathbf{f}$ from the given inexact point spread function $\mathbf{h}_{l_1 l_2}^z$ ($l_1 = 0, 1, \cdots, L_1 - 1$, $l_2 = 0, 1, \cdots, L_2 - 1$, $z \in \{x, y\}$) and the observed noisy signal vector $\mathbf{g}$.

The constrained TLS formulation for multisensors is considered. Using

$$H_L(\bar{\epsilon})\mathbf{f} = \sum_{l_1=0}^{L-1} \sum_{l_2=0}^{L-1} D_{l_1 l_2} H_{l_1 l_2}(\bar{\epsilon}_{l_1, l_2})\mathbf{f} = \bar{\mathbf{g}} = \mathbf{g} + \delta \mathbf{g}$$

(from (10.34)) and (10.35) and (10.36), the preceding equation can be reformulated as follows:

$$\left[\sum_{l_1=0}^{L-1}\sum_{l_2=0}^{L-1}D_{l_1l_2}\left(H_{l_1l_2}^x(\epsilon_{l_1,l_2})\otimes H_{l_1l_2}^y(\epsilon_{l_1,l_2})\right)\right]\mathbf{f}-\mathbf{g}+$$

$$\left[\sum_{l_1=0}^{L-1}\sum_{l_2=0}^{L-1}\delta\epsilon_{l_1l_2}^x\left(D_{l_1l_2}E\otimes H_{l_1l_2}^y(\epsilon_{l_1,l_2})\right)\right]\mathbf{f}+$$

$$\left[\sum_{l_1=0}^{L-1}\sum_{l_2=0}^{L-1}\delta\epsilon_{l_1l_2}^y\left(D_{l_1l_2}H_{l_1l_2}^x(\epsilon_{l_1,l_2})\otimes E\right)\right]\mathbf{f}+$$

$$\left[\sum_{l_1=0}^{L-1}\sum_{l_2=0}^{L-1}\delta\epsilon_{l_1l_2}^x\delta\epsilon_{l_1l_2}^y\left(D_{l_1l_2}E\otimes E\right)\right]\mathbf{f}-\delta\mathbf{g}=0$$

or

$$\left[\sum_{l_1=0}^{L-1}\sum_{l_2=0}^{L-1}D_{l_1l_2}\left(H_{l_1l_2}^x(\epsilon_{l_1,l_2})\otimes H_{l_1l_2}^y(\epsilon_{l_1,l_2})\right)\right]\mathbf{f}-\mathbf{g}+\sum_{l_1=0}^{L-1}\sum_{l_2=0}^{L-1}\delta\epsilon_{l_1l_2}^x\mathbf{f}_{l_1l_2}^y+$$

$$\sum_{l_1=0}^{L-1}\sum_{l_2=0}^{L-1}\delta\epsilon_{l_1l_2}^y\mathbf{f}_{l_1l_2}^x+\left[\sum_{l_1=0}^{L-1}\sum_{l_2=0}^{L-1}\delta\epsilon_{l_1l_2}^x\delta\epsilon_{l_1l_2}^y\left(D_{l_1l_2}E\otimes E\right)\right]\mathbf{f}-\delta\mathbf{g}=0,$$

$$(10.37)$$

where

$$E=\frac{1}{L}\begin{pmatrix}\overbrace{0\cdots\quad0}^{L/2\ \text{zeros}}-1&&0\\\vdots&\ddots&\ddots&\ddots&\ddots\\0&\ddots&\ddots&\ddots&\ddots&-1\\1&\ddots&\ddots&\ddots&\ddots&0\\&\ddots&\ddots&\ddots&\ddots&\vdots\\0&&1&0&\cdots&0\end{pmatrix}+\frac{1}{L}\begin{pmatrix}\overbrace{0\cdots\quad0}^{L/2-1\ \text{zeros}}1&&0\\\vdots&\ddots&&\ddots\\0&\ddots&&\ddots&-1\\1&&&&0\\&\ddots&&\ddots&\ddots&\vdots\\0&-1&0&\cdots&0\end{pmatrix}$$

$$\underbrace{\qquad}_{L/2\ \text{zeros}}\qquad\qquad\underbrace{\qquad}_{L/2-1\ \text{zeros}}$$

and

$$\mathbf{f}_{l_1l_2}^x=D_{l_1l_2}\left(H_{l_1l_2}^x(\epsilon_{l_1,l_2})\otimes E\right)\mathbf{f},\qquad\mathbf{f}_{l_1l_2}^y=D_{l_1l_2}\left(E\otimes H_{l_1l_2}^y(\epsilon_{l_1,l_2})\right)\mathbf{f}.$$

The constrained TLS formulation amounts to minimizing the norms of vectors associated with

$$\{\delta\epsilon_{l_1l_2}^x,\delta\epsilon_{l_1l_2}^y\}_{l_1,l_2=0}^{L-1}\quad\text{and}\quad\delta\mathbf{g}$$

as explained below such that (10.37) is satisfied.

However, it is first noted that because $\delta\epsilon_{l_1 l_2}^z$ $(l_1, l_2 = 0, 1, \cdots, L-1)$ should be very small, the quantity $|\delta\epsilon_{l_1 l_2}^x \delta\epsilon_{l_1 l_2}^y|$ can be assumed to be negligible, and, therefore, the nonlinear term

$$\left[\sum_{l_1=0}^{L-1} \sum_{l_2=0}^{L-1} \delta\epsilon_{l_1 l_2}^x \delta\epsilon_{l_1 l_2}^y \left(D_{l_1 l_2} E \otimes E \right) \right] \mathbf{f} \tag{10.38}$$

can be ignored in (10.37). In this case, (10.37) reduces to a linear system involving $\{\delta\epsilon_{l_1 l_2}^x, \delta\epsilon_{l_1 l_2}^y\}_{l_1, l_2=0}^{L-1}$ and $\delta\mathbf{g}$.

For simplicity, denote the $2L^2 \times 1$ vector

$$\triangle = [\delta\epsilon_{00}^x \; \delta\epsilon_{00}^y \; \delta\epsilon_{01}^x \; \delta\epsilon_{01}^y \; \cdots \; \delta\epsilon_{L-1L-2}^x \; \delta\epsilon_{L-1L-2}^y \; \delta\epsilon_{L-1L-1}^x \; \delta\epsilon_{L-1L-1}^y]^t.$$

Mathematically, the constrained TLS formulation can be expressed as

$$\min_{\mathbf{f}} \left\| \begin{bmatrix} \triangle \\ \delta\mathbf{g} \end{bmatrix} \right\|_2^2$$

subject to

$$H_L(\epsilon)\mathbf{f} - \mathbf{g} + \sum_{l_1=0}^{L-1} \sum_{l_2=0}^{L-1} \delta\epsilon_{l_1 l_2}^x \cdot \mathbf{f}_{l_1 l_2}^y + \sum_{l_1=0}^{L-1} \sum_{l_2=0}^{L-1} \delta\epsilon_{l_1 l_2}^y \cdot \mathbf{f}_{l_1 l_2}^x - \delta\mathbf{g} = 0. \tag{10.39}$$

Image reconstruction problems are in general ill-conditioned inverse problems and reconstruction algorithms can be extremely sensitive to noise [28]. Regularization can be used to achieve stability. Using classical Tikhonov regularization [27], stability is attained by introducing a regularization operator P and a regularization parameter μ to restrict the set of admissible solutions. More specifically, the regularized solution $\mathbf{f}$ is computed as the solution to

$$\min_{\mathbf{f}} \left\| \begin{bmatrix} \triangle \\ \delta\mathbf{g} \end{bmatrix} \right\|_2^2 + \mu\|P\mathbf{f}\|_2^2 \tag{10.40}$$

subject to (10.39). The term $\mu\|P\mathbf{f}\|_2^2$ is added in order to regularize the solution. The regularization parameter μ controls the degree of regularity (i.e., degree of bias) of the solution. In many applications [28, 30], $\|P\mathbf{f}\|_2$ is chosen to be $\|\mathbf{f}\|_2$ or $\|R\mathbf{f}\|_2$ where R is a first-order difference operator matrix.

The theorem below characterizes the regularized constrained total least squares (RCTLS) formulation of the high-resolution reconstruction problem with multisensors.

THEOREM 10.7 *The regularized constrained total least squares solution can be obtained as the $\mathbf{f}$ that minimizes the functional:*

$$J(\mathbf{f}) = (H_L(\epsilon)\mathbf{f} - \mathbf{g})^t \{[Q(\mathbf{f}) | -\mathbf{I}_{M_1 M_2}][Q(\mathbf{f}) | -\mathbf{I}_{M_1 M_2}]^t\}^{-1} (H_L(\epsilon)\mathbf{f} - \mathbf{g}) + \mu\mathbf{f}^t P^t P\mathbf{f}, \tag{10.41}$$

$$Q(\mathbf{f}) = [\,\mathbf{f}_{00}^y \mid \mathbf{f}_{00}^x \mid \mathbf{f}_{01}^y \mid \mathbf{f}_{01}^x \mid \cdots \mid \mathbf{f}_{L-1L-2}^y \mid \mathbf{f}_{L-1L-2}^x \mid \mathbf{f}_{L-1L-1}^y \mid \mathbf{f}_{L-1L-1}^x\,]$$

and $\mathbf{I}_{M_1 M_2}$ is the $M_1 M_2$-by-$M_1 M_2$ identity matrix.

The proof of the above theorem is similar to Theorems 10.1 and 10.5. The resulting objective function (10.41) to be minimized is nonconvex and nonlinear. In the next section, an iterative algorithm that takes advantage, computationally, of the fast solvers for image reconstruction problems with known displacement errors [49, 50] is developed.

By (10.39) and (10.40),

$$\min_{\mathbf{f},\triangle} \; J(\mathbf{f},\triangle) \equiv \min_{\mathbf{f},\triangle} \left\{ \|\triangle\|_2^2 + \right.$$

$$\left. + \left\| H_L(\epsilon)\mathbf{f} - \mathbf{g} + \sum_{l_1=0}^{L-1}\sum_{l_2=0}^{L-1} \delta\epsilon_{l_1 l_2}^x \cdot \mathbf{f}_{l_1 l_2}^y + \sum_{l_1=0}^{L-1}\sum_{l_2=0}^{L-1} \delta\epsilon_{l_1 l_2}^y \cdot \mathbf{f}_{l_1 l_2}^x \right\|_2^2 + \mu\|P\mathbf{f}\|_2^2 \right\}.$$

$$(10.42)$$

It is noted that the above objective function is equivalent to (10.41).

Before solving for $\mathbf{f}$ in the RCTLS formulation, it is first noted that for a given $\mathbf{f}$, the function $J(\mathbf{f}, \cdot)$ is convex with respect to $\triangle$, and for a given $\triangle$, the function $J(\cdot, \triangle)$ is also convex with respect to $\mathbf{f}$. Therefore, with an initial guess $\triangle_0$, one can minimize (10.42) by first solving

$$J(\mathbf{f}_1, \triangle_0) \equiv \min_{\mathbf{f}} J(\cdot, \triangle_0)$$

and then

$$J(\mathbf{f}_1, \triangle_1) \equiv \min_{\triangle} J(\mathbf{f}_1, \cdot).$$

Therefore, an alternating minimization algorithm is developed in which the function value $J(\mathbf{f}_n, \triangle_n)$ always decreases as n increases. More precisely, the algorithm is stated as follows:

Assume that $\triangle_{n-1}$ is available:

- Determine $\mathbf{f}_n$ by solving the following least squares problem:

$$\min_{\mathbf{f_n}} \left\{ \left\| H_L(\epsilon)\mathbf{f_n} - \mathbf{g} + \sum_{l_1=0}^{L-1}\sum_{l_2=0}^{L-1} \delta\epsilon_{n-1\ l_1 l_2}^x \mathbf{f}_{n\ l_1 l_2}^y + \right.\right.$$

$$\left.\left. + \sum_{l_1=0}^{L-1}\sum_{l_2=0}^{L-1} \delta\epsilon_{n-1\ l_1 l_2}^y \mathbf{f}_{n\ l_1 l_2}^x \right\|_2^2 + \mu\|P\mathbf{f_n}\|_2 \right\}.$$

$$(10.43)$$

Here,

$$\mathbf{f}^x_{n \, l_1 l_2} = D_{l_1 l_2} \left(H^x_{l_1 l_2}(\epsilon_{l_1,l_2}) \otimes E \right) \mathbf{f}_n \text{ and } \mathbf{f}^y_{n \, l_1 l_2} = D_{l_1 l_2} \left(E \otimes H^y_{l_1 l_2}(\epsilon_{l_1,l_2}) \right) \mathbf{f}_n. \tag{10.44}$$

As we have noted that the nonlinear term (10.38) can be assumed to be negligible, we can add this term in the objective function (10.43). Therefore, the least squares solution $\mathbf{f}_n$ can be found by solving the following linear system:

$$\left[H_L(\epsilon + \triangle_{n-1})^t H_L(\epsilon + \triangle_{n-1}) + \mu P^t P \right] \mathbf{f}_n = H_L(\epsilon + \triangle_{n-1})^t \mathbf{g}. \tag{10.45}$$

- Determine $\triangle_n$ by solving the following least squares problem:

$$\min_{\triangle_n} \left\{ \| \triangle_n \|_2^2 + \Big\| H_L(\epsilon)\mathbf{f}_n - \mathbf{g} + \sum_{l_1=0}^{L-1} \sum_{l_2=0}^{L-1} \delta\epsilon^x_{l_1 l_2} \mathbf{f}^y_{n \, l_1 l_2} + \right.$$

$$\left. + \sum_{l_1=0}^{L-1} \sum_{l_2=0}^{L-1} \delta\epsilon^y_{l_1 l_2} \mathbf{f}^x_{n \, l_1 l_2} \Big\|_2^2 \right\}. \tag{10.46}$$

Using (10.44) and (10.41), the above equation can be rewritten as

$$\min_{\triangle_n} \left\{ \| \triangle_n \|_2^2 + \| H_L(\epsilon)\mathbf{f}_n - \mathbf{g} + Q(\mathbf{f}_n)\triangle_n \|_2^2 \right\}, \tag{10.47}$$

where $Q_n(\mathbf{f})$ is defined as in Theorem 10.7. The cost function in (10.47) becomes

$$\begin{aligned} J(\mathbf{f}_n, \triangle_n) &= \triangle_n^t \triangle_n + (H_L(\epsilon)\mathbf{f}_n - \mathbf{g} + Q(\mathbf{f}_n)\triangle_n)^t (H_L(\epsilon)\mathbf{f}_n - \mathbf{g} + Q(\mathbf{f}_n)\triangle_n) \\ &= \triangle_n^t \triangle_n + (H_L(\epsilon)\mathbf{f}_n - \mathbf{g})^t (H_L(\epsilon)\mathbf{f}_n - \mathbf{g}) + \\ &\quad + (H_L(\epsilon)\mathbf{f}_n - \mathbf{g})^t Q(\mathbf{f}_n) \triangle_n + (Q(\mathbf{f}_n)\triangle_n)^t (H_L(\epsilon)\mathbf{f}_n - \mathbf{g}) + \\ &\quad + (Q(\mathbf{f}_n)\triangle_n)^t (Q(\mathbf{f}_n)\triangle_n). \end{aligned}$$

Here $\mathbf{f}_n$ is fixed. The gradient of this cost function with respect to $\triangle_n$ is equal to

$$2\triangle_n + 2Q(\mathbf{f}_n)^t (H_L(\epsilon)\mathbf{f}_n - \mathbf{g}) + 2Q(\mathbf{f}_n)^t Q(\mathbf{f}_n)\triangle_n,$$

and therefore the minimum 2-norm least squares solution $\triangle_n$ is given by

$$\triangle_n = [\mathbf{I}_{2L^2} + Q(\mathbf{f}_n)^t Q(\mathbf{f}_n)]^{-1} Q(\mathbf{f}_n)^t (\mathbf{g} - H_L(\epsilon)\mathbf{f_n}). \tag{10.48}$$

The costly stage in the algorithm is the inversion of the matrix

$$H_L(\epsilon + \triangle_{n-1})^t H_L(\epsilon + \triangle_{n-1}) + \mu P^t P.$$

The preconditioned conjugate gradient method can be applied to solve the linear system

$$[c(H_L(\epsilon + \triangle_{n-1}))^t c(H_L(\epsilon + \triangle_{n-1})) + \mu c(P^t P)]^{-1} \cdot$$
$$[H_L(\epsilon + \triangle_{n-1})^t H_L(\epsilon + \triangle_{n-1}) + \mu P^t P] \mathbf{f}_n =$$
$$[c(H_L(\epsilon + \triangle_{n-1}))^t c(H_L(\epsilon + \triangle_{n-1})) + \mu c(P^t P)]^{-1} H_L(\epsilon + \triangle_{n-1})^t \mathbf{g}. \quad (10.49)$$

It is noted that when the regularization matrix P is equal to R, the first-order difference operator matrix $c(R^t R) = R^t R$ and the matrix $R^t R$ can be diagonalized by the discrete cosine transform matrix. In [50], it has been shown that the singular values of the preconditioned matrices in (10.49) are clustered around 1 for sufficiently small subpixel displacement errors. Hence when the conjugate gradient method is applied to solve the preconditioned system (10.49), fast convergence is expected. Numerical results in [50] have shown that the cosine transform preconditioners can indeed speed up the convergence of the method.

For each PCG iteration, one needs to compute a matrix–vector product

$$\left[H_L(\epsilon + \triangle_n)^t H_L(\epsilon + \triangle_{n-1}) + \mu R^t R \right] \mathbf{v}$$

for some vector $\mathbf{v}$. Since $H_L(\epsilon + \triangle_n)$ has only $(L + 1)^2$ nonzero diagonals and $R^t R$ has at most five nonzero entries in each row, the computational complexity of this matrix–vector product is $O(L^2 M_1 M_2)$. Thus the cost per each PCG iteration is $O(M_1 M_2 \log M_1 M_2 + L^2 M_1 M_2)$ operations. Hence the total cost for finding the solution in (10.45) is $O(M_1 M_2 \log M_1 M_2 + L^2 M_1 M_2)$ operations.

Besides solving the linear system in (10.45), one also needs to solve the least squares problem (10.46) or determine the solution from (10.48). From (10.48), it is noted that one needs to compute

$$Q(\mathbf{f}_n) = [\mathbf{f}_{n00}^y \| \mathbf{f}_{n00}^x \| \mathbf{f}_{n01}^y \| \mathbf{f}_{n01}^x \| \cdots \| \mathbf{f}_{nL-1L-2}^y \| \mathbf{f}_{nL-1L-2}^x \| \mathbf{f}_{nL-1L-1}^y \| \mathbf{f}_{nL-1L-1}^x]$$

and then solve a $2L^2$-by-$2L^2$ linear system. Since the matrix–vector product $H_L(\epsilon + \triangle_n)\mathbf{v}$ for any vector $\mathbf{v}$ can be computed in $O(L^2 M_1 M_2)$ operations, the cost of computing $Q(\mathbf{f}_n)$ is $O(L^4 M_1 M_2)$. The cost of solving the corresponding $2L^2$-by-$2L^2$ linear system is $O(L^6)$ by using Gaussian elimination. Hence the total cost for finding the solution in (10.46) is $O(L^6 + L^2 M_1 M_2)$ operations.

10.7.3 Simulation Results

In the computer simulation, a 256×256 image (Figure 10.9 (left)) is taken to be the original high-resolution image. A (2×2) sensor array with sub-pixel

FIGURE 10.9:   Original image of size 256×256 (left) and observed blurred and noisy image 256×256; PSNR = 24.20 dB (right).

FIGURE 10.10:   Reconstructed image by RLS; PSNR = 25.64 dB. (left) and Reconstructed image by RCTLS; PSNR = 25.88 dB (right).

displacement errors retrieves four 128×128 blurred and undersampled images, which are corrupted by white Gaussian noise with a SNR of 30 dB. The image interpolated from these low-resolution images is shown in Figure 10.9 (right). The parameters $\bar{\epsilon}^x_{l_1 l_2}$ and $\bar{\epsilon}^y_{l_1 l_2}$ are random values chosen between $\frac{1}{2}$ and $-\frac{1}{2}$.

In the first simulation, the estimated subpixel displacement errors, $\epsilon^x_{l_1 l_2}$ and $\epsilon^y_{l_1 l_2}$, are set to 85 percent of the real subpixel displacement errors. In the proposed RCTLS algorithm, the choice of "proper" regularization parameter, μ, is very important. Initially, the proposed RCTLS algorithm was implemented

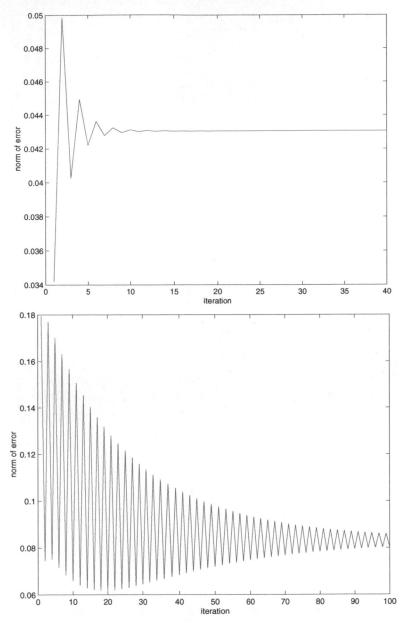

FIGURE 10.11: Norm of error with $\mu = 10^{-4}$ (upper) and norm of error with $\mu = 1$ (lower).

with various values of μ. Define

$$\text{the norm of error at the } n\text{th iteration } = \ \|\bar{\epsilon} - (\epsilon + \triangle_n)\|_2$$

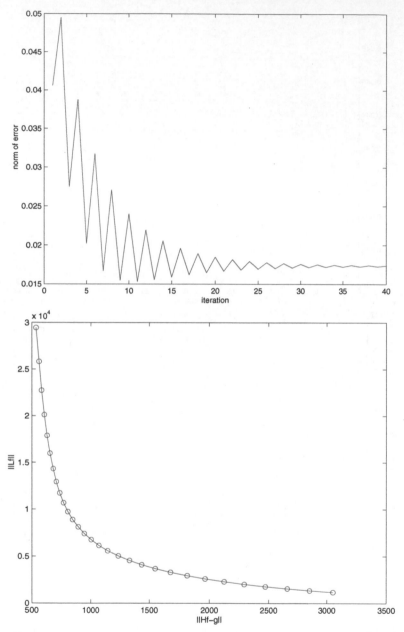

FIGURE 10.12: Norm of error with $\mu = \mu_{opt}$ (upper) and L-curve (lower).

and plot it for different values of μ (Figures 10.11 and 10.12). The speed of convergence decreases as μ increases. In the case when $\mu = 10^{-4}$ and

TABLE 10.2: Estimation errors.

Errors (x-direction)	$\ell_1 = 0, \ell_2 = 0$	$\ell_1 = 1, \ell_2 = 0$	$\ell_1 = 0, \ell_2 = 1$	$\ell_1 = 1, \ell_2 = 1$
$\bar{\epsilon}^x_{l_1 l_2}$	0.2	-0.2	0.1	0.1
$\epsilon^x_{l_1 l_2}$	-0.17	-0.17	0.085	0.085
$\epsilon_{l_1 l_2} + \delta\epsilon^x_{l_1 l_2}$ with $\mu = 10^{-4}$	0.1871	-0.1806	0.0915	0.0856
$\epsilon_{l_1 l_2} + \delta\epsilon^x_{l_1 l_2}$ with $\mu = 1$	0.2343	-0.2218	0.1290	0.1262
$\epsilon_{l_1 l_2} + \delta\epsilon^x_{l_1 l_2}$ with $\mu_{opt} = 0.005092$	0.1987	-0.1946	0.0967	0.0957

Errors (x-direction)	$\ell_1 = 0, \ell_2 = 0$	$\ell_1 = 1, \ell_2 = 0$	$\ell_1 = 0, \ell_2 = 1$	$\ell_1 = 1, \ell_2 = 1$
$\bar{\epsilon}^y_{l_1 l_2}$	0.1	0	-0.3	-0.2
$\epsilon^y_{l_1 l_2}$	0.085	0	-0.255	-0.17
$\epsilon_{l_1 l_2} + \delta\epsilon^y_{l_1 l_2}$ with $\mu = 10^{-4}$	0.0950	-0.0067	-0.2720	-0.1863
$\epsilon_{l_1 l_2} + \delta\epsilon^y_{l_1 l_2}$ with $\mu = 1$	0.1034	-0.0235	-0.3373	-0.2361
$\epsilon_{l_1 l_2} + \delta\epsilon^y_{l_1 l_2}$ with $\mu_{opt} = 0.005092$	0.1038	-0.0011	-0.2857	-0.1952

$\mu = 1$, the norm of error after convergence is greater than that in the first iteration (Figure 10.11). The "inappropriate" value of μ makes the RCTLS solution fall into a local minimum. The L-curve method proposed in [51] may be used to get the optimum value of the regularization parameter. The L-curve method to estimate the "proper" μ for RCTLS was used here. The L-curve plot is shown in Figure 10.12 (right). With μ_{opt} retrieved by the L-curve, the RCTLS converges to a better minimum point (the norm of error is significantly smaller than those obtained by choosing $\mu = 10^{-4}$ and $\mu = 1$). Table 10.2 shows the estimated subpixel displacement errors by RCTLS with different values of μ. It is seen that the updated subpixel displacement errors with μ_{opt} are closer to the real subpixel displacement errors than the updated subpixel displacement errors with arbitrarily chosen values for regularization parameters μ. It implies better use of the more precise blurring matrix for image reconstruction when the "proper" regularization parameter is chosen.

To show the advantage of the proposed RCTLS algorithm over the conventional RLS (regularized least squares) algorithm, the two methods are compared. The reconstructed image by RLS and the reconstructed image by RCTLS are shown in Figure 10.10 (right), respectively. The reconstructed high-resolution image using RCTLS shows improvement both in image quality and PSNR. It is remarked that the optimal regularization parameter for RLS is also determined by L-curve for use in the RLS algorithm.

10.8 Concluding Remarks and Current Work

In summary, we have presented a new approach to image restoration by using regularized, constrained total least squares image methods, with Neumann boundary conditions. Numerical results indicate the effectiveness of the method. An application for the approach proposed here is regularized constrained total least squares (RCTLS) reconstruction from an image sequence captured by multisensors with subpixel displacement errors that produces a high-resolution output, whose quality is enhanced by a proper choice of the regularization parameter. The novelty of this application lies in the treatment of the nonsymmetric PSF estimation problem arising from the multisensor image acquisition system.

High-resolution image reconstruction is largely known as a technique whereby multiframe motion is used to overcome the inherent resolution limitations of a low-resolution imaging system. High-resolution image reconstruction methods consist of three basic components: (i) motion compensation; (ii) interpolation; and (iii) blur and noise removal. The performance of high-resolution image reconstruction algorithms will depend on the effectiveness of motion estimation and modeling. For analysis, algorithms, and applications of blind high-resolution imaging, see the papers in the recent special issues [52–54].

Finally, we mention some ongoing work on the development of high-resolution algorithms and software for array imaging systems. Compact, multilens cameras can provide a number of significant advantages over standard single-lens camera systems, but the presence of more than one lens poses a number of difficult problems. Modern work on the design and testing of array imaging systems began with seminal papers on the TOMBO (Thin Observational Module for Bound Optics) system described by Tanida et al., in [55,56]. High-resolution reconstruction methods based on interpolation and projection of pixel values were developed by the authors.

In [57], Chan, Lam, Ng, and Mak describe a new superresolution algorithm and apply it to the reconstruction of a high-resolution image from low-resolution images from a simulated compound eye (array) imaging system. They explore several variations of the imaging system, such as the incorporation of phase masks to extend the depth of focus. Software simulations with a virtual compound-eye camera are used to verify the feasibility of the proposed architecture and its variations. They also report on the tolerance of the virtural camera system to variations of physical parameters, such as optical aberrations.

In [58], the team consisting of Barnard, Chung, van der Gracht, Nagy, Pauca, Plemmons, Prasad, Torgersen, Mathews, Mirotznik, and Behrmann investigate the fabrication and use of novel multilens imaging systems for iris recognition as part of the PERIODIC (Practical Enhanced-Resolution Integrated Optical-Digital Imaging Camera) project. The authors of [58]

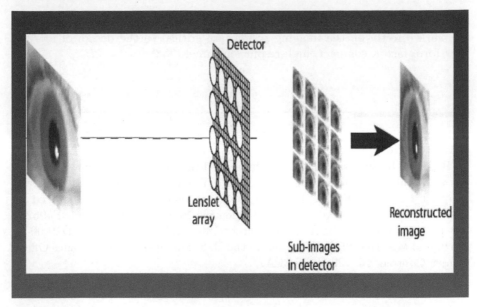

FIGURE 10.13: Lenslet array imaging for biometric identification.

use established, conventional techniques to compute registration parameters and leverage appropriate reconstruction of high-resolution images from low-resolution imagery. To verify that the generated high-resolution images have comparable fidelity to those captured by single-lens camera systems they apply the reconstruction methods to iris recognition biometric problems for personnel identification and verification. Figure 10.13 illustrates the collection of an eye image for this biometric application with an array imaging camera.

In [58] the authors also report on image registration and reconstruction results using both simulated and real laboratory multilens image data from prototype array cameras fabricated by the PERIODIC research team. The chief goals of the PERIODIC project are to analyze, optimize, simulate, design, and fabricate beta prototype, integrated, optical-digital, low-profile, array-based imaging systems. The approaches to this work include research and development studies involving mathematical and engineering technology to optimize and fabricate a subpixel-shift-based superresolving array camera system with multiple diversity channels, low cost, and enhanced reconfigurability. Current work on superresolution construction methods includes not only the use of subpixel displacement methods, but also of individual lenslet image diversities such as dynamic range, wavelength, and polarization. Array cameras based on novel diffractive lenslets are also being investigated. For near real-time computational capabilities, Graphics Processing Units (GPUs) are being used, leading to on-chip processing capability. The recent development of a number of mathematical libraries for GPUs containing advanced linear algebra, fast Fourier transform (FFT), and image/signal processing functions to work directly on the imaging system are critical to the design of a truly low-form-factor, compact, high-resolution camera [58].

Acknowledgments

Research by the first author was supported in part by Hong Kong Research Grants Council Numbers 7046/03P, 7035/04P, and 7035/05P and Hong Kong Baptist University FRGs. Research by the second author was supported in part by the U. S. Air Force Office of Scientific Research under grant F49620-02-1-0107, by the U. S. Army Research Office under grants DAAD19-00-1-0540 and W911NF-05-1-0402, and by the U. S. Disruptive Technologies Office under Contract No. 2364-AR03-A1.

References

[1] M. Roggemann and B. Welsh, *Imaging Through Turbulence*, CRC Press, Boca Raton, FL, 1996.

[2] K. Castleman, *Digital Image Processing*, Prentice–Hall, Upper Saddle River, NJ, 1996.

[3] M. Banham and A. Katsaggelos, Digital Image Restoration, *IEEE Signal Processing Magazine,* March 1997, pp. 24–41.

[4] D. Kundur and D. Hatzinakos, Blind Image Deconvolution, *IEEE Signal Processing Magazine,* May 1996, pp. 43–64.

[5] J. Bardsley, S. Jefferies, J. Nagy, and R. Plemmons, A Computational Method for the Restoration of Images with an Unknown, Spatially-varying Blur, *Optics Express,* 14, no. 5 (2006), pp. 1767–1782.

[6] C. Vogel, T. Chan, and R. Plemmons, Fast Algorithms for Phase Diversity-based Blind Deconvolution, in *Adaptive Optical System Technologies, Proceedings of SPIE,* Edited by D. Bonaccini and R. Tyson, 3353 (1998), pp. 994–105.

[7] M. Ng, R. Chan, and T. Wang, A Fast Algorithm for Deblurring Models with Neumann Boundary Conditions, *SIAM Journal on Scientific Computing,* 21 (1999), pp. 851–866.

[8] H. Andrews and B. Hunt, *Digital Image Restoration*, Prentice-Hall, Englewood Cliffs, NJ, 1977.

[9] T. Bell, Electronics and the Stars, *IEEE Spectrum* 32, no. 8 (1995), pp. 16–24.

[10] J. Hardy, *Adaptive Optics for Astronomical Telescopes*, Oxford Press, New York, 1998.

[11] J. Nelson, Reinventing the Telescope, *Popular Science*, 85 (1995), pp. 57–59.

[12] V. Mesarović, N. Galatsanos, and A. Katsaggelos, Regularized Constrained Total Least Squares Image Restoration, *IEEE Transactions on Image Processing*, 4 (1995), pp. 1096–1108.

[13] T. Chan and C. Wong, Total Variation Blind Deconvolution, IEEE *Transactions on Image Processing*, 7, no. 3 (1998), pp. 370–375.

[14] S. Jefferies and J. Christou, Restoration of Astronomical Images by Iterative Blind Deconvolution, *Astrophysics Journal*, 415 (1993), pp. 862–874.

[15] M. Ng, R. Plemmons, and S. Qiao, Regularized blind deconvolution using recursive inverse filtering. In G. H. Golub, S. H. Lui, F. T. Luk, and R. J. Plemmons, editors, *Proceedings of the Hong Kong Workshop on Scientific Computing*, (1997), pp. 110–132.

[16] M. Ng, R. Plemmons, and S. Qiao, Regularization of RIF Blind Image Deconvolution, *IEEE Transactions on Image Processing*, 9, no. 6 (2000), pp. 1130–1134.

[17] Y. You and M. Kaveh, A Regularization Approach to Joint Blur Identification and Image Restoration, *IEEE Transactions on Image Processing*, 5 (1996), pp. 416–427.

[18] H. Fu and J. Barlow, A Regularized Structured Total Least Squares Algorithm for High Resolution Image Reconstruction, *Linear Algebra and its Applications*, 391, 1 (2004), pp. 75–98.

[19] G. Golub, P. Hansen, and D. O'Leary, Tikhonov Regularization and Total Least Squares, *SCCM Research Report*, Stanford, 1998.

[20] J. Kamm and J. Nagy, A Total Least Squares Method for Toeplitz Systems of Equations, *BIT*, 38 (1998), pp. 560–582.

[21] M. Ng, R. Plemmons, and F. Pimentel, A New Approach to Constrained Total Least Squares Image Restoration, *Journal of Numerical linear algebra with applications*, 316 (2000), pp. 237–258.

[22] A. Prussner and D. O'Leary, Blind Deconvolution Using a Regularized Total Least Norm Algorithm, *SIAM Journal on Matrix Analysis and Applications*, 24, 4 (2003), pp. 1018–1037.

[23] W. Zhu, Y. Wang, Y. Yao, J. Chang, H. Graber, and R. Barbour, *Iterative Total Least Squares Image Reconstruction Algorithm for Optical Tomography by the Conjugate Gradient Algorithm*, Journal of the Optical Society of America A, 14 (1997), pp. 799–807.

[24] W. Zhu, Y. Wang, and J. Zhang, Total Least Squares Reconstruction with Wavelets for Optical Tomography, *Journal of the Optical Society of America A*, 15 (1997), pp. 2639–2650.

[25] G. Golub and C. Van Loan, An Analysis of Total Least Squares Problems, *SIAM Journal on Numerical Analysis*, 17 (1980), pp. 883–893.

[26] S. Van Huffel and J. Vandewalle, *The Total Least Squares Problem*, SIAM Press, Philadelphia, 1991.

[27] H. Engl, M. Hanke, and A. Neubauer, *Regularization of Inverse Problems*, Kluwer Academic Publishers, Dordrecht, The Netherlands, 1996.

[28] R. Gonzalez and R. Woods, *Digital Image Processing*, Addison Wesley, New York, 1992.

[29] R. Chan, M. Ng, and R. Plemmons, Generalization of Strang's Preconditioner with Applications to Toeplitz Least Squares Problems, *J. Numer. Linear Algebra Appls.*, 3 (1996), pp. 45–64.

[30] A. Jain, Fundamentals of Digital Image Processing, Prentice-Hall, Englewood Cliffs, NJ, 1989.

[31] F. Luk and D. Vandevoorde, Reducing boundary distortion in image restoration, *Proceedings SPIE 2296*, Advanced Signal Processing Algorithms, Architectures and Implementations VI, 1994.

[32] R. Lagendijk and J. Biemond, *Iterative Identification and Restoration of Images*, Kluwer Academic Publishers, Boston, 1991.

[33] J. Lim, *Two-dimensional Signal and Image Processing*, Prentice Hall, Englewood Cliffs, NJ, 1990.

[34] G. Strang, The Discrete Cosine Transform, *SIAM Review*, 41 (1999), pp. 135–147.

[35] M. Hanke and J. Nagy, Restoration of Atmospherically Blurred Images by Symmetric Indefinite Conjugate Gradient Techniques, *Inverse Problems*, 12 (1996), pp. 157–173.

[36] R. Lagendijk, A. Tekalp, and J. Biemond, Maximum Likehood Image and Blur Identification: a Unifying Approach, *Optical Engineering*, 29 (1990), pp. 422–435.

[37] K. Rao and P. Yip, *Discrete Cosine Transform: Algorithms, Advantages, Applications*, Academic Press, Boston, 1990.

[38] J. Dennis and R. Schnabel, *Numerical Methods for Unconstrained Optimalization and Nonlinear Equation*, Prentice-Hall, Upper Saddle River, NJ, 1983.

[39] T. Kelley, *Iterative Methods for Optimization*, SIAM Press, Philadelphia, 1999.

[40] S. Nash and A. Sofer, *Linear and Nonlinear Programming*, The McGraw-Hill Companies, Inc., New York, 1996.

[41] Y. Saad, *Iterative Methods for Sparse Linear Systems*, PWS, Boston, 1996.

[42] M. Hestenes and E. Steifel, Methods of Conjugate Gradient for Solving Linear Systems, *Journal of Research of the National Bureau of Standards,* 49 (1952), pp. 409–436.

[43] D. Luenberger, *Introduction to Linear and Nonlinear Programming*, Addison-Wesley, New York, 1973.

[44] O. Axelsson and V. Barker, *Finite Element Solution of Boundary Value Problems, Theory and Computation*, Academic Press, Orlando, FL, 1984.

[45] R. Chan and M. Ng, Conjugate Gradient Methods for Toeplitz Systems, *SIAM Review,* 38 (1996), pp. 427–482.

[46] M. Ng and R. Plemmons, Fast Recursive Least Squares Using FFT-based Conjugate Gradient Iterations, *SIAM Journal on Scientific Computing,* 17 (1996), pp. 920–941.

[47] N. K. Bose, H. C. Kim, and H. M. Valenzuela, Recursive Total Least Squares Algorithm for Image Reconstruction from Noisy, Undersampled Frames, *Multidimensional Systems and Signal Processing,* 4 (1993), pp. 253–268.

[48] S. Van Huffel and J. Vandewalle, *The Total Least Squares Problem: Computational Aspects and Analysis*, SIAM Publications, Philadelphia, 1991.

[49] N. K. Bose and K. J. Boo, High-resolution Image Reconstruction with Multisensors, *International Journal of Imaging Systems and Technology,* 9 (1998), pp. 294–304.

[50] M. Ng and A. Yip, A Fast MAP Algorithm for High-Resolution Image Reconstruction with Multisensors, *Multidimensional Systems and Signal Processing,* 12, 2 (2001), pp. 143–164.

[51] P. Hansen and D. O'Leary, The Use of the L-curve in the Regularzation of Discrete Ill-posed Problems, *SIAM Journal on Scientific Computing,* 14 (1993), pp. 1487–1503.

[52] M. Ng, N. Bose, and R. Chan, Guest editorial of high-resolution image reconstruction, *International Journal of Imaging System and Technology,* 14 (2004), pp. 35-89.

[53] M. Ng, T. Chan, M. Kang, and P. Milanfar, Guest editorial of Special Issue on Super-resolution imaging: analysis, algorithms, and applications, *EURASIP Journal on Applied Signal Processing,* Volume 2006 (2006), pp. 1–2.

[54] M. Ng, E. Lam, and C. Tong, Guest editorial of Special Issue on the International Conference on Superresolution imaging, *Multidimensional Systems and Signal Processing,* 2007, to appear.

[55] J. Tanida, T. Kumagai, K. Yamada, S. Miyatake, K. Ishida, T. Morimoto, N. Kondou, D. Miyazaki, and Y. Ichioka, Thin Observation Module by Bound Optics (TOMBO): Concept and Experimental Verification, *Applied Optics,* 40 (2001), pp. 1806–1813.

[56] J. Tanida, R. Shogenji, Y. Kitamura, K. Yamada, M. Miyamoto, and S. Miyatake, Color Imaging with an Integrated Compound Imaging System, *Optics Express,* 11 (2003), pp. 2109–2117.

[57] W. Chan, E. Lam, M. Ng, and G. Mak, *Super-resolution reconstruction in a Computational Compound-eye Imaging System,* Multidimensional Systems and Signal Processing (2007), to appear.

[58] R. Barnard, J. Chung, J. van der Gracht, J. Nagy, P. Pauca, R. Plemmons, S. Prasad, T. Torgersen, S. Mathews, M. Mirotznik, and G. Behrmann, High-Resolution Iris Image Reconstruction From Low-Resolution Imagery, *Proceedings of SPIE Annual Conference, Symposium on Advanced Signal Processing Algorithms, Architectures, and Implementations XVI,* Vol. 6313, (2006), on CD, Section D, pp. 1–13.

Index

T - #0026 - 101024 - C4 - 229/152/26 [28] - CB - 9780849373671 - Gloss Lamination